Progress in
Medicinal Chemistry 35

Progress in Medicinal Chemistry 35

Editors:

G.P. ELLIS, D.SC., PH.D., F.R.S.C.

Department of Chemistry
University of Wales
P.O. Box 912, Cardiff, CFI 3TB
United Kingdom

D.K. LUSCOMBE, B.PHARM., PH.D., F.I.BIOL., F.R.PHARM.S

Welsh School of Pharmacy
University of Wales
P.O. Box 13, Cardiff, CFI 3XF
United Kingdom

and

A.W. OXFORD, M.A., D.PHIL.

Consultant in Medicinal Chemistry
P.O. Box 151
Royston SG8 5YQ
United Kingdom

1998

ELSEVIER
AMSTERDAM•LAUSANNE•NEW YORK•OXFORD•SHANNON•SINGAPORE•TOKYO

Elsevier Science BV
P.O. Box 211
1000 AE Amsterdam
The Netherlands

RM30
. P7
vol. 35

Library of Congress Cataloging in Publication Data:

Please refer to card number 62-2712 for this series.

ISBN 0-444-82909-1
ISBN Series 0-7204-7400-0

Printed in The Netherlands

Preface

Five topics are reviewed in this volume. Chapter 1 traces the biochemical and medical significance of Vitamin D and its derivatives from the first quarter of this century, a field which continues to promise further valuable results. In contrast, the relatively recent development of neurokinin antagonists, especially NK1-selective compounds, is surveyed in Chapter 2. These compounds have potential for the treatment of pain, migraine, emesis and asthma. A plethora of types of compound structures have recently been found to possess selective antagonism for each of the neurokinin receptors.

Chapter 3 surveys recent progress in our understanding of opioid antagonists and of their therapeutic potentials. Bacterial resistance to antibacterial agents and especially chemotherapeutic drugs is of increasing importance. Current knowledge of the mechanisms involved in resistance together with ways of combating this threat to successful chemotherapy is discussed in Chapter 4.

Much progress has recently been made in our understanding of structure-activity relationships of cannabinoids. In view of the current interest in the clinical usefulness of these compounds as analgesics, Chapter 5 gives a timely survey of present scientific knowledge of these drugs.

We are grateful to our authors for condensing the vast literature of these topics and to owners of copyright material for permission to reproduce, and to our publishers for their encouragement.

September 1997

G. P. Ellis
D. K. Luscombe
A. W. Oxford

Contents

Progress in Medicinal Chemistry – Vol. 35
Edited by G.P. Ellis, D.K. Luscombe and A.W. Oxford

1 Modern View of Vitamin D_3 and its Medicinal Uses

MATTHEW J. BECKMAN, PH.D. AND HECTOR F. DELUCA, PH.D.

Department of Biochemistry, College of Agricultural and Life Sciences, University of Wisconsin-Madison, 420 Henry Mall, Madison, Wisconsin 53706

INTRODUCTION

The involvement of vitamin D as an accessory food substance in the health of individuals has been well-known since the turn of the century stemming mainly from studies investigating the cause of rickets. The disease rickets is characterized by the failure of mineralization of the bone matrix and/or growth plate and is marked by dramatic skeletal deformities. The documentation of this disease extends back to ancient times, but the sharp rise in the incidence of rickets and osteomalacia (the adult form of rickets) in the United States and Northern Europe at the approximate time of the Industrial Revolution stimulated a surge of determined interest in clarifying the aetiology of rickets [1]. The three most profound determinants contributing to the cause of rickets were undernourishment, pollution and the implementation of long workdays in factories especially by young growing children.

Perhaps the most significant of the early classical studies was that of Sir Edward Mellanby in 1919 experimenting with dogs. He induced rickets by housing them indoors and feeding them oatmeal [2]. Around this time, McCollum and his co-workers had discovered a potent substance in cod liver oil that he termed vitamin A which was active in preventing diseases of deficiency such as xeropthalmia and night blindness [3]. Mellanby demonstrated that cod liver oil was also effective in curing rickets in dogs, but he considered that vitamin A was the important nutritional factor responsible for healing rickets. McCollum *et al.* destroyed vitamin A in cod liver oil by heating and oxidation, but found that the antirachitic factor remained [4]. The antirachitic factor was given the name vitamin D by McCollum. In addition, other observations provided the rationale indicating that UV irradiation is beneficial in the healing of rickets. Steenbock and Black tied both the nutritional and environmental observations together by demonstrating the formation of the antirachitic factor in skin and the fat-soluble fraction of foods by UV irradiation [5]. Hess and Weinstock confirmed Steenbock and Black's results and this led eventually to the isolation and identification of ergocalciferol (vitamin D_2) from plants [6]. At first, vitamin D_2 was thought to be the same antirachitic factor of fish oils until it was found to be less active in curing rickets in chickens [7, 8]. In 1936, Windaus *et al.* synthesized 7-dehydrocholesterol and demonstrated its conversion via UV irradiation to cholecalciferol (vitamin D_3), the natural form of vitamin D in animal species [9]. Vitamin D_3 differs from vitamin D_2 by not having a C-24 methyl group and a double bond at C-22. From the discovery of Steenbock and Black, effective methods were developed to introduce vitamin D into foods, eliminating rickets as a major medical problem in developed countries throughout the northern hemisphere.

The relationship of serum calcium and phosphate with rickets was discovered by Howland and Kramer [10]. They found that blood from normal rats could mineralize rachitic rat cartilage, whereas blood from rachitic rats could not. They also provided evidence that a low serum calcium and phosphate status caused rickets. Orr *et al.* [11] demonstrated that UV irradiation stimulated calcium absorption. This study was largely unappreciated for 30 years until Nicolaysen and Eeg-Larsen [12] and Schachter and Rosen [13] demonstrated evidence for vitamin D-induced intestinal absorption of calcium by an active transport process.

Up to 1960, there had been several years of study dedicated to the metabolism of vitamin D_3 with practically no breakthroughs resulting in the detection of further biologically active metabolites of vitamin D_3. Following this period a new era of important advances led to a virtual flood of information regarding vitamin D physiology and metabolic activation. These advances included the chemical synthesis of radioactive vitamin D_3 of high specific activity [14, 15], the development of a number of new chromatographic systems [16–18] and technological advances in high resolution mass spectrometry and nuclear magnetic resonance spectrometry that allowed for the accurate identification of metabolite structures and facilitated chemical synthesis of both the metabolites and their analogues. These notable advances gave way to new investigations of vitamin D_3 metabolism at truly physiological levels, and allowed for the detailed isolation, separation, and characterization of many novel metabolites of vitamin D_3 (*Figure 1.1*).

The first indication that vitamin D_3 might require further metabolic alteration before it could be active came from experiments showing a lag between the time of vitamin D_3 administration and the first observed biological response [19]. The lag time could be shortened by intravenous administration (9 h) compared with oral administration (18 h) of vitamin D_3, but not completely eliminated [20]. A major polar metabolite fraction

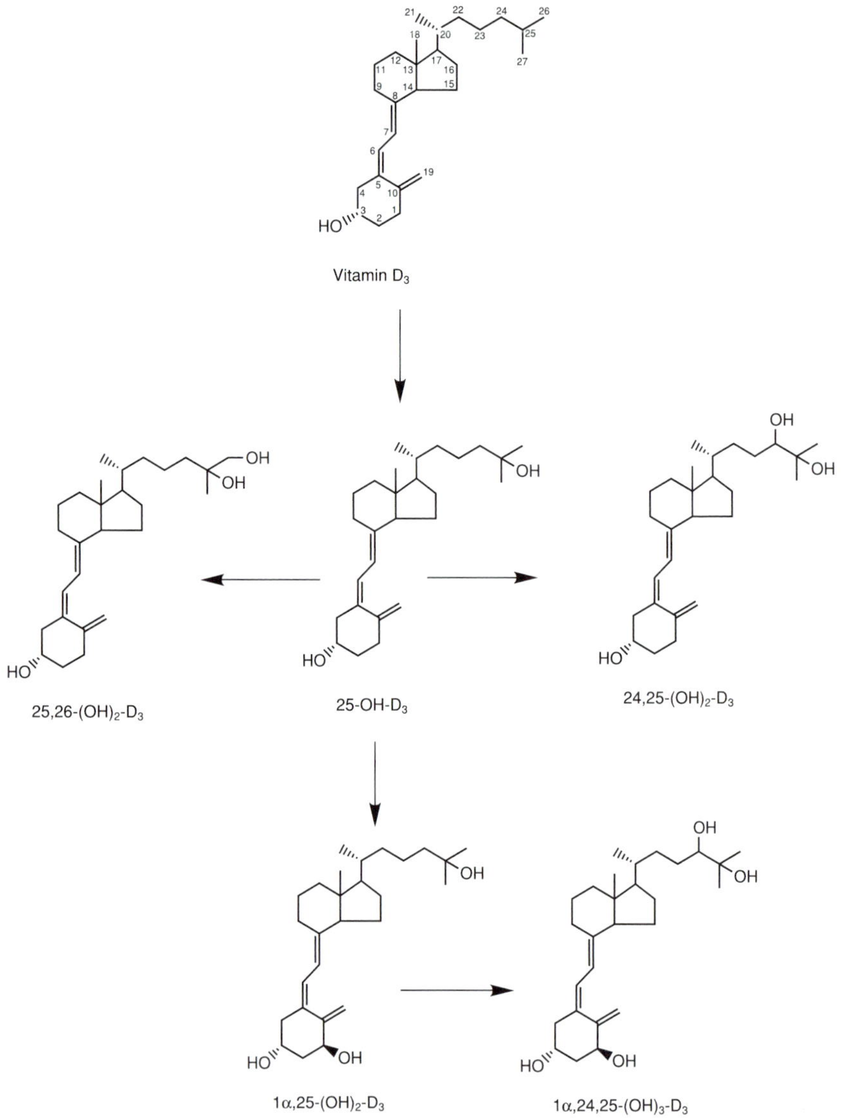

Figure 1.1. Numbering scheme of the vitamin D_3 molecule and early structural identification of several vitamin D_3 metabolites.

was then separated from vitamin D_3 in target tissue extracts and found to have significantly greater antirachitic activity and acted more rapidly than vitamin D_3 [21, 22]. This metabolite fraction was further resolved into two fractions and the major component purified and identified as 25-hydroxyvitamin D_3 (25-OH-D_3) [23]. Despite the initial belief that 25-OH-D_3 was the metabolically active form [24], it soon was discovered that ^{3}H-25-OH-D_3 could rapidly be metabolized to more polar metabolites [25, 26]. A second polar metabolite was isolated from chick intestine that possessed potent activity [27], but this metabolite appeared only briefly then and disappeared [28] which made the isolation of this metabolite very difficult. Eventually, this metabolite too was isolated to purity [29], its structure identified as 1,25-$(OH)_2D_3$ [30], and was shown to be produced in the kidney [31].

It is now well established that 1,25-$(OH)_2D_3$ is the active hormonal form of vitamin D_3 [32]. The production of 1,25-$(OH)_2D_3$ in the kidney is regulated by dietary calcium and phosphate and also by changes in serum calcium and parathyroid hormone, which clearly highlight the hormonal nature of this compound. Functionally, the three classical actions of 1,25-$(OH)_2D_3$ are to stimulate intestinal calcium and independently phosphate absorption, the mobilization of calcium from bone, and increase renal reabsorption of calcium. The focus of this review will be to explore the most recent concepts of vitamin D in regard to its metabolism and physiology, and with respect to the medicinal applications of vitamin D_3 metabolites and analogues.

METABOLISM OF VITAMIN D_3

BIOSYNTHESIS OF VITAMIN D_3 IN SKIN

Vitamin D_3 is not an essential exogenous micronutrient as such because it is made endogenously from a precursor in skin, 7-dehydrocholesterol (provitamin D_3), by exposure to the high-energy ultraviolet B (UVB)* photons (290–315 nm) of the solar spectrum [33]. The photons penetrate the epider-

---

*The following abbreviations are used in this review: CaBP, calcium-binding protein; DBD, DNA-binding domain; DBP, Vitamin D-binding protein; DNFB, 2,4-dinitrofluorobenzene; ER, oestrogen receptor; GR, glucocorticoid receptor; HPDR, hypophosphataemic vitamin D-resistant rickets; LBD, ligand-binding domain; NAF, nuclear accessory factor; NOD, non-obese diabetes; OH-D_3, hydroxy-vitamin D_3; $(OH)_2D_3$, dihydroxy-vitamin D_3; OHase, hydroxylase; PCR, polymerase chain reaction; PCT, proximal tubular cells; PKA, protein kinase A; PKC, protein kinase C; PTH, parathyroid hormone; RAF, receptor auxilliary factor; RAR, retinoic acid receptor; RARE, retinoic acid receptor response elements; RDA, recommended dietary allowance; RXR, retinoid X receptor; TR, thyroid hormone receptor; TRE, thyroid hormone receptor response elements; UVB, ultraviolet B; VDDR, vitamin D-dependency rickets; VDR, 1,25-dihydroxy vitamin D_3 receptor; VDRE, 1,25-dihydroxy vitamin D_3-responsive element.

mis and photolyze 7-dehydrocholesterol causing rupture of the B-ring followed by a 1,7-sigmatropic shift, forming the previtamin D_3 intermediate. During initial exposure to sunlight, provitamin D_3 is efficiently converted to previtamin D_3. Once formed, previtamin D_3 undergoes a thermally-dependent rearrangement of its double bonds at $37°C$ to form vitamin D_3 that takes 2–3 days to reach completion [34]. This slow thermal conversion of previtamin D_3 to vitamin D_3 equilibrates to reach a mixture of 96% vitamin D_3 and 4% previtamin D_3 [35]. No more than 10–20% of the initial provitamin D_3 concentrations ultimately end up as previtamin D_3. Continued exposure to sunlight causes isomerization of previtamin D_3, principally to a biologically dead-end lumisterol (*Figure 1.2*). Therefore, prolonged exposure to sunlight does not result in formation of toxic levels of vitamin D_3.

Transport of vitamin D_3 away from the dermal junction of skin is accomplished by a 52 kDa serum vitamin D-binding protein (DBP). Serum DBP is a member of the α-fetoprotein-albumin superfamily [36]. DBP has high affinity for vitamin D_3, but does not bind to its precursors or the products of previtamin D_3 side-reactions, lumisterol and tachysterol [37]. Accumulation of 7-dehydrocholesterol in skin occurs in sebaceous glands at the malpighian layer of the epidermis, mostly in the stratum spinosum and stratum basal

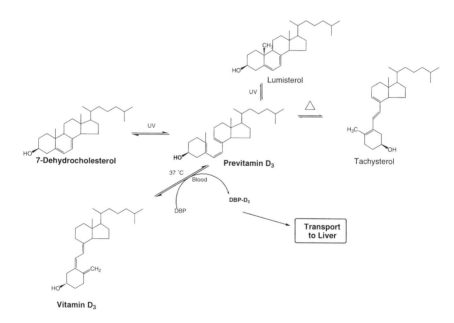

Figure 1.2. Skin biosynthesis of vitamin D_3.

[38]. A significant amount of 7-dehydrocholesterol is also found in the deeper dermis layer, but because most UVB photons are absorbed by the epidermis, very little production of vitamin D_3 occurs in the dermal layer. There is some evidence that formation of 7-dehydrocholesterol is under the regulation of the vitamin D_3 endocrine system [39]; however, details of this regulation are not fully understood.

Aging, sunscreens, seasonal changes, time of day, and latitude also significantly affect the cutaneous production of this vitamin-hormone [37, 40, 41]. More darkly pigmented skin interferes with the efficiency of vitamin D_3 formation because melanin effectively competes with provitamin D_3 for UV irradiation [38]. Persons living at latitudes more northerly than 34 degrees N produce little to no previtamin D_3 in the cutaneous layer during the months of November-February, and at latitudes as high as 52 degrees N, individuals may experience an extended 'vitamin D winter' [42]. Because vitamin D_3 insufficiency or deficiency exacerbates osteoporosis, causes osteomalacia and increases the risk of fracture, dietary supplementation of vitamin D_3 for individuals living at high latitudes may be advisable. For example, an intake of 5 μg vitamin D_3 to postmenopausal women living at 42 degrees N was found to be sufficient to limit bone loss from the spine and whole body, but was not adequate to minimize bone loss from the femoral neck [43].

The recommended dietary allowance (RDA) for vitamin D is 5 μg (200 IU)/d, but this value includes casual exposure to sunlight without regard to lifestyle or climate. For individuals who do not receive adequate sunlight because of illness, advanced age, or are otherwise shut in, the actual daily requirement for vitamin D_3 could be as much as 15 μg/d. Vitamin D_3 deficiency also may occur with chronic biliary obstruction and steatorrhoea limiting intestinal absorption of vitamin D_3 and lead to osteomalacia [44]. Another factor is the increased use of sunscreens among Western societies, that limits the penetration of UV photons to the site of vitamin D_3 synthesis [41, 45, 46].

Serum levels of calcitropic hormones are often used as predictors of the seasonal changes in sunlight-activated vitamin D_3 synthesis. In one such study, Rosen *et al.* [47] measured bone mineral density and serum 25-OH-D_3 of elderly women in the New England area (latitude 45.5) and examined the seasonal variations of these factors. In the 24-month observational study, significant seasonal changes were demonstrated. During the winter months, 25-OH-D_3 decreased in correlation with the loss in bone mineral content. Increasing vitamin D intake during the second year of the study elevated 25-OH-D_3 serum concentration slightly. The difference in serum

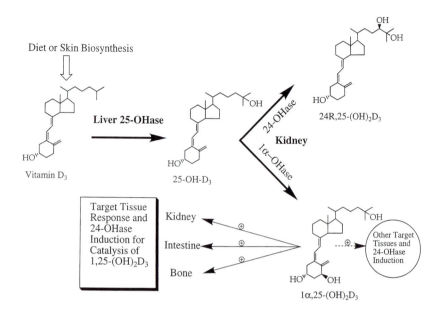

Figure 1.3. Scheme of the functional metabolism of vitamin D_3.

25-OH-D_3 between the first and second winters was the strongest predictor of bone accretion during the second year of the study.

FUNCTIONAL METABOLISM OF VITAMIN D_3

Critical to vitamin D_3 action is its further metabolic conversion to more active compounds (*Figure 1.3*). Via its transport by DBP, vitamin D_3 accumulates in the liver [48]. In rats, as much as 60–80% of an injected or oral dose of vitamin D_3 locates to the liver [49–51]. Intestinal absorption of vitamin D_3 is in association with the chylomicron fraction via the lymphatic system. Vitamin D_3 is delivered to the liver in blood from the thoracic duct only a few hours post ingestion [44]. A specific portion of hepatic vitamin D_3 in the rat is converted to 25-OH-D_3 by a 25-hydroxylase system in the endoplasmic reticulum of hepatocytes [52, 53]. This enzyme (Km 10^{-8} M) is regulated to an extent by 25-OH-D_3 and its metabolites. Higher concentrations of vitamin D_3 are handled by a second 25-hydroxylase located in liver mitochondria [54]. This enzyme, also known as CYP27, 27-hydroxylates cholesterol and thus appears less discriminating than the microsomal 25-OHase which does not use cholesterol as substrate [55, 56]. In humans, however, the mitochondrial form is the predominant liver form of 25-OHase [57]. Furthermore, the mitochondrial enzyme (Km for vitamin $D_3, 10^{-6}$ M) does not appear to be regulated by end-products. Thus, the circulating concentration of 25-OH-D_3 reflects more accurately vitamin D_3 status of an organism than serum vitamin D_3 itself.

The affinity of 25-OH-D_3 for DBP is much greater than it is for vitamin D_3, and whereas vitamin D_3 accumulates in liver cells, its metabolites are cleared completely [58, 59]. The half-life of 25-OH-D_3 in serum is two to three weeks [60], indicating very slow metabolism of this metabolite by catabolic enzymes. Serum 25-OH-D_3 is the major circulating metabolite of vitamin D_3 and the source from which the active hormonal form of vitamin D_3, 1,25-$(OH)_2D_3$ is synthesized. The C-1 hydroxylation of 25-OH-D_3 to 1,25-$(OH)_2D_3$ is a highly regulated step that completes the hormonal activation of vitamin D_3. The proximal renal tubule appears to be the exclusive site of 25-OH-D_3-1α-hydroxylation under normal conditions [61]. A physiologically relevant 1α-hydroxylase, however, also develops in placental tissue of pregnant rats [62]. A firmly established exception to these is in the case of chronic granulomatous disorders and some lymphomas [63, 64]. Elevations in serum 1,25-$(OH)_2D_3$ from these disorders is not the result of a defect in the renal 1α-OHase because elevated 1,25-$(OH)_2D_3$ persists even in sarcoidotic patients with end-stage renal disease or nephrectomy [65]. In humans, the amount of 1,25-$(OH)_2D_3$ produced per day is in the range of 1–1.5 μg which maintains a circulating concentration of 25–45 pg/ml. Secretion of 1,25-$(OH)_2D_3$ is a continual process that is subject to regulation by many endocrine factors and certain physiological stresses that will be discussed in detail in this review.

As an endocrine factor itself, 1,25-$(OH)_2D_3$ localizes in target sites and mediates its biological actions by interacting with a specific nuclear receptor protein that, in turn, responds as a transcriptional factor for the activation or deactivation of specific genes [66, 67]. The expression of the 1,25-$(OH)_2D_3$-vitamin D receptor (VDR) in tissues identifies these tissues as target sites of 1,25-$(OH)_2D_3$ action. One important biological action of 1,25-$(OH)_2D_3$ is the potent induction of mitochondrial 1,25-$(OH)_2D_3$-24-hydroxylase (24-OHase) [68]. C-24-Hydroxylation is a catabolic step that results in diminished action of 1,25-$(OH)_2D_3$ on target genes and decreases the cellular half-life of 1,25-$(OH)_2D_3$ as well [69, 70]. In addition, 24-OHase catalyzes C-24 hydroxylation of 25-OH-D_3 to form 24,25-$(OH)_2D_3$ [71]. This is specifically regulated in the kidney and is induced by 1,25-$(OH)_2D_3$ [72]. There are as many as 30 different metabolites of vitamin D_3 that exist naturally. Many of these metabolites are derivatives of either 25-OH-D_3 or 1,25-$(OH)_2D_3$ (*Figure 1.4*). The only known metabolite of vitamin D_3 of any sig-

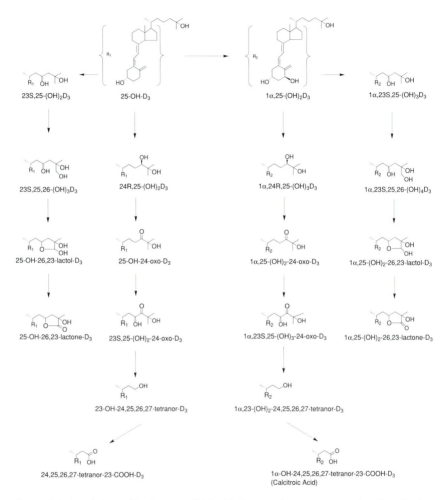

Figure 1.4. Metabolism of 25-OH-D$_3$ and 1,25-(OH)$_2$D$_3$ to their respective end products by the C-23/C-24-catabolic and C-23/C-26-lactone pathways.

nificant biological importance, however, is 1,25-(OH)$_2$D$_3$ [73]. Therefore, most vitamin D$_3$ metabolites are intermediates in the catabolism of the vitamin D$_3$ molecule, and pathways which produce these metabolites are induced by 1,25-(OH)$_2$D$_3$ [74]. It is also interesting that expression of the catabolic pathways appear in all classical target sites of 1,25-(OH)$_2$D$_3$ (bone, kidney, and intestine), but in particular 24-hydroxylase activity or mRNA has also been found in many of the nonclassical target tissues of 1,25-$(OH)_2D_3$. The extrarenal levels of 24-OHase are very low normally, but increase in a time and dose-dependent fashion in the presence of 1,25-$(OH)_2D_3$ [75]. As such, it is thought that 1,25-$(OH)_2D_3$ plays a role in its own destruction at the level of the target cell.

Over the last 3 decades, several independent investigators contributed to an enormous investigation of the metabolism of vitamin D_3. Collectively, these studies have identified at least two catabolic pathways of relevance for 1,25-$(OH)_2D_3$, the C-23/C-24 pathway and the C-23/C-26 lactone pathway (*Figure 1.4*) [76]. Serum 25-OH-D_3-lactone does bind five times more avidly to DBP, but because the lactone end products form primarily as an escape from supraphysiological administration of vitamin D_3 [77, 78], they are not broadly considered as biologically active, despite some reports to the contrary [79, 80]. Therefore, the principal physiological pathway of interest is the C-23/C-24 catabolic pathway, consisting of C-24-hydroxylation, C-24-keto-oxidation, C-23-hydroxylation and side-chain cleavage to 24,25,26,27-tetranor-23-OH-D_3 [78, 81]. Some side reactions also take place, but it is not known how these reactions influence the overall metabolism of vitamin D_3. Calcitroic acid is the final excretion product of vitamin D_3 [82]. It is a highly polar substance that is hydrophilic enough to associate with the aqueous fluids of the body and finally be excreted in bile [83]. Hydroxylation of C-26 on 25-OH-D_3 can also occur under physiological conditions forming 25,26-$(OH)_2D_3$ [84]. This metabolite can be further metabolized to 1,25,26-$(OH)_3D_3$ by renal 1α-OHase [85], and in this form, mimic actions of 1,25-$(OH)_2D_3$. The caveat here is that C-26 hydroxylation occurs only if serum calcium is normal or increased due to excess vitamin D_3 [86]. Because this circumstance is reciprocal to 1α-OHase activity, the formation of 1,25,26-$(OH)_3D_3$ *in vivo* is negligible and, therefore, probably not of meaningful importance.

Further evidence that C-25 and C-1 hydroxylation are the activation steps of vitamin D_3, and that C-24/C-23 and C-23/C-26-lactone metabolic conversions do not produce physiologically important products is with the use of side-chain fluoridated analogues of 25-OH-D_3 [87, 88]. Studies with these analogues were prompted by assertions that 24-hydroxylated or lactone metabolites are involved or required for such biological actions as: mineralization of bone [89], suppression of parathyroid hormone secretion [90], cartilage metabolism [91], and embryonic development in the chick [92]. It is well established that plasma 24,25-$(OH)_2D_3$ concentrations (2–5 ng/ml) are approximately 50 times greater than those of 1,25-$(OH)_2D_3$. Even so, 24,25-$(OH)_2D_3$ has little affinity for VDR. At pharmacological concentrations, though, 24,25-$(OH)_2D_3$ has been demonstrated to compete for VDR binding and stimulate transcriptional activation of a vitamin D-responsive gene-reporter system [93]. Perhaps this explains the studies in which 24,25-$(OH)_2D_3$ has been implicated as having a specific effect but only at extremely high (10^{-6} M) concentrations [94, 95]. Other studies have been directed toward the search for a 24,25-$(OH)_2D_3$-binding protein or receptor [96]; however, this work is far from conclusive. The synthesis and use of 24,24-difluoro-25-OH-D_3 [97] and 26,26,26,27,27,27-hexafluoro-25-hydroxyvitamin D_3 [88] have provided research tools for examining the functional consequence of the absence of 24- and 26-hydroxylation. Both of these compounds can be further activated by 1α-hydroxylation, but the presence of fluorine groups at C-24 and C-26 blocks hydroxylation at these sites [88]. In experiments that were controlled by the use of 25-OH-D_3 versus the synthetic compounds in rats and chickens, dosed with either of the synthetic compounds, the animals were completely normal with respect to intestinal calcium transport, the mobilization of calcium from bone and the mineralization of vitamin D-deficient bone [88, 98]. Furthermore, animals supported on these fluoridated analogues were normal with respect to their reproduction, growth, and development [99, 100]. As for parathyroid hormone secretion, it is now well established that the calcium receptor isolated and cloned by Hebert and Brown [101] is the sensing signal that responds to low serum calcium concentration and triggers parathyroid hormone release from parathyroid gland cells [102]. Any role for 24,25-$(OH)_2D_3$ in this process is not convincing.

REGULATION OF VITAMIN D_3 METABOLISM

HEPATIC VITAMIN D_3-25-HYDROXYLASE

Under physiological conditions, there is no clear evidence of extrahepatic 25-hydroxylation of vitamin D_3. In avians, however, some evidence exists for functional 25-OHase activity in intestine and kidney [103]. In addition to its major role of 27-hydroxylation of cholesterol [104, 105], mitochondrial CYP27 catalyzes 25-hydroxylation of vitamin D_3 and also can catalyze 1α-hydroxylation of 25-OH-D_3, thus forming the active 1,25-$(OH)_2D_3$ metabolite [106]. Following the cDNA cloning of CYP27, the enzyme was found to be expressed in both liver and kidney [107]. These findings have promoted speculation that CYP27 might be the activation enzyme of the vitamin D_3 system because a mitochondrial 1α-OHase has not been purified and cloned [108], but there are concrete reasons for opposing this theory. Firstly, CYP27 is predominantly a cholesterol enzyme and has very low specificity for vitamin D_3 (Km, 10^{-6}) [53, 109] and even far less specificity for 25-OH-D_3 [109]. Therefore, the kinetics would alone prohibit the *in vivo* formation of 1,25-$(OH)_2D_3$ by this enzyme. Secondly, there is not sufficient evidence that renal CYP27 is regulated in conjunction with vitamin D_3 or calcium status, because it is still expressed in the vitamin D-deficient rat kidney [107]. Furthermore, the metabolizing action of CYP27 on the vitamin D_3 molecule occurs only when the level of vitamin D_3 ingestion is excessive and beyond the capacity of the microsomal 25-OHase.

Excessive ingestion of vitamin D_3 leads to toxicity by hypercalcaemia. Under conditions of vitamin D_3 toxicity, 25-OH-D_3 is probably the causative agent, because unlike 1,25-$(OH)_2D_3$, the production of 25-OH-D_3 is generally unregulated [110]. High concentrations of vitamin D_3 inhibit microsomal 25-OHase [53, 111], but this regulation does not affect CYP27 and the 25-hydroxylation function of CYP27 allows for a much greater production of 25-OH-D_3 than microsomal 25-OHase. A 50- to 100-fold increase in dietary vitamin D_3 in rats results in a 10- to 20-fold increase in blood 25-OH-D_3, from 25 ng/ml to 300–500 ng/ml [112]. This increase in 25-OH-D_3 is in the face of reduced 1,25-$(OH)_2D_3$ production [113]; yet, it leads to increased VDR concentrations in kidney and intestine [113, 114]. It is well known that 25-OH-D_3 competitively binds VDR at 10^{-6} M [115], and 25-OH-D_3 acts on bone at high concentrations [116, 117]. Furthermore, vitamin D toxicity has been demonstrated in an anephric child, in which 1α-hydroxyl activation of 25-OH-D was prohibited [118].

The fact that vitamin D_3 toxicity results from primarily uncontrolled intestinal calcium absorption suggests that it is dietary calcium and not vitamin D_3 that exacerbates the hypervitaminosis D_3 toxicity effect [119]. This was tested by the interaction of excess vitamin D_3 and calcium restriction [113]. Rats fed a calcium-deficient diet and given 25,000 IU of vitamin D_3 three times/week for 2.5 weeks did not succumb to overt hypervitaminosis D_3. Simple calcium restriction increased intestinal but not renal 24-OHase activity, presumably because of the absence of parathyroid hormone regulation in the intestine [113]. Coupled with vitamin D_3, excess intestinal 24-OHase increased several fold more. However, when dietary calcium was adequate, vitamin D_3 excess increased intestinal 24-OHase activity only slightly because of a suppressive mechanism regulated in part by increased blood calcitonin [120].

Another potential regulator of the 25-OHase system is 1,25-$(OH)_2D_3$; however, this has not been firmly established. Studies do show that blood 25-OH-D_3 concentrations rise and fall in accordance with a variety of physiological stresses that influence the 1α-OHase system [121]. Low blood 1,25-$(OH)_2D_3$ and high 25-OH-D_3 concentrations are found in postmenopausal osteoporosis. Long-term treatment with 1,25-(OH)$_2$D$_3$ corrected this imbalance [122]. Calcium restriction, in otherwise normal vitamin D rats, leads to a profound decrease in serum 25-OH-D$_3$ [123] and 24,25-(OH)$_2$D$_3$ [113, 124] concentrations. Because 25-OH-D$_3$ clearance is not significantly increased by calcium restriction, it is probable that the endogenous increase in blood 1,25-(OH)$_2$D$_3$ is responsible for slowing 25-OH-D$_3$ production. The opposite occurs with exogenous administration of 1,25-(OH)$_2$D$_3$, which decreases 25-OH-D$_3$ via metabolic clearance [125]. Vitamin D-deficiency leads to increased 25-OHase activity in liver homogenates, but small doses of vitamin D$_3$, and not 25-OH-D$_3$ or 1,25-(OH)$_2$D$_3$, given to vitamin D-deficient rats, reduces this effect markedly [126]. Other studies have shown that a rise in intracellular calcium, mediated by 1,25-(OH)$_2$D$_3$, signals down-regulation of 25-OH-D$_3$ in hepatic cells [127, 128]. This effect was also produced by increasing intracellular calcium with the use of calcium ionophores and blocked in the presence of the calcium chelator, EGTA [127]. Despite these varied results, reliable *in vivo* data showing the direct mechanistic effect of 1,25-(OH)$_2$D$_3$ on hepatic 25-OHase are still lacking.

Developmentally, microsomal 25-OHase activity is low 3 days prior to birth in rat foetuses [129]. After birth, 25-OH-D$_3$ production steadily increases for the first two weeks of neonatal life and then increases six-fold more in the third week. Maternal 25-OH-D$_3$ production increases slightly the day of parturition and 22 days postpartum. Therefore, the microsomal 25-OHase appears to be fully activated in conjunction with the time of weaning.

RENAL 24- AND 1α-HYDROXYLASES: RECIPROCAL REGULATION

In the kidney, 25-OH-D$_3$ enters either a pathway of activation by 1α-hydroxylation or catabolism by 24-hydroxylation (*Figure 1.5*). The regulation of these two pathways are under reciprocal control of several integrated calcitropic factors and hormones. The two key enzymes central to this regulation checkpoint are 1α-OHase and 24-OHase. Both are mitochondrial cytochrome P450 mono-oxygenases that require NADPH, flavoprotein (ferredoxin reductase) and iron-sulphur protein (ferredoxin) as essential cofactors [74]. Because of their similarities, it has been suggested that the renal 1α- and 24-hydroxylases are closely linked or even the same enzyme [130]. On the contrary, however, the enzymes are quite distinct [75, 131]. The two factors that link these renal enzymes are parathyroid hormone (PTH) and 1,25-(OH)$_2$D$_3$. The role of PTH in the reciprocal regulation of 25-OH-D$_3$ metabolism was first noted by Garabedian *et al.* [132]. PTH can also be influenced by manipulating dietary calcium concentration [133]. At high percent-

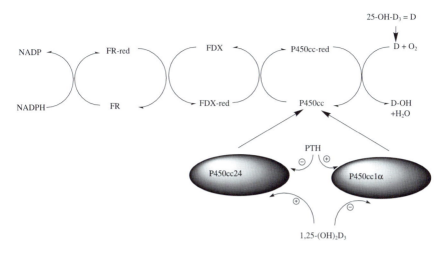

Figure 1.5. Reciprocal regulation of 1α-OHase and 24-OHase by parathyroid hormone and 1,25-(OH)₂D₃ and organization of the mitochondrial P-450 complex.

ages of dietary calcium, parathyroid hormone secretion is at a minimum, thus 1α-OHase is not stimulated and 24-OHase is activated [72, 134]. This situation reverses as blood calcium dips below the typical set-point value of 10 mg/dl [134]. Blood calcium is the trigger that sets off a sensing mechanism in parathyroid glands for either stimulation or suppression of PTH secretion [101, 102]. Receptors for PTH, located in renal proximal tubular cells (PCT) receives the message of hypocalcaemia and extends the signal in two ways; first by causing potent down-regulation of 24-OHase, and second by stimulating 1α-OHase [135]. Several reports demonstrate quite conclusively that c-AMP acts as the second messenger in this reciprocal regulation [75, 134, 136], but the actual molecular mechanism is still the focus of investigation.

The half-life of 1,25-(OH)₂D₃ in humans is 2–6 h, whereas the half-life of 24,25-(OH)₂D₃ is several days [60]. Newly synthesized 1,25-(OH)₂D₃ then localizes in its target sites and elicits its three major actions aimed at reaching a balance of blood calcium and phosphate to meet the integrated requirements for normal bone mineralization and neuromuscular function. As the calcium demand is normalized, the parathyroid gland calcium-sensing trigger is reversed and PTH secretion is slowed [101]. In renal PCT, the PTH suppression on 24-OHase is lifted and these cells revert back to a target-type cell of 1,25-(OH)₂D₃. In this state, VDR increases and potently activates the transcription of PCT genes, among which is the 24-OHase [68, 75,

137]. The increase in 24-OHase by 1,25-$(OH)_2D_3$ is further sensitized by phorbol esters [138], indicating the involvement of protein kinase C [139, 140]. However, it is unclear whether protein kinase C functions in conjunction with 1,25-$(OH)_2D_3$ by a modification of the ferredoxin subunit [139] or by a direct modification of the expression of the 24-OHase cytochrome P450 [138, 140]. In addition, 1,25-$(OH)_2D_3$ serves a negative feedback regulation on the 1α-OHase system [72]. The molecular details of this regulation are completely unknown to date.

The difficulty in studying the reciprocal phenomenon of renal vitamin D hydroxylases has been homing-in on the specific cell type of 1,25-$(OH)_2D_3$ production, and secondly, developing a system in which one can examine all of the calcitropic signals independently. Kawashima *et al.* [141] performed microdissection of rat nephrons and used defined nephron segments in the measurement of [^{3}H]-24,25-$(OH)_2D_3$ or [^{3}H]-1,25-$(OH)_2D_3$ production. Both 24,25-$(OH)_2D_3$ and 1,25-$(OH)_2D_3$ are produced in proximal tubules. Production of 1,25-$(OH)_2D_3$ was detected only in PCT of vitamin D-deficient rats, and production of 24,25-$(OH)_2D_3$ was detected in either normal or 1,25-$(OH)_2D_3$-treated rats but not in vitamin D-deficient rats. Thyroparathyroidectomy of vitamin D-deficient rats abolishes 1α-OHase activity [132], and administration of PTH restores diminished 1α-OHase activity [135]. There is a time lag in bringing about both the stimulation and suppression of 1α-OHase that is independent of the rapid changes in serum calcium and phosphate in response to these manipulations [72]. Because of this, it is likely that the mechanisms involve regulation at the level of transcription or protein synthesis.

An intriguing aspect unique to renal physiology in regard to the vitamin D endocrine system is VDR regulation. A supraphysiological dose of 1,25-$(OH)_2D_3$ leads to homologous up-regulation of VDR protein in many target tissues [142] due to ligand-induced stabilization of the VDR [143]. This is in contrast to the case *in vivo* where endogenous production of 1,25-$(OH)_2D_3$, stimulated by nutritional hyperparathyroidism, occurs in the face of renal VDR down-regulation [144]. Hypocalcaemia results in a marked downregulation of VDR [113, 144,145]. Recently, Iida *et al.* [137] developed a polymerase chain reaction- (PCR) based technique for quantifying gene expression from defined microdissections of the rat nephron. With this technique, they demonstrated (using models of enhanced 1,25-$(OH)_2D_3$ production) that VDR down-regulation in PCT is permissive of increased 1α-OHase activity. In these same rat models, expression of VDR and 24-OHase genes in cortical collecting ducts was not suppressed. The working hypothesis proposed from this work is that down-regulation of VDR plays a critical role in production of 1,25-$(OH)_2D_3$ in renal proximal tubules. Because this regulation is unique to PCT, it is the PCT cells that can be considered the endocrine cells of the vitamin D system.

What causes the VDR down-regulation in PCT is still the subject of great curiosity. Reinhardt and Horst [146] obtained evidence that PTH mediates this regulation, while other work, including that of this laboratory, has suggested that calcium is required for normal VDR regulation in kidney [145, 147]. The same may also be true for the intestine [144]. Clearly, the down-regulation of renal VDR is in response to an endocrine signal because of the specific nature of this effect and its occurrence only in PCT where the synthesis of $1,25\text{-}(OH)_2D_3$ takes place [137]. The decrease in renal VDR but not intestinal VDR by sex steroids [137], which also are known to regulate $1,25\text{-}(OH)_2D_3$ production, is further evidence that renal VDR down-regulation is under endocrine control. The challenge now will be to address the molecular mechanism by which reciprocal regulation of vitamin D hydroxylases in response to PTH occurs. This must be developed in a system of proximal kidney tubules responsive to PTH, which has been a practical problem in most renal cell culture lines that lose PTH receptors as the cells are immortalized.

OTHER FACTORS REGULATING 1α-OHASE

Hypophosphataemia is also an important regulator of 1α-OHase [148, 149]. The major difference in hypophosphataemic and hypocalcaemic induction of 1α-OHase is that PTH secretion is not stimulated in hypophosphataemia [150]. In thyroparathyroidectomized rats that are otherwise normal, no $1,25\text{-}(OH)_2D_3$ production can occur and the renal 24-hydroxylation pathway is activated [148]. Lowering the dietary phosphate in this circumstance leads to increased 1α-OHase and decreased 24-OHase [148, 151]. The increased $1,25\text{-}(OH)_2D_3$ produced by phosphate restriction results in increased utilization of phosphate from bone to restore blood phosphate. It also increases phosphate absorption from the intestine when available. Thus, $1,25\text{-}(OH)_2D_3$ is an important regulator of phosphate metabolism as well as calcium. An examination of nutritional hypophosphataemia and VDR regulation in the intestine and kidney by Sriussadaporn *et al.* [152] revealed a tissue-specific and time-dependent decrease in renal VDR mRNA starting at day 3 and continuing until day 10, which was commensurate with the rise in serum $1,25\text{-}(OH)_2D_3$ concentration. Contrasting this is intestinal VDR mRNA that rises to its highest level on day 2 of the low phosphate diet and then declines to control levels by day 5. However, it should be noted that the mechanism by which low phosphate regulates 1α-OHase is unknown. One proposed mechanism is by a growth hormone-mediated pathway [153, 154].

The negative feedback regulation by 1,25-(OH)$_2$D$_3$ is another form of major regulation imposed on 1α-OHase. This regulation is observed when dietary calcium and phosphate are adequate and 1,25-(OH)$_2$D$_3$ is given exogenously. At the same time, 1,25-(OH)$_2$D$_3$ induces 24-OHase. Both of these effects are nuclear mediated because they can be blocked in the presence of actinomycin D [155, 156]. It is also of some interest that exogenous 1,25-(OH)$_2$D$_3$ acts as a stimulus of 1α-OHase induction in calcium-restricted rats as long as they remain hypocalcaemic [157]. Because the production of 1,25-(OH)$_2$D$_3$ in these rats can far exceed production capacity ordinarily measured by calcium restriction alone, there must be a role for 1,25-(OH)$_2$D$_3$ in combination with PTH and cAMP to induce 1α-OHase further.

Other hormones have been reported over the years to control more minor aspects of 1α-OHase regulation. Included in this list are calcitonin [158, 159], insulin [160], prolactin [161], growth hormone and the sex steroid oestrogen [162, 163]. The physiological relevance of many of these hormones and 1α-OHase regulation is circumspect, but clearly, the absence of oestrogen in the post menopausal state leads to the release of calcium and phosphate from bone that raises blood calcium and suppresses PTH secretion, and hence, 1α-OHase activity. Administration of oestrogen corrects this symptom partly by stimulating 1α-OHase again [164, 165].

Certain physiological stresses also play a role in the function of 1α-OHase. During growth, pregnancy and lactation, the body's requirement for calcium increases. In response, the circulating concentration of 1,25-(OH)$_2$D$_3$ increases to facilitate increased intestinal calcium absorption. During pregnancy and lactation, the increase in plasma 1,25-(OH)$_2$D$_3$ may be related to an oestrogen-mediated increase in liver DBP production and to effects of prolactin on 1α-OHase [166, 167].

Because the kidney is the only significant source of 1α-OHase, inadequate formation of 1,25-(OH)$_2$D$_3$ occurs in renal failure [168]. Not only is the mass of kidney tissue and therefore of enzyme decreased, but also with renal failure, phosphate excretion is reduced and serum phosphate rises. Increased phosphate inhibits 1α-OHase so that little 1,25-(OH)$_2$D$_3$ is formed. Acidosis, a frequent result of renal failure, also impairs 1α-OHase activity [169, 170]. Deficiency of the active form of vitamin D causes osteomalacia, a prominent feature of renal osteodystrophy. Therapy is directed toward use of 1,25-(OH)$_2$D$_3$, reduction of serum phosphate, and correction of acidosis, so that residual 1α-OHase can be expressed.

A rare disease associated with renal 1α-OHase is vitamin D-dependency rickets type I (VDDR-I), which is an autosomal recessive defect of the renal

1α-OHase which impairs the synthesis of 1,25-(OH)$_2$D$_3$. The molecular details of this defect await the cloning of the 1α-OHase, but the disease locus has been mapped to human chromosome 12q14–13 [171], which is near the chromosomal location for human VDR also located on chromosome 12 [172].

BIOLOGICAL EFFECTS OF 1,25-(OH)$_2$D$_3$

The intestine is a principal target of 1,25-(OH)$_2$D$_3$, although the hormone also affects bone and to a lesser extent kidney. Like other steroid hormones, 1,25-(OH)$_2$D$_3$ binds with high affinity (10^{-10} to 10^{-11} M) and specificity to its nuclear receptor, VDR, which in turn functions to modulate gene expression. The net effect of 1,25-(OH)$_2$D$_3$ is to increase intestinal absorption of calcium. At a normal calcium intake of 1000 mg about 15% of calcium absorption occurs by passive diffusion. The amount of calcium absorbed via this mechanism is not sufficient to sustain calcium balance, even when the intake of calcium is significantly increased. The major component of calcium absorption occurs by active transport, a process that is activated solely by 1,25-(OH)$_2$D$_3$ principally in the duodenum and jejunum [173]. The low intracellular concentration of calcium (μM) relative to extracellular calcium (mM) means that calcium entering the cell must be sequestered as it is transported from the mucosal surface to the luminal surface where it is extruded out of the cell into the extracellular compartment. Calcium is believed to be sequestered by a cytosolic calcium-binding protein (CaBP). The CaBP was shown by Wasserman *et al.* [174, 175] to be up-regulated by 1,25-(OH)$_2$D$_3$. This protein, also known as calbindin, has a molecular weight of 28,000 daltons (CaBP-28k) in the avian species and 8,000–11,000 daltons (CaBP-9k) in mammalian species [176]. There are four specific calcium-binding sites (10^{-6} M) for each molecule of CaBP-28K and two binding sites for CaBP-9K [177, 178]. The suggested role of these proteins is to set up an intracellular gradient that shuttles calcium across the enterocyte [179, 180]. In addition, 1,25-(OH)$_2$D$_3$ stimulates alkaline phosphatase [181, 182] and a calcium-dependent ATPase [179, 180]. The manner in which all of these components come together to increase calcium uptake, transcellular transport and extrusion of calcium is still a matter of investigation, and other processes, induced by 1,25-(OH)$_2$D$_3$, are likely to occur as well. 1,25-(OH)$_2$D$_3$ also increases intestinal absorption of phosphate by an active transport process that is less tightly regulated than that of calcium [183, 184].

The second major target for 1,25-(OH)$_2$D$_3$ is bone. The most firmly established role of 1,25-(OH)$_2$D$_3$ in bone is the mobilization of calcium to support serum calcium levels, first described by Carlsson [19]. This process requires both PTH and 1,25-(OH)$_2$D$_3$ [185]. Whether it is mediated by osteoclasts is unknown, and it may be a mechanism involving calcium movement by the osteoblasts and bone-lining cells [186]. Overall, 1,25-(OH)$_2$D$_3$ stimulates bone mineralization.

Mineralization of bone is a complex process involving both simple and complex molecules in addition to calcium and phosphate. The collagen matrix at the growing end of long bones is synthesized and secreted by chondrocytes. Influx of calcium and phosphate, as well as other substances to the extracellular matrix, is regulated via these cells. Mineralization takes place in the extracellular matrix, and crystal growth then fills the matrix with mineral. Blood vessels invade the calcified cartilage matrix and this is resorbed and remodeled into trabecular bone. Bone remodeling is a process which continues throughout life, long after epiphyseal fusion and cessation of linear growth [187]. Bone remodeling, which consists of osteoclastic-mediated bone resorption followed by osteoblastic-mediated bone formation, occurs continually, and about one-fifth of bone is replaced each year.

The primary role 1,25-(OH)$_2$D$_3$ plays with respect to bone formation is to provide supersaturating levels of calcium and phosphate to form new matrix and induce ossification [188–191]. Osteoblasts are the primary cells concerned with synthesis of new bone. These cells, which cover bone-forming surfaces, produce osteoid which will subsequently undergo calcification. In addition, osteoblasts are the bone's target cells of 1,25-(OH)$_2$D$_3$ action [192], and 1,25-(OH)$_2$D$_3$ up-regulates several osteoblast-related genes including osteocalcin (bone-specific *gla*-protein) [193], osteopontin [194, 195] and regulates αI-procollagen at the level of transcription [196]. Osteocalcin is expressed only in the osteoblast cell-type in response to 1,25-(OH)$_2$D$_3$ in a dose-dependent manner [197]. The function of osteocalcin is unknown except that it is thought to be involved in the recruitment of osteoclasts. Osteocalcin is a 49-residue γ-carboxyglutamic acid containing peptide that accounts for 10–20% of the soluble protein during the mineralization of bone [197]. In addition, osteocalcin is excreted in urine so it is commonly used as a marker of bone resorptive activity. Osteocytes are mature bone cells which are less active than osteoblasts. Osteoclasts are multinucleated cells derived from macrophages which function to resorb bone. 1,25-(OH)$_2$D$_3$ is a potent stimulus of osteclastic bone resorption. However, because osteoclasts lack VDR, 1,25-(OH)$_2$D$_3$-mediated effects on osteoclastic resorption must be through an indirect activation pathway(s). A primary action of 1,25-(OH)$_2$D$_3$ in bone is to induce the differentiation of promyelocytes to monocytes and further to form osteoclasts from the promyelocytes [198, 199]. In conjunction with PTH, 1,25-(OH)$_2$D$_3$ stimulates osteoclastic progenitors to fuse to form multinucleate osteoclast cells [200]. 1,25-(OH)$_2$D$_3$ also acti-

Table 1.1.   1,25-$(OH)_2D_3$ TARGET CELLS

Proven	Putative
Intestinal enterocyte	Pancreatic islet cell (β-cell)
Bone (osteoblasts / chondroblasts)	Stomach
Distal renal cells	Pituitary cells (somatomammotroph)
Parathyroid gland (chief cells)	Placenta
Skin ketatinocytes	Epididymis
Bone marrow (promyelocytes, monocytes)	Brain (hypothalamus)
Lymphocytes	Myoblasts (developing)
Colon enterocytes	Mammary epithelium
Avian shell gland	Aortic endothelial cells
Chick chorioallantoic membrane	Skin fibroblasts
Thymus (reticular / T lymphocytes)	Ovarian cells
Thyroid (C-cell)	Testis (sertoli / seminiferous cells)
Breast (epithelial cells)	Muscle (myoblast)
Parotid (acinar)	Uterus

vates mature preformed osteoclasts, possibly via osteoblast-derived resorption factors. It also is a potent immunomodulatory molecule in that 1,25-$(OH)_2D_3$ suppresses T-cell proliferation and IL-2 production from cells with monocytic characteristics [201].

In the kidney, 1,25-$(OH)_2D_3$ facilitates calcium reabsorption in the distal nephron, an effect that is potentiated by PTH [202]. Also, there are several tissues that translocate calcium, and for which, the presence of functional VDR is well documented, including skin, mammary gland, placenta, and the avian shell gland. A current list of 1,25-$(OH)_2D_3$ target sites is provided in *Table 1.1*. Although the precise role of 1,25-$(OH)_2D_3$ in many of these 'nonclassical' tissues is uncertain, effects on calcium and phosphate transport and feedback effects on overall calcium homeostasis are likely. In addition, 1,25-$(OH)_2D_3$ plays a specific role in cellular differentiation [203, 204]. Cells of the epidermis and cells of the immune system have VDR [66, 205], alluding to as yet unidentified functions in these systems. A paracrine function for 1,25-$(OH)_2D_3$ has been suggested from *in vitro* experiments with macrophages but whether this occurs *in vivo* is unknown [206].

EXTRARENAL METABOLISM OF 1,25-DIHYDROXYVITAMIN D_3

A great deal of effort has been expended to examine vitamin D_3 metabolism as it relates to active metabolites. The kidney, being the site of 1,25-$(OH)_2D_3$ and 24,25-$(OH)_2D_3$ synthesis, is one of the most extensively studied organs. The inducibility of 24-OHase by 1,25-$(OH)_2D_3$ and the natural formation of 1,24,25-$(OH)_3D_3$ in intestinal homogenates [207] and other ex-trarenal tissues [208] encourages the idea that the 24-OHase serves as a mod-ulator of biological potency of 1,25-$(OH)_2D_3$ in its target sites [73]. It is now known that the 24-OHase gene is one of the most strongly induced of all 1,25-$(OH)_2D_3$-responsive genes [209, 210]. In addition, it has become clear that the tissue distribution of 24-OHase extends to nearly every tissue where 1,25-$(OH)_2D_3$ localizes and has VDR-mediated biological actions [211]. However, in studies using intestine and bone tissues, PTH does not ap-pear to affect 24-OHase regulation as it does in kidney [75, 212]. This is not surprising in the intestine which lacks PTH responsiveness, but it is in bone cells that show an increase in cAMP production in response to PTH without a reduction in 24-hydroxylase [212].

Purification of rat kidney 24-OHase was made possible because of the great inducibility of the enzyme by 1,25-$(OH)_2D_3$. Classification of 24-OHase as a cytochrome P450 enzyme was confirmed by the carbon monox-ide difference spectrum and UV spectrophotometry at 450 nm of the purified protein [213]. In addition, the enzyme could be reconstituted in the presence of cofactors obtained from bovine adrenal glands; adrenodoxin, adrenodox-in reductase and NADPH [214]. Following this, antibodies to rat 24-OHase were produced and used to clone the cDNA, providing the primary structure of 24-OHase [215]. The human 24-OHase was cloned by homology and expressed [216]. The chromosomal location of human 24-OHase is 20q13.2→q13.3 [214], and from all that is presently known of 24-OHase, it exists as only one form with wide tissue distribution and tissue-specific regu-lation.

The promoter region of the 24-OHase gene is unique among 1,25-$(OH)_2D_3$-responsive genes in that it consists of multiple response elements that recognize VDR [209, 210, 217, 218]. The rat 24-OHase promoter has been the more studied of the 24-OHase genes (*Figure 1.6*). Zierold *et al.* first cloned a 1,25-$(OH)_2D_3$-responsive element (VDRE) at −262/−238 [209]. Ohyama *et al.* [210] found a VDRE at −151/−137. This response element is comprised of two direct repeat half-sites, AGGTGAgtgAGGGCG that bind VDR and a nuclear accessory factor from COS cells. Zierold *et al.* [209, 218] further characterized the rat 5′-flanking sequence of 24-OHase as having two functional VDREs at bases −154/−125 and −262/−238, respec-tively. The proximal VDRE (−154/−125) consists of three half-sites each separated by three base pairs and was similar to the VDRE discovered by Ohyama *et al.* [210]. The more distal VDRE (−262/−238) consists of two half-sites in direct repeat orientation. The intervening 93 base pair sequence is not important for full functionality of the 24-OHase promoter, but re-moval of one of the five half-site compromises transcriptional activation as

25-OH-Vitamin D3-24-hydroxylase gene structure

Rat gene: 15kb, 12 exons

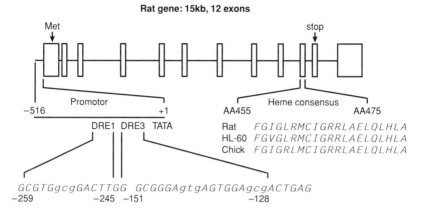

Figure 1.6. Organization of the rat 24-OHase gene.

determined by a reporter construct [218]. Similar results have likewise been obtained for the promoter regions of human [219, 220] and chicken 24-OHase genes (Jehan and DeLuca, unpublished results).

Studies of the kinetics of 24-OHase show that the enzyme preferentially 24-hydroxylates $1,25\text{-}(OH)_2D_3$ (Km, 10^{-8}) versus $25\text{-}OH\text{-}D_3$ (Km, 10^{-7}) [221]. Low affinity of $1,25\text{-}(OH)_2D_3$ for plasma DBP and high affinity for nuclear VDR makes free and not DBP-bound $1,25\text{-}(OH)_2D_3$ more readily available to its target cells [222]. Therefore, in vitamin D_3 target cells, $1,25\text{-}(OH)_2D_3$ is the likely natural metabolite for 24-hydroxylation, whereas in PCT of kidney, the conversion of $25\text{-}OH\text{-}D_3$ to $24,25\text{-}(OH)_2D_3$ predominates. This is also supported by nephrectomy studies that demonstrate the contribution of kidney in the production of $24,25\text{-}(OH)_2D_3$ [223, 224]. So one might envisage extrarenal 24-hydroxylation as serving a modulating role for $1,25\text{-}(OH)_2D_3$ action. Ironically, the sites of richest 24-OHase inducibility are intestine, bone and kidney – the so-called classical target sites of vitamin D_3 because of their abundance of VDR and involvement in calcium homeostasis.

Recently the rat 24-OHase cDNA was recombinantly overexpressed in a bacterial cell line and reconstituted with its required cofactors [225]. Cell fractionation revealed the 24-OHase to be primarily localized in membranes, with minimum inclusion body formation, and that upon reconstitution with adrenodoxin, adrenodoxin reductase, and NADPH, 24-OHase activity was apparent. Interestingly, this preparation produced three products with each substrate analyzed by reverse phase HPLC. These products were identified as 24-hydroxy-, 24-oxo-, and 24-oxo-23-hydroxy-modified side-chains, all members of the C-24/C-23 catabolic pathways for vitamin D. In a similar study [226] using recombinant human 24-OHase expressed in *Spodoptera frugiperda* (*Sf* 21) insect cells, several further oxidized products of 25-OH-D$_3$ were isolated, purified and characterized as: 24,25-(OH)$_2$D$_3$; 23,25-(OH)$_2$D$_3$: 24-oxo-25-OH-D$_3$; 24-oxo-23,25-(OH)$_2$D$_3$; and 24,25,26,27-tetranor-23-OH-D$_3$. The latter product is the immediate precursor of 24,25,26,27-tetranor-23-COOH-D$_3$, which is one of the final excretion products of vitamin D$_3$ metabolism. This system differed from the previous bacterial system in that Sf21 cells are eukaryotic, and 24-OHase activity was reconstituted and restored from solubilized mitochondrial proteins. In the case of wild-type infected Sf21 cells, no metabolism of 25-OH-D$_3$ took place. Therefore, it would appear that the high inducibility of 24-OHase by 1,25-(OH)$_2$D$_3$, its wide-spread tissue distribution, and its apparent multi-catalytic activity are strong evidence that extrarenal 24-OHase regulates the cellular half-life of 1,25-(OH)$_2$D$_3$, as well as converting it to at least one of its final excretion products.

MOLECULAR MECHANISM OF 1,25-DIHYDROXYVITAMIN D$_3$ ACTION

THE 1,25-(OH)$_2$D$_3$-RECEPTOR (VDR)

The mechanism of 1,25-(OH)$_2$D$_3$ action was not well understood until it was learned that it localizes in a specific manner in the nuclei of various target site cells [208]. This supported the theory that 1,25-(OH)$_2$D$_3$ participates as a hormone analogous to other steroid hormones like oestrogen and gluco-corticoids [227]. The first breakthrough came with the clear demonstration of a 1,25-(OH)$_2$D$_3$ receptor in intestine [115, 228]. The chick protein was used to provide the first reproducible and accurate radioreceptor assay for 1,25-(OH)$_2$D$_3$ in serum [229]. Eventually, the soluble factor, termed 1,25-(OH)$_2$D$_3$-receptor (VDR), was purified from chicken intestine [230]. Limited proteolysis of intact VDR demonstrated the existence of at least two distinct functional domains, an N-terminal DNA-binding domain and a C-terminal ligand-binding domain [231, 232]. These findings provided further evidence of the steroid hormone receptor-like nature of VDR. The use of sulphydryl-modifying reagents demonstrated the importance of divalent cations and cysteines for the structural and functional integrity of VDR [208]. Finally, the development of monoclonal antibodies [233, 234] to intestinal VDR provided the tools for isolation of a partial length avian VDR cDNA [235]. Following this, the full length VDR cDNAs for human and rat species were like-wise obtained [236, 237]. The full length human VDR of 4605 bp was consistent with the mRNA transcript size of 4.6 Kb [236]. The open reading frame consisted of 1,281 bp, encoding for 427 amino acids and yielding a protein of 48,000 daltons. In addition, there is a 115 bp leader sequence leaving 3,209 bp of 3' noncoding sequence [236]. Sequence alignment of the human, rat, and avian VDR cDNAs demonstrates a region in the N-terminal DNA-binding domain with 95% identity and another region in the C-terminal ligand binding domain with 93% identity [237]. Even more striking than this, however, is the amount of sequence identity the DNA-binding domain of VDR shares with other members of the steroid hormone receptor superfamily [227]. Furthermore, it was shown that transient expression of the human VDR in CV-1 green monkey cells conferred $1,25\text{-}(OH)_2D_3$-responsiveness thus confirming the essential role of VDR in the transcriptional effects of $1,25\text{-}(OH)_2D_3$ [238]. These results unequivocally classified VDR as another member of the steroid hormone/thyroid receptor superfamily [236, 237]. The biological significance of a functional VDR is demonstrated by the disease, vitamin D-dependency rickets type II, wherein mutations in the ligand or DNA binding domains cause $1,25\text{-}(OH)_2D_3$ resistance [239, 240].

STRUCTURE AND FUNCTION OF VDR

Originally, the domain classification scheme for VDR consisted of three domains designated as C1, C2 and C3. The C1 domain was comprised of a 70 amino acid portion of the N-terminus of VDR rich in cysteine, lysine and arginine residues. The C2 and C3 domains represented small regions in the C-terminal domain of VDR that contain the ligand-binding domain (C3) and other regions of high similarity for potential protein-protein interactions (C2). The C2 domain also is a site of high antigenicity [238] and is often referred to as the hinge region bridging the DNA-binding (DBD) and ligand-binding (LBD) domains (*Figure 1.7*). These domains were more finely mapped by the use of deletion mutational analysis of VDR cDNA at both its 5' and 3' ends [238].

The LBD has been defined as residues 114–427 at the C-terminal end of VDR [241]. Residues 403–427 are apparently important for maximal ligand affinity, whereas the other 200–300 bases are involved in ligand recognition [241]. Forman and Samuels [242] performed a sequence comparison of the extreme most portion of the C-terminal domain for several steroid receptors and found that this sequence is unique to each receptor, which is indicative

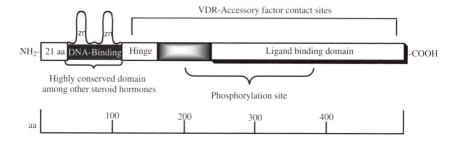

Figure 1.7. Structural and functional domains of the VDR.

of the preference each receptor has for its specific ligand. There are also several intradomain tandem repeats of a common structural motif known as heptad repeats [242]. They consist of leucine residues at positions 1 and 8, and have a hydrophobic amino acid or a charged amino acid with a hydrophobic side-chain at position 5 in the repeat. These classic heptad repeats are common among transcription factors that form protein-protein contacts through dimerization interfaces, similar to leucine zipper and helix-loop-helix motifs. In all, there are nine heptad repeats in the LBD of VDR and at least heptads 4 and 9 are qualified as authentic sites of protein-protein contacts [242].

The mapping of DBD confines this domain to the first 113 amino acids by C-terminal deletion mutational analysis [238]. In addition, Sone *et al.* [243] have shown that deletion of the first 21 amino acids from the N-terminal end is not deleterious to DNA-binding, delineating the minimum domain for DNA-binding between residues 22–113. Within the DBD, there are 9 cysteine residues that are absolutely conserved among all steroid hormone/thyroid receptors. Eight of the cysteines are tetrahedrally co-ordinated about two zinc atoms forming the zinc fingers that comprise the functional portion of the DBD which interacts with response elements of 1,25-(OH)₂D₃-responsive genes [239]. Site-directed mutagenesis of each respective cysteine showed that mutations at any of the first eight cysteines disrupted the structural motif of the zinc finger, impaired VDR interaction with DNA, and decreased transcriptional activation of the human osteocalcin promoter in response to 1,25-(OH)₂D₃ [243]. Mutagenesis of the ninth cysteine residue was not detrimental to the functionality of the DBD, delineating the first eight cysteine residues as the precise residues that comprise DBD [243].

Delineation of the hinge region between the LBD and the DBD was determined to begin at Met 90 and extend to the region between Glu 114-His

166. Internal deletion of amino acids 103–114 abolishes VDR/DNA interaction and, therefore, appears to be of vital importance [243]. This site overlaps with the epitope for 9A7γ (anti-VDR) monoclonal antibody [238]. The 9A7γ epitope is highly conserved among human, rat, and chicken species, extending from amino acids 89–105, which probably explains interspecies cross-reactivity of 9A7γ. The epitope is located immediately adjacent to the second α-helix situated at the base of the more C-terminal zinc finger motif. For the retinoid X receptor-alpha (RXRα) and retinoid X receptor-beta (RXRβ) homologues, the corresponding region to amino acids 103–114 of VDR has been implicated in RXR homodimer formation [244], and for VDR, the possibility exists that this domain is likewise a protein contact site.

MOLECULAR BIOLOGY OF VDR FUNCTION 1,25-(OH)$_2$D$_3$ (RESPONSE ELEMENTS)

The basic theme concerning the mechanism of action of steroid/thyroid hormone receptors is that they mediate the biological actions of the cognate hormone by regulating the rate of gene transcription within their respective target cells. The genomic actions of 1,25-(OH)$_2$D$_3$ are ligand-induced and the result of the ligand/receptor complex interacting with specific DNA sequences or response elements (VDRE) within the promoters of 1,25-(OH)$_2$D$_3$-responsive genes. Several 1,25-(OH)$_2$D$_3$-responsive genes have now been cloned and promoter regions of these genes have been characterized (*Table 1.2*). It is too early to decipher a consensus response element se-

Table 1.2   SEQUENCE COMPARISON OF THE VITAMIN D DESPONSIVE ELEMENTS (VDRE)

Promotor	VDRE				
human PTH			GGTTCA	AAG	CAGACA
rat calbindin D-9K			GGGTGT	CGG	AAGCCC
rat osteocalcin			GGGTGA	ATG	AGGACA
human osteocalcin			GGGTGA	ACG	GGGGCA
mouse osteopontin			GGTTCA	CGA	GGTTCA
rat 24-hydroxylase distal			GGTTCA	GCG	GGTGCG
human 24-hydroxylas distal			AGTTCA	CCG	GGTGTG
rat 24-hydroxylase proximal	GAGTCA	GCG	AGGTGA	GTG	AGGGCG
human 24-hydroxylase proximal	GAGTCA	GCG	AGGTGA	GCG	AGGGCT
chicken carbonic anhydrase-II			GGGGGA	AAA	AGTCCA
mouse calbindin D-28K			AGGTGA	TGA	AAGTCA
human calbindin D-9K			TGCCCTTCCTTATGGGGTTCA		
mouse calbindin D-28K		CTGGGGGGATGTGAGGAGAAATGAGTCTGAGC			
mouse *c-fos*		AGGTGAAAGATGTATGCCAAGACGGGGGTTGAAAG			

quence because not all of the $1,25\text{-}(OH)_2D_3$-regulated genes are available for analysis. But unlike the response elements of oestrogen and glucocorticoid receptors, which are palindromic or inverted repeats of two six-base half-sites, VDREs appear to comprise at least two imperfect half-sites in direct repeat with three bases of spacing between them [67, 245, 246]. This is analogous to retinoic acid receptor response elements (RARE) and thyroid hormone receptor response elements (TRE) [247].

Although among the various known VDREs, there is some flexibility in precise sequence mediating $1,25\text{-}(OH)_2D_3$-responsiveness, specific changes in the bases of the half-sites can dramatically affect VDR-VDRE interactions [245]. In addition, mutations in the spacer sequence of VDRE can also disrupt VDR binding [245]. Of great interest is that sequence specificity of a VDRE is generally similar to that of a TRE or RARE except that the intervening spacing of half-sites is 4 bases for a TRE and 5 bases for an RARE [227]. It is now understood that a general theme of VDR, RAR and TR is one of heterodimer formation with other transcription factors [246], unlike the co-operative homodimer formation of oestrogen and glucocorticoid receptors. Recent studies, however, with synthetic elements with larger spacer regions and inverted arrangements can be demonstrated to confer $1,25\text{-}(OH)_2D_3$-responsiveness [247]. Also, there are a few reported natural VDREs for which the three-base spacing rule is not obeyed [245–249], but the biological relevance of these response elements is not known. Therefore, it is possible that as yet unidentified VDREs may fall into a dissimilar category of the more common theme. One example of this is the VDRE in the promoter of the human preproparathyroid hormone gene, which is repressed by $1,25\text{-}(OH)_2D_3$. This VDRE is a 25 bp region from -125 to -101 and was characterized by Demay *et al.* [250]. The PTH VDRE consists of one half-site that exactly matches the GGTTCA half-site of mouse osteopontin but in the antisense orientation [195]. When inserted into a reporter system in front of a heterologous promoter, the PTH VDRE conferred suppression in response to $1,25\text{-}(OH)_2D_3$ in a cell-type specific manner [250]. This indicates that an unidentified factor found in parathyroid gland cells, but not in certain other types of cells, plays an interactive role in the suppression mechanism.

Studies are now underway to elucidate the mechanism of VDR interaction with the VDRE. Liao *et al.* [251] provided the first evidence that an accessory factor is essential for VDR binding to VDRE, by showing that recombinant VDR does not bind to human osteocalcin VDRE without an accessory factor from nuclear extract. This result was also found by two other independent laboratories [252, 253]. The accessory factor, termed nuclear accessory factor (NAF) or receptor auxiliary factor (RAF), is a 55 kilodalton (Kd) protein from human cells [254] and somewhat larger from porcine intestine [253]. RAF confers high affinity of VDR to VDRE, and with a dissociation constant of 0.24 nM for the VDR-RAF-VDRE complex [243]. The VDR-RAF interaction in solution can occur in the absence of DNA, and the VDR-RAF-VDRE association is markedly enhanced by 1,25-$(OH)_2D_3$ [243]. Thus, 1,25-$(OH)_2D_3$ either initiates heterodimer formation or a tighter interaction of the heterodimer with the VDRE [255]. Concomitantly, a purified nuclear factor from HeLa cell extracts that promoted the specific interaction of RAR and TR with their corresponding response elements was identified as RXRβ [256]. It is now realized that other homologues of RXR (RXRα and RXRγ) are also capable accessory factors to RAR, TR and VDR [257]. Antibodies to RXRβ were found to be immunoreactive with RAF and to interfere with VDR-RAF heterodimer formation [258]. Also, supershifts of specific gel shift complexes consisting of VDRE and VDR and highly purified porcine intestinal RAF occur in the presence of both anti-VDR and anti-RXR monoclonal antibodies [259]. Thus, it is likely if not certain that RAF is an RXR but which one is not certain.

REGULATION OF 1,25-$(OH)_2D_3$ RESPONSIVE GENES

Several integrated cell-type specific and gene specific factors function in the modulation of 1,25-$(OH)_2D_3$ responsive genes. These factors have become too numerous and complex to be comprehensively reviewed here, so the interested reader is referred to more specific reviews [247, 260]. This review will focus attention on more general themes that have arisen over the last few years and have enhanced the mechanistic understanding of the way in which 1,25-$(OH)_2D_3$-responsive genes function.

VDR uptake

It is well known that occupied VDR is predominantly in the nucleus, but how the VDR localizes to the nucleus has long been a controversial issue and is still poorly understood. Cell fractionation studies show the VDR to be exclusively nuclear [228]. The presence of VDR in both nuclear and cytosolic fractions, however, has been reported with the use of physiological ionic strength buffers or by gentle fractionation methods [261, 262]. The work of Barsony *et al.* [263] has used immunocytochemical detection to study the intracellular compartmentalization of VDR. They report the transcellular movement of VDR from the cytoplasm to the nucleus in microwave-fixed cells. In the presence of 1,25-$(OH)_2D_3$, VDR becomes organized on cytoplasmic fibres that resemble or are microtubules [264] and are oriented in the direction of the nuclear envelope [263, 265]. Within 1–5 min following 1,25-(OH)$_2$D$_3$-treatment, the organized bundles of VDR translocate into the nucleus in correlation with 3′,5′-cGMP production [266]. By disrupting the integrity of monocyte microtubules with colchicine, Kamimura *et al.* [267] showed that induction of 24-hydroxylase mRNA and activity were abolished without affecting either total 1,25-(OH)$_2$D$_3$ uptake or maximal 1,25-(OH)$_2$D$_3$-VDR binding. Also, in a case study of hereditary vitamin D-resistant rickets type II, a subclass of this disease was identified as having no mutations in the VDR open reading frame and functional 1,25-(OH)$_2$D$_3$-VDR binding. However, both skin fibroblasts and peripheral blood lymphocytes from the afflicted patient showed defective VDR translocation to the nucleus [268]. All these studies suggest a specific mechanism involving cytoskeletal matrices to facilitate movement of VDR to the nucleus for its more specialized role as a transcription factor. However, these require additional investigation before they can be accepted.

Phosphorylation

One of the more intriguing aspects of steroid hormone receptor study is deciphering the role phosphorylation plays in signalling nuclear events mediated by steroid receptors. Phosphorylation of the VDR contributes significantly to its functional activity, but the specific mechanisms that mediate this regulation are not well understood. Phosphorylation may influence DNA binding, ligand binding, and protein-protein interactions, including heterodimerization and/or transcriptional activation functions [66, 67]. The essential elements of the VDR phosphorylation event are that 1,25-(OH)$_2$D$_3$ controls the process by activating cellular pathways involving protein kinases. The use of phosphatase inhibitors, like okadaic acid, enhance this, whereas agents that specifically block either protein kinase C (PKC) or protein kinase A (PKA) have been found to decrease or abolish 1,25-(OH)$_2$D$_3$-mediated phosphorylation [67]. In the case of PKC, 1,25-(OH)$_2$D$_3$ causes translocation of PKC-β to the plasma membrane and PKC-α to the nucleus [269–271]. Both PKC isotypes are calcium-dependent, and 1,25-(OH)$_2$D$_3$ does not affect the calcium-independent PKC-ζ [271]. Brown and DeLuca [272] demonstrated that VDR phosphorylation is rapid, occurring within 1 h following treatment of 1,25-(OH)$_2$D$_3$ of embryonic chick duodenal organ culture. In addition, phosphorylation occurred before calcium uptake and the 1,25-(OH)$_2$D$_3$-dependent increase in calcium-binding protein mRNA. Proteolytic digestion of the 1,25-(OH)$_2$D$_3$-dependent phosphorylation of porcine VDR was mapped and shown to be localized to a 23 Kd fragment of the C-terminal domain [273].

A rough delineation of a phosphorylation site was also mapped out by Jones *et al.* [274] by transfecting internally deleted mutants of the human VDR into ROS 17/2.8 cells in the presence of labelling with [^{32}P]-orthophosphate. This site was assigned to a fragment between Met 197 and Val 234, and was previously shown to consist of a putative recognition site for phosphorylation by casein kinase II [275], a regulatory enzyme of significance in the function of nuclear proteins [275]. Highly purified casein kinase II was later found to phosphorylate human VDR [276], and the serine at 208 was determined to be the substrate site for casein kinase II as identified by site-directed mutagenesis [277]. However, despite experiments showing a 1,25-(OH)$_2$D$_3$-dependent phosphorylation of VDR [272], the phosphorylation by casein kinase II is 1,25-(OH)$_2$D$_3$-independent [277]. Hilliard *et al.* [278] also have elegantly mapped a casein kinase II recognition site in human VDR which under their nomenclature is Ser 205. However, blocking phosphorylation of this site did not preclude phosphorylation at other sites by casein kinase II nor did it affect transcriptional activation.

Other work has shown that Ser 51 in the DNA-binding domain is another potential site of phosphorylation, in this case, a PKC-β dependent phosphorylation [279]. Because Ser 51 is between two zinc fingers, modification of this residue or flanking residues can alter VDR binding to the human osteocalcin VDRE [280]. In contrast, changing Ser 51 to an alanine, which is the residue in the corresponding position of most other steroid receptors, eliminated PKC-β phosphorylation but left intact the fully functional specific DNA binding activity and transcriptional activation capacity of hVDR [280].

Despite these results, it is clear that phosphorylation of VDR plays some underlying role in the transcriptional activation 1,25-(OH)$_2$D$_3$-responsive genes [281]. The signalling pathway by which phosphorylation of VDR is accomplished may occur either by protein kinase C or protein kinase A transduction [282, 283] or by an alternative kinase system [284].

Molecular model of 1,25-(OH)$_2$D$_3$-dependent gene transcription

Regulation of gene function by 1,25-(OH)$_2$D$_3$ occurs at several distinct stages involving facets of ligand and protein-protein interactions and probably physical modifications of VDR either by a change in its conformation or by phosphorylation [246]. The latter two aspects are less well characterized, but indirect evidence exists showing that ligand binding to the oestrogen and progesterone receptors causes conformational changes that in turn influence transcriptional activation [285, 286]. Ligand binding to VDR stabilizes the receptor, a function that is important during periods of increased

Model for Vitamin D–influenced Target Gene Expression

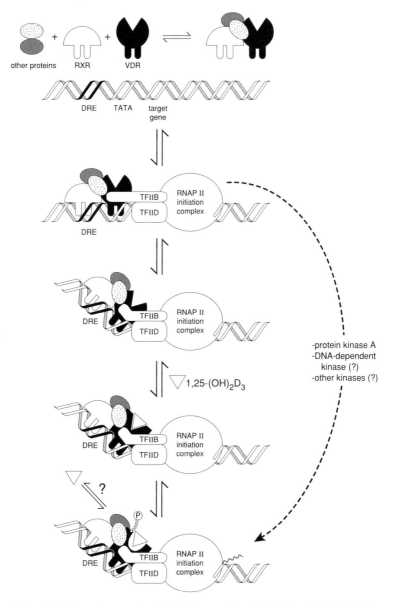

Figure 1.8. Current proposed model for the genomic mode of action of 1,25-(OH)$_2$D$_3$.

1,25-$(OH)_2D_3$ production [144, 287]. The affinity of 1,25-$(OH)_2D_3$ for VDR is extremely high, exhibiting a dissociation constant in the range of 10^{-10} to 10^{-11} M [115, 288]. Measurement of cellular 1,25-$(OH)_2D_3$ based on this assumption shows that only 5–7% of VDR is occupied under normal circumstances, and physiological increases in 1,25-$(OH)_2D_3$ raises this value to about 15% [70]. At least some VDR binds to genomic sites in the absence of ligand, but 1,25-$(OH)_2D_3$ is also important for enhancing the affinity of VDR for VDRE and in signalling of gene transcriptional activation [255]. The work of Barsony *et al.* [263] suggests that 1,25-$(OH)_2D_3$ has an involvement in transcellular movement of VDR to the nucleus, and other work shows that 1,25-$(OH)_2D_3$ functions in the modification of VDR by events such as phosphorylation and possibly conformational changes that allow for interactions with other proteins [260]. Upon entry into target cells, 1,25-$(OH)_2D_3$ elicits rapid acting transduction pathways and finds unoccupied VDR either nuclear or cytoplasmic and enables a conformational change such that the heterodimer VDR-RXR complex can form (*Figure 1.8*). In the nucleus, these heterodimers as larger complexes bind to VDREs of 1,25-$(OH)_2D_3$-responsive genes. Ligand is then believed to bind to the complexes and phosphorylation occurs in an analogous manner to the progesterone receptor [289]. The use of agents such as okadaic acid that block phosphatase activity and allow for basal VDR phosphorylation [283], would support this theory. The recent evidence that TFIIB, a basal transcription factor that aids in the assembly and bridging of the pre-initiation complex to many gene promoters, binds to a contact site on VDR in the presence of RXR [290] is suggestive of an even more complex model of 1,25-$(OH)_2D_3$-induced transcriptional activation. Regions in the long carboxyl-terminus of VDR are essential for the TFIIB interaction [290]. Future work will undoubtedly reveal other such interacting factors that bind to contact sites on VDR or RXR. Indeed, new work has already surfaced demonstrating the involvement of a ligand-independent co-activator (SRC-1) for certain steroid receptors [291] and co-repressors (N-CoR and SMRT) for RAR and TR [292–294]. These transcriptional regulators were identified and tested using dihybrid systems sensitive to protein-protein contact sites of the interacting proteins [295], and similar work is also underway for the VDR and RXR components.

TREATMENT OF CALCIUM AND PHOSPHORUS DISORDERS WITH VITAMIN D$_3$ METABOLITES AND ANALOGUES

RENAL OSTEODYSTROPHY

Renal osteodystrophy is a disease resulting from renal failure. The hallmarks of this disease are low plasma 1,25-(OH)$_2$D$_3$, hyperparathyroidism, and hyperphosphataemia. The loss of kidney function results in phosphate retention, and in addition, renal 1α-OHase is dysfunctional, causing low serum 1,25-(OH)$_2$D$_3$ concentrations. Low 1,25-(OH)$_2$D$_3$ results in little to no intestinal calcium absorption for soft tissue needs. Thus, blood calcium can be low and secondary hyperparathyroidism develops. The inability to regulate renal handling of phosphate by PTH leads to phosphate-mediated repression of ionized calcium. Serious osteodystrophic lesions occur because of high PTH activity on bone resorption resulting in osteitis fibrosa and osteosclerosis.

The understanding that parathyroid glands are a bona fide target tissue of 1,25-(OH)$_2$D$_3$ led to intensive investigations of 1,25-(OH)$_2$D$_3$ action in this tissue. It is well known that 1,25-(OH)$_2$D$_3$ decreases PTH secretion but more recently it has been demonstrated that 1,25-(OH)$_2$D$_3$ also decreases expression of preproparathyroid gene [296, 297]. This result was confirmed by Demay *et al.* [250] who showed that a VDRE located in the 5′-flanking sequence of the preproparathyroid hormone gene mediates negative transcriptional regulation in response to 1,25-(OH)$_2$D$_3$ when linked to a heterologous promoter/reporter construct. Further, 1,25-(OH)$_2$D$_3$ probably inhibits parathyroid gland proliferation. It is now clear why 1,25-(OH)$_2$D$_3$ is so effective in the treatment of renal osteodystrophy [298]. Intravenous administration of 1,25-(OH)$_2$D$_3$ met with excellent success in controlling hyperparathyroidism, and the same can be said for use of these compounds in renal osteodystrophy [298, 299]. The use of 1,25-(OH)$_2$D$_3$ in some patients may be limiting due to its potent hypercalcaemic tendency; therefore, the employment of less calcaemic analogues of 1,25-(OH)$_2$D$_3$ may provide more benefit to the patient.

FANCONI SYNDROME

Fanconi syndrome develops from a generalized defect in the proximal renal tubule limiting the transport of glucose, several minerals, and organic acids, as well as low molecular weight proteins. The syndrome represents global manifestations in proximal tubulopathy, and therefore, has many associated disorders including defective phosphate reabsorption causing phosphaturic hypophosphataemia and associated skeletal defects. Treatment with 1α-OH-D_3 often corrects the skeletal symptoms related to fanconi syndrome [300].

X-LINKED HYPOPHOSPHATAEMIA VITAMIN D-RESISTANT RICKETS

X-Linked hypophosphataemia, hypophosphataemic vitamin D resistant rickets (HPDR), is an X-linked dominant trait [300] with the mutant gene being located in the distal part of the short arm of the X-chromosome, Xp22.31-p21.3 [301, 302]. A similar mutation has been identified in the mouse (Hyp) [303], which has provided a useful model to study this disease in greater detail. The hypophosphataemia of HPDR is a consequence of a primary inborn error of phosphate transport at the location of the proximal tubules [304]. One hypothesis in regard to HPDR is that the proximal tubular Na(+)-Pi transporter plays a role in the murine form of X-linked hypophosphataemic vitamin D-resistant rickets [305]. The gene expression of the Na(+)-Pi transporter is up-regulated by phosphate restriction in normal mice but displays abnormally low transcription in Hyp mice [306]. The molecular mechanisms for this response, however, are unclear. Other interesting features in the Hyp model is the uncoupling of renal 25-OH-D_3 metabolism in response to dietary phosphate restriction where there is the inappropriate increase in 24-OHase and no increase in 1α-OHase activity [307, 308]. It is not known whether the same defects in renal enzymes occur in human HPDR, but a higher than normal catabolism of 1,25-$(OH)_2D_3$ is suspected [309]. Treatment of HPDR can be accomplished by supplementation with phosphate salts combined with the administration of oral 1,25-$(OH)_2D_3$ (1–2 μg/day), which both suppresses PTH secretion and combats hypocalcaemia caused by the phosphate supplementation [309].

VITAMIN D-DEPENDENCY RICKETS TYPE I

Vitamin D-dependency rickets type I (VDDR-I) is an autosomal recessive inheritance that was first described by Prader *et al.* [310]. The disease develops owing to a defect in the 1α-OHase. Though molecular details await purification and/or cloning of human 1α-OHase, the phenotype for VDDR-I has been temporarily mapped to chromosome 12q14–13 [171, 172]. A variant of VDDR-I was recently identified and reported in a strain of pigs which may provide an animal model for the study of VDDR-I [311]. Treatment strategy is to provide the missing vitamin D hormone at 0.5–1 μg dose/day [312]. In the past, large doses of vitamin D brought about high levels of 25-OH-D_3 that acted as an analogue of 1,25-$(OH)_2D_3$ and were used for treatment but management was more difficult [313].

VITAMIN D-DEPENDENCY RICKETS TYPE II

Vitamin D-dependency rickets type II (VDDR-II) is a rare heritable end-organ resistance disorder [314] caused by one of several genetic defects in the VDR [315]. There are five major classifications of VDR defects including hormone binding [316], binding capacity [317], binding affinity [318], deficient nuclear localization [319], and decreased affinity of the hormone/VDR complex for heterologous DNA [239, 240]. The disease is not vitamin D-dependent as such because patients are not responsive to any form of vitamin D_3 therapy. In more severe cases, total alopecia occurs [320], indicating a potential role of $1,25\text{-}(OH)_2D_3$ in hair growth. In these patients, remission even with extremely high doses of vitamin D_3 is not likely. Remission of non-alopecic patients is carried out by administering massive regimens of vitamin D_3 (6 μg/kg body weight or a total of 30 to 60 μg for a duration of 3–5 months) for sufficient new bone mineralization to occur. In addition, supplemental calcium up to 3 g of elemental calcium per day can augment calcium absorption by passive diffusion. Patients are best treated by calcium and phosphorus infusion [321].

OSTEOPOROSIS

Osteoporosis is a general decrease in bone mass. In America alone, 25 million individuals, most women, are affected to some degree by this bone-thinning disease. Unfortunately, if more attention is not given to communicating appropriate preventative strategies, osteoporosis and its related costs will continue to escalate. The three most important interacting factors are diet, exercise and oestrogen. For women the primary risk factors include Northern European descent or Caucasian, petite body size, cigarette smoking, a sedentary life-style, a family history of osteoporosis, life-long low calcium intake, the first 5–10 years following menopause, and oophorectomy [322]. Throughout the world, the prevalence of osteoporosis is increasing as the population and life-expectancy both increase, and is becoming a substantial burden to society in general.

Bone size and mass are genetically determined, but can also be influenced by a range of calcitropic factors that are most critical throughout the first three decades of life. Peak bone mass is attained by age 30–40. The two major forms of osteoporosis, senile involutional and postmenopausal, occur in women over the age of 50, with a significant proportion of the male population being afflicted by the age-dependent form of osteoporosis. In addition to the sedentary aspects brought on by advanced age for some, such as limb disuse and less than appropriate exposure to sunlight, certain biochemical functions also decline. The best example of this is intestinal calcium absorption which becomes less responsive to 1,25-$(OH)_2D_3$ around the sixth decade of life [323]. The same pattern can be demonstrated in the rat model and this trend appears to correlate with decreased VDR numbers in the intestine [324]. The reason for this tissue resistance is unclear, but may relate to a lower calcium demand following the attainment of peak bone mass. Also in rats, there have been some reports of lower renal 1α-OHase with advancing age and poor localization of 1,25-$(OH)_2D_3$ in bone [325], but this is not clearly established in adult humans [326]. In cases where there is a deficiency of vitamin D or calcium, secondary hyperparathyroidism accelerates the rate of bone loss from cortical zones and eventually hip fractures develop. Treatment with 1,25-$(OH)_2D_3$ has potential usefulness in the management of hyperparathyroidism, whereas supplementation with calcium and vitamin D_3 can be important for preventing cortical bone loss [327].

During menopause, rapid bone loss occurs in the first five years in direct correlation with lost synthesis of oestrogen [328, 329]. The same effect can be generated by oophorectomy in the rat [330] which has been an effective model for examining the efficacy of treatment modalities [331]. Oestrogen replacement is effective in avoiding profound loss of trabecular bone, which, if untreated, leads to vertebral crush fractures [332]. Oestrogen therapy is also useful for restoring lost bone, and the administration of oestrogen is frequently in combination with progestins to lower the risk of endometrial cancer [333], a potential side-effect of long- term treatment with oestrogen.

Glucocorticoid treatment for arthritis or other ailments can very quickly produce a form of osteoporosis caused by the inhibition of bone formation [334]. In such cases, the decrease in bone mass may be as much as 10–20%, but examination of trabecular bone reveals a much greater (30–40%) decrease in this component of bone [335]. Combination therapies with vitamin D and bisphosphonates, calcitonin or fluoride can be effective [336]. Therapy employing vitamin D or 1,25-$(OH)_2D_3$, the latter being highly calcaemic, should also include serum calcium monitoring and the use of thiazide diuretics as appropriate.

Polymorphisms in the VDR gene may be related to bone mineral density and predispose an individual to osteoporosis [337]. The use of restriction length fragment polymorphism (RFLP) analysis and other markers of VDR polymorphisms could be an additional early assessment parameter for the future risk of developing osteoporosis. Identification of the defective allele also referred to as BB or bb, depending on the laboratory from which the studies first originated, has been correlated to an increased risk of primary and secondary osteoporosis and primary hyperparathyroidism. But not all studies agree with these conclusions [338], so more work is necessary on the relationship of VDR polymorphisms to the incidence of osteoporosis. The b allele is potentially linked to a decreased transcriptional activity or mRNA stability leading to reduced VDR expression, which over the long term may have impending effects on bone density [339]. However, the power equations used to determine these relationships are still under scrutiny. Despite these findings, all individuals should be urged to recognize the importance of regular dietary intake of recommended levels of calcium and vitamin D$_3$.

The use of 1,25-(OH)$_2$D$_3$ and 1α-OH-D$_3$ has had considerable success in the treatment of osteopenia associated with osteoporosis. In Japan, 1α-OH-D$_3$ has been a highly successful treatment of osteoporosis [340, 341]. Further, 1,25-(OH)$_2$D$_3$ has been successfully applied to the treatment of osteoporosis in Japan and Italy [342]. Tilyard et al. [343] have also provided strong evidence that 1,25-(OH)$_2$D$_3$ reduces the fracture rate in an extensive clinical trial in New Zealand. Other successes have been reported by Gallagher [344] and Aloia [345]. Ott and Chesnut failed to find efficacy of 1,25-(OH)$_2$D$_3$ when given at 0.4 μg/day [346] but reported efficacy at higher doses [0.5 μg/day] [347]. Because of low calcium intakes in Japan, larger doses of 1,25-(OH)$_2$D$_3$ or 1α-OH-D$_3$ are possible, accounting for their higher success level. There is little doubt that 1,25-(OH)$_2$D$_3$ and its analogues are useful in the treatment of osteoporosis. However, in high calcium-intake countries like the United States, it is likely that their use will wait the development of an analogue with a greater window of efficacy to toxicity.

VITAMIN D$_3$ ANALOGUES OF CLINICAL IMPORTANCE FOR OTHER TYPES OF DISEASES

As has been discussed, 1,25-(OH)$_2$D$_3$ carries out a broad range of actions in many nonclassical target tissues, including immune cells and several leukaemia cell lines in addition to its central role in bone and calcium metabolism. Also, 1,25-(OH)$_2$D$_3$ has been demonstrated to have important general effects on cell proliferation and differentiation including cancer cells, epidermal keratinocytes, and activated lymphocytes.

PSORIASIS

In skin, 1,25-(OH)$_2$D$_3$ inhibits epidermal proliferation and promotes epidermal differentiation. The life cycle of a normal skin cell is about 28 days, compared with 4 days for a psoriatic cell. Sunlight of the UVB range is a known deterrent to skin cell proliferation, but the reason for this is not understood, even though it is a commonly used treatment in the healing of psoriasis. Because epidermal cells in skin also contain significant numbers of VDR, the application of $1,25\text{-}(OH)_2D_3$ to treat psoriatic skin was conceived. Clinical evaluation of $1,25\text{-}(OH)_2D_3$ and its synthetic analogues $1,24\text{-}(OH)_2D_3$ and MC-903 (calcipotriol) have proven efficacious in slowing chronic epidermal proliferation and healing psoriasis via topical application [348, 349]. $1,25\text{-}(OH)_2D_3$ can cause hypercalcaemia at the levels sometimes used to treat psoriasis [350]. MC-903, on the other hand, has equal affinity for VDR but once it enters the blood stream it is rapidly metabolically cleared [351], and thus is far less calcaemic than $1,25\text{-}(OH)_2D_3$. The mechanism by which $1,25\text{-}(OH)_2D_3$ and its analogues heal psoriasis remains unknown.

IMMUNOBIOLOGY

Receptors for $1,25\text{-}(OH)_2D_3$ are found throughout the immune system in many different cell types including monocytes and activated (but not resting) lymphocytes and also in specific subsets of thymocytes [352, 353]. An intensive study of $1,25\text{-}(OH)_2D_3$ function in these cells over the past 10 years has revealed many important immunomodulatory properties of $1,25\text{-}(OH)_2D_3$ *in vitro* and *in vivo* [354]. In particular, $1,25\text{-}(OH)_2D_3$ has potent ability to stimulate differentiation of monocytic cells of both normal myeloid and malignant lineage to become macrophages [355, 356]. Both humoral and cell-mediated immune effects can be demonstrated by $1,25\text{-}(OH)_2D_3$ action. In T-cells, $1,25\text{-}(OH)_2D_3$ decreases IL-2 mRNA [201] and secretion [357], and suppresses interferon-γ gene expression [201, 358]. In the T-helper cell population, $1,25\text{-}(OH)_2D_3$ selectively decreases cell proliferation; however, VDR is equally represented in both helper/inducer and suppressor/cytotoxic T-lymphocyte subsets [359]. Furthermore, in the presence of $1,25\text{-}(OH)_2D_3$, T-helper cell induction of immunoglobulin production by B-lymphocytes is blocked [360]. Other effects of $1,25\text{-}(OH)_2D_3$ include increasing the antigen-presenting and phagocytic activities of macrophages [361, 362]. $1,25\text{-}(OH)_2D_3$ also inhibits activated macrophage growth in response to colony-stimulating factor-1 [363, 364], thus representing a possible autocrine/paracrine action of $1,25\text{-}(OH)_2D_3$ within the immune system.

Supraphysiological doses of $1,25\text{-}(OH)_2D_3$ produce immunosuppressive effects by causing depletion of thymocytes and decreased immunoglobulin production in mice [365]. A further study of these effects by $1,25\text{-}(OH)_2D_3$ showed that hypercalcaemia, induced by $1,25\text{-}(OH)_2D_3$ treatment (20 ng/d for 4 days) markedly decreased thymus weight, thymocyte number, and in particular caused atrophy of immature thymocyte subpopulations which reside in the cortical region of the thymus gland [366]. This same dose of 1,25-$(OH)_2D_3$ in calcium restricted mice had relatively no effect on cortical T-cell populations. Further evidence of the vitamin D role in immune function comes from studies of vitamin D deficiency. In humans, vitamin D deficiency has been associated with reduced levels of serum immunoglobulins [367], and impaired phagocytic activity [368], neutrophil migration [369], and a delayed inflammatory response [370, 371]. In some cases, patients suffered recurrent bouts by infectious agents causing bronchitis, bronchopneumonia or tuberculosis [372]. In a recent study, HIV-infected patients with abnormally low serum $1,25-(OH)_2D_3$ concentrations (< 25 pg/ml) suffered shorter survival times than other HIV-patients. In this case, however, low serum $1,25-(OH)_2D_3$ was independent of vitamin D status [373]. In vitamin D-deficient mice, characterized by severe hypocalcaemia and undetectable blood 25-$OH-D_3$, a reduced delayed hypersensitivity response to the skin contact antigen, dinitrofluorobenzene (DNFB), was observed in comparison with vitamin D-sufficient animals [374]. This indicates impairment to cell-mediated immune mechanisms. Prolonged treatment of these vitamin D-deficient mice with $1,25-(OH)_2D_3$ restored a normal hypersensitivity response to DNFB, and the ability of T-lymphocytes to respond to mitogenic treatment [374].

The immunosuppressive effects of $1,25-(OH)_2D_3$ in such diseases as diabetes in non-obese diabetic (NOD) mice [375], and the autoimmune diseases murine lupus [376] and encephalomyelitis [377] reveal potentially important and novel uses for vitamin D treatment. However, it is clear that noncalcemic immunopotent vitamin D analogues will have to be employed to prevent hypercalcaemia caused by $1,25-(OH)_2D_3$.

DIFFERENTIATION

One of the more recent major developments in the vitamin D field is that $1,25-(OH)_2D_3$ plays an important role in regulating cellular growth and differentiation of cells of myeloid lineage [378]. This now has become an important area of study because of the potential benefit that the use of $1,25-(OH)_2D_3$ or an analogue of $1,25-(OH)_2D_3$ might be in suppressing cancerous growth by causing differentiation [379]. This concept has been tested *in vivo* where $1,25-(OH)_2D_3$ treatment markedly decreased the growth of breast cancer tumours [380, 381], and *in vitro* in a host of normal and malignant cell lines [382], in which the effect of $1,25-(OH)_2D_3$ is both antiproliferative and prodifferentiative. These results agree well with a number of epidemiological studies that suggest a relationship exists between solar radiation exposure and the prevention of cancer diseases [383], as well as, the importance of adequate dietary vitamin D_3 and sufficient blood levels of calcium and vitamin D metabolites in reducing the incidence of colon and prostate cancers [384, 385].

Both genomic and nongenomic pathways have been explored in the regulation of 1,25-$(OH)_2D_3$ effects on the processes of cellular growth and differentiation; however, the evidence supporting the genomic pathway is more conclusive. In many cell lines, the degree of 1,25-$(OH)_2D_3$-induced differentiation or growth inhibition correlates extremely well with the level of nuclear VDR [386, 387]. Also, in proliferating cells that do not express VDR, 1,25-$(OH)_2D_3$ is neither capable of suppressing growth nor differentiating these cells, whereas, recombinant expression of VDR into cells of this type is permissive of a 1,25-$(OH)_2D_3$ effect on differentiation [388]. In experiments with prostatic cells, where functional VDR expression is blocked by antisense RNA, the effect of 1,25-$(OH)_2D_3$ on differentiation is impaired [388]. In addition, the end-result of the differentiation of mononuclear phagocytes and preosteoclasts by 1,25-$(OH)_2D_3$, is a disappearance of VDR with the emergence of the mature osteoclasts [389]. All of these data are indicative of VDR expression being crucial to carrying out the effects of 1,25-$(OH)_2D_3$ on the regulation of cell growth and differentiation. Other cellular events involving fluxes in membrane fluidity [390], membrane coupled G-protein signal transduction [391], changes in the intracellular Ca^{2+} gradient [392] and protein phosphorylation cascades have been considered [393].

The most widely studied model for examining the mechanism of 1,25-$(OH)_2D_3$-induced differentiation and inhibition of cell proliferation has been the human promyeloid leukaemia cell line (HL-60), characterized by deficiency of the tumor suppressor gene p53. 1,25-$(OH)_2D_3$ rapidly suppresses the c-myc protooncogene and the retinoblastoma gene products [394]. Both of these oncogenes are known to be positively associated with cell proliferation. In the case of c-myc, 1,25-$(OH)_2D_3$ decreases the mRNA stability of the c-myc transcript [395]. Other oncogenes, c-fos and c-fes, are influenced by 1,25-$(OH)_2D_3$ and may also play a role in the differentiation process mediated by 1,25-$(OH)_2D_3$ [396]. In addition, 1,25-$(OH)_2D_3$ upregulates c-fms receptor expression with subsequent induction of tyrosine auto-phosphorylation of c-fms and amelioration of HL-60 differentiation [397]. Gene expression of myeloblastin, a serine protease associated with decreased proliferation and increased differentiation, is also stimulated by 1,25-$(OH)_2D_3$ [398]. Other 1,25-$(OH)_2D_3$-regulated factors related to HL-60 cell differentiation include up-regulation of markers of cell surface attachment, vitronectin, $\alpha V\beta V$ integrin, and talin [399], as well as, increased synthesis of thromboxane [400] and increased secretion of plasminogen activator [401].

The HL-60 VDR protein has been studied by Goto *et al.* [402] and found to be identical with VDR from human intestine and T47-D breast carcinoma cells based on nucleotide structure and 100 per cent immunoprecipitation of [^{3}H]-1,25-(OH)$_2$D$_3$ binding to HL-60 nuclear extracts. This demonstrates the unlikelihood of a receptor isotype of VDR mediating 1,25-(OH)$_2$D$_3$ effects in HL-60 cells. More specifically, the affinity of vitamin D$_3$ analogues to HL-60 VDR parallels the ability of each analogue to induce differentiation [403]. Furthermore, 1,25-(OH)$_2$D$_3$-mediated differentiation of HL-60 cells was demonstrated to be calcium-independent [404]. Other differentiation agents such as dimethyl sulphoxide [405] and retinoic acid [406] induce HL-60 cell differentiation to granulocytes, while phorbol esters induce a monocytic pathway of differentiation [407]. In contrast, 1,25-(OH)$_2$D$_3$-mediated differentiation of HL-60 cells is assessed by induction of phagocytic activity, nitro blue tetrazolium reduction, and an increase in nonspecific esterase activity [408].

The use of 1,25-(OH)$_2$D$_3$ in the treatment of cancer *in vivo* is not appropriate because of calcaemic consequences at the dose level required for efficacy. In light of this, a major pharmaceutical effort, over the past several years, has generated and tested nearly 400 analogues [409] of 1,25-(OH)$_2$D$_3$ in a quest to identify the structural basis for the calcaemic actions of 1,25-(OH)$_2$D$_3$ and form super analogues that are noncalcaemic but retain an equal or improved potency in effecting cellular differentiation and inhibiting cell growth [410]. From this collective work only a handful of synthetic analogues have been developed that retain potentially useful properties of 1,25-(OH)$_2$D$_3$ *in vitro*, but exhibit less calcemic activity *in vivo* or in organ culture. The testing of these compounds for their therapeutic utility must be considered at several levels of competence in comparison with 1,25-(OH)$_2$D$_3$, including binding affinity to plasma DBP, metabolic clearance, cellular uptake, affinity for VDR, induction of metabolism, as well as ability to inhibit cancer growth and stimulate differentiation.

CONCLUDING REMARKS

An enormous volume of information is now available concerning the action and involvement of 1,25-(OH)$_2$D$_3$ in biological systems. Much headway has been made with the use of 1,25-(OH)$_2$D$_3$ and its analogues in the management of several bone diseases. However, the greatest advances in their uses are yet to come. The presence of high concentrations of VDR in cancerous tissues and the initial success in curing these abnormalities using 1,25-$(OH)_2D_3$ or its analogues underscores a renewed and vital interest in the development of vitamin D-based anti-cancer agents. Finally, there also appears to be great potential for treating several immune-related diseases and in exploring heretofore unanticipated functions of 1,25-$(OH)_2D_3$.

ACKNOWLEDGMENTS

This work was supported in part by a program project grant (no. DK14881) from the National Institutes of Health, a fund from the National Foundation for Cancer Research, and a fund from the Wisconsin Alumni Research Foundation.

REFERENCES

1. Hess, A. (1929) in Rickets, Including Osteomalacia and Tetany: The History of Rickets, Including Osteomalacia and Tetany, (Hess, A., ed.), pp. 22–37, Lee and Febiger, Philadelphia.
2. Mellanby, E. (1919) Lancet 1, 407–412.
3. McCollum, E.V., Simmonds, N. and Pitz, W. (1916) J. Biol. Chem. 27, 33–43.
4. McCollum, E.V., Simonds, N., Becker, J.E. and Shipley, P.G. (1922) Bull. Johns Hopkins Hosp. 33, 229–230.
5. Steenbock, H. and Black, A. (1924) J. Biol. Chem. 61, 405–422.
6. Hess, S.F. and Weinstock, M. (1924) Proc. Soc. Exp. Biol. Med. 22, 5–6.
7. Massengale, O.N. and Nussmeier, M. (1930) J. Biol. Chem. 87, 423–425.
8. Steenbock, H. and Kletzien, S.W.F. (1932) J. Biol. Chem. 97, 249–264.
9. Windaus, A., Schenck, F. and von Werder, F. (1936) Hoppe-Seylers Z. Physiol. Chem. 241, 100–103.
10. Howland, J. and Kramer, B. (1921) Am. J. Dis. Child. 22, 105–119.
11. Orr, W.J., Holt, L.E.J., Wilkins, L. and Boone, F.H. (1923) Am. J. Dis. Child. 26, 362–372.
12. Nicolaysen, R. and Eeg-Larsen, N. (1953) Vitam. Horm. 11, 29–60.
13. Schachter, D. and Rosen, S.M. (1959) Am. J. Physiol. 196, 357–362.
14. DeLuca, H.F., Weller, M., Blunt, J.W. and Neville, P.F. (1968) Arch. Biochem. Biophys. 124, 122–128.
15. Neville, P.F. and DeLuca, H.F. (1966) Biochemistry 5, 2201–2207.
16. Holick, M.F. and DeLuca, H.F. (1971) J. Lipid Res. 12, 460–465.
17. Norman, A.W. and DeLuca, H.F. (1963) Anal. Chem. 35, 1247–1250.
18. Suda, T., DeLuca, H.F., Schnoes, H.K., Ponchon, G., Tanaka, Y. and Holick, M.F. (1970) Biochemistry 9, 2917–2922.
19. Carlsson, A. (1952) Acta Physiol. Scand. 26, 212–220.
20. DeLuca, H.F. (1967) Vitam. Horm. 25, 315–367.
21. Lund, J. and DeLuca, H.F. (1966) J. Lipid Res. 7, 739–744.
22. Morii, H., Lund, J., Neville, P.F. and DeLuca, H.F. (1967) Arch. Biochem. Biophys. 120, 508–512.

23. Blunt, J.W., Tanaka, Y. and DeLuca, H.F. (1968) Proc. Natl. Acad. Sci. U.S.A. 61, 714–718.
24. DeLuca, H.F. (1969) Arch. Intern. Med. 124, 442–450.
25. Cousins, R.J., DeLuca, H.F., Chen, T., Suda, T. and Tanaka, Y. (1970) Biochemistry 9, 1453–1459.
26. DeLuca, H.F. (1970) in The Fat Soluble Vitamins (DeLuca, H.F. and Suttie, J.W., eds.), pp. 3–20, The University of Wisconsin Press, Madison.
27. Cousins, R.J., DeLuca, H.F. and Gray, R.W. (1970) Biochemistry 9, 3649–3652.
28. Haussler, M.R., Boyce, D.W., Littledike, E.T. and Rasmussen, H. (1971) Proc. Natl. Acad. Sci. U.S.A. 68, 177–181.
29. Holick, M.F., Schnoes, H.K. and DeLuca, H.F. (1971) Proc. Natl. Acad. Sci. U.S.A. 68, 803–804.
30. Holick, M.F., Schnoes, H.K., DeLuca, H.F., Suda, T. and Cousins, R.J. (1971) Biochemistry 10, 2799–804.
31. Fraser, D.R. and Kodicek, E. (1970) Nature (London) 228, 764–766.
32. DeLuca, H.F. (1988) FASEB J. 2, 224–236.
33. Holick, M.F., Frommer, J.E., McNeill, S.C., Richtand, N.M., Henley, J.W. and Potts, J.T., Jr. (1977) Biochem. Biophys. Res. Commun. 76, 107–114.
34. Holick, M.F. and Clark, M.B. (1978) Fed. Proc. 37, 2567–2574.
35. Holick, M.F., MacLaughlin, J.A., Clark, M.B., Holick, S.A., Potts, J.T., Jr., Anderson, R.R., Blank, I.H., Parrish, J.A. and Elias, P. (1980) Science (Washington, D.C.) 210, 203–205.
36. Cooke, N.E. and Haddad, J.G. (1989) Endocrine Rev. 10, 294–307.
37. Holick, M.F., MacLaughlin, J.A. and Doppelt, S.A. (1981) Science (Washington, D.C.) 211, 590–593.
38. Loomis, W.F. (1967) Science (Washington, D.C.) 157, 501–506.
39. Esvelt, R.P., DeLuca, H.F., Wichmann, J.K., Yoshizawa, S., Zurcher, J., Sar, M. and Stumpf, W.E. (1980) Biochemistry 19, 6158–6161.
40. MacLaughlin, J. and Holick, M.F. (1985) J. Clin. Invest. 76, 1536–8.
41. Matsuoka, L.Y., Wortsman, J., Hanifan, N. and Holick, M.F. (1988) Arch. Dermatol. 124, 1802–4.
42. Holick, M.F. (1995) Am. J. Clin. Nutr. 61, 638S-645S.
43. Dawson-Hughes, B., Harris, S.S., Krall, E.A., Dallal, G.E., Falconer, G. and Green, C.L. (1995) Am. J. Clin. Nutr. 61, 1140–1145.
44. Lo, C.W., Paris, P.W., Clemens, T.L., Nolan, J. and Holick, M.F. (1985) Am. J. Clin. Nutr. 42, 644–649.
45. Fine, R.M. (1988) Int. J. Dermatol. 27, 300–301.
46. Kligman, E.W., Watkins, A., Johnson, K. and Kronland, R. (1989) Fam. Prac. Res. J. 9, 11–19.
47. Rosen, C.J., Morrison, A., Zhou, H., Storm, D., Hunter, S.J., Musgrave, K., Chen, T., Wei, W. and Holick, M.F. (1994) Bone Miner. 25, 83–92.
48. Holick, M.F. and Clark, M.B. (1978) Fed. Proc. 37, 2567–2574.
49. Ponchon, G., Kennan, A.L. and DeLuca, H.F. (1969) J. Clin. Invest. 48, 2032–2037.
50. Ponchon, G. and DeLuca, H.F. (1969) J. Clin. Invest. 48, 1273–1279.
51. Ponchon, G., Deluca, H.F. and Suda, T. (1970) Arch. Biochem. Biophys. 141, 397–408.
52. Bhattacharyya, M.H. and DeLuca, H.F. (1974) Arch. Biochem. Biophys. 160, 58–62.
53. Madhok, T.C. and DeLuca, H.F. (1979) Biochem. J. 184, 491–499.
54. Bjorkhem, I. and Holmberg, I. (1978) J. Biol. Chem. 253, 842–849.
55. Guo, Y.D., Strugnell, S., Back, D.W. and Jones, G. (1993) Proc. Natl. Acad. Sci. U.S.A. 90, 8668–8672.
56. Usui, E., Noshiro, M., Ohyama, Y. and Okuda, K. (1990) FEBS Lett. 274, 175–177.

57. Saarem, K. and Pedersen, J.I. (1985) Biochim. Biophys. Acta 840, 117–126.
58. Haddad, J.G., Jr. (1979) Clin. Orthop. Relat. Res. 142, 249–261.
59. Imawari, M., Kida, K. and Goodman, D.S. (1976) J. Clin. Invest. 58, 514–523.
60. Mawer, E.B., Backhouse, J., Holman, C.A., Lumb, G.A. and Stanbury, S.W. (1972) Clin. Sci. 43, 413–431.
61. DeLuca, H.F. (1981) in Subcellular Biochemistry: Subcellular Mechanisms Involving Vitamin D Subcellular Biochemistry, (Roodyn, D.B., ed.), pp. 251–272, Plenum Publishing, New York.
62. Tanaka, Y., Halloran, B., Schnoes, H.K. and DeLuca, H.F. (1979) Proc. Natl. Acad. Sci. U.S.A. 76, 5033–5035.
63. Adams, J.S., Singer, F.R., Gacad, M.A., Sharma, O.P., Hayes, M.J., Vouros, P. and Holick, M.F. (1985) J. Clin. Endocr. Metab. 60, 960–966.
64. Fetchick, D.A., Bertolini, D.R., Sarin, P.S., Weintraub, S.T. and Mundy, G.R. (1986) J. Clin. Invest. 78, 592–596.
65. Maesaka, J.K., Batuman, V., Pablo, N.C. and Shakamuri, S. (1982) Arch. Int. Med. 142, 1206–1207.
66. Darwish, H. and DeLuca, H.F. (1993) Crit. Rev. Euk. Gene Expr. 3, 89–116.
67. Ross, T.K., Darwish, H.M. and DeLuca, H.F. (1994) Vitam. Horm. 49, 281–326.
68. Tanaka, Y. and DeLuca, H.F. (1974) Science (Washington, D.C.) 183, 1198–1200.
69. Horst, R.L., Reinhardt, T.A., Ramberg, C.F., Koszewski, N.J. and Napoli, J.L. (1986) J. Biol. Chem. 261, 9250–9256.
70. Reinhardt, T.A. and Horst, R.L. (1989) J. Biol. Chem. 264, 15917–15921.
71. Holick, M.F., Schnoes, H.K., DeLuca, H.F., Gray, R.W., Boyle, I.T. and Suda, T. (1972) Biochemistry 11, 4251–4255.
72. Tanaka, Y., Lorenc, R.S. and DeLuca, H.F. (1975) Arch. Biochem. Biophys. 171, 521–526.
73. Brommage, R. and DeLuca, H.F. (1985) Endocr. Rev. 6, 491–511.
74. DeLuca, H.F. (1986) Adv. Exp. Med. Biol. 196, 361–375.
75. Shinki, T., Jin, C.H., Nishimura, A., Nagai, Y., Ohyama, Y., Noshiro, M., Okuda, K. and Suda, T. (1992) J. Biol. Chem. 267, 13757–13762.
76. Kumar, R. (1984) Physiol. Rev. 64, 478–504.
77. Tanaka, Y., Wichmann, J.K., Paaren, H.E., Schnoes, H.K. and DeLuca, H.F. (1980) Proc. Natl. Acad. Sci. U.S.A. 77, 6411–6414.
78. Horst, R.L., Wovkulich, P.M., Baggiolini, E.G., Uskokovic, M.R., Engstrom, G.W. and Napoli, J.L. (1984) Biochemistry 23, 3973–3979.
79. Ishizuka, S., Ishimoto, S. and Norman, A.W. (1984) J. Steroid Biochem. 20, 611–615.
80. Wilhelm, F., Mayer, E. and Norman, A.W. (1984) Arch. Biochem. Biophys. 233, 322–329.
81. Makin, G., Lohnes, D., Byford, V., Ray, R. and Jones, G. (1989) Biochem. J. 262, 173–180.
82. Esvelt, R.P. and DeLuca, H.F. (1981) Arch. Biochem. Biophys. 206, 403–413.
83. Onisko, B.L., Esvelt, R.P., Schnoes, H.K. and DeLuca, H.F. (1980) Biochemistry 19, 4124–4130.
84. Suda, T., DeLuca, H.F., Schnoes, H.K., Tanaka, Y. and Holick, M.F. (1970) Biochemistry 9, 4776–4780.
85. Tanaka, Y., Schnoes, H.K., Smith, C.M. and DeLuca, H.F. (1981) Arch. Biochem. Biophys. 210, 104–109.
86. Horst, R.L., Shepard, R.M., Jorgensen, N.A. and DeLuca, H.F. (1979) J. Lab. Clin. Med. 93, 277–285.
87. Halloran, B.P., DeLuca, H.F., Barthell, E., Yamada, S., Ohmori, M. and Takayama, H. (1981) Endocrinology 108, 2067–2071.
88. Tanaka, Y., Pahuja, D.N., Wichmann, J.K., DeLuca, H.F., Kobayashi, Y., Taguchi, T. and Ikekawa, N. (1982) Arch. Biochem. Biophys. 218, 134–141.

89. Ornoy, A., Goodwin, D., Noff, D. and Edelstein, S. (1978) Nature (London) 276, 517–519.
90. Henry, H.L., Taylor, A.N. and Norman, A.W. (1977) J. Nutr. 107, 1918–1926.
91. Corvol, M.T., Dumontier, M.F., Maroteaux, P., Rappaport, R., Guyda, M. and Posner, B. (1978) Arch. Fr. Pediatr. 35, 57–64.
92. Henry, H.L. and Norman, A.W. (1978) Science (Washington, D.C.) 201, 835–837.
93. Uchida, M., Ozono, K. and Pike, J.W. (1994) J. Bone Miner. Res. 9, 1981–1987.
94. Cantley, L.K., Russell, J.B., Lettieri, D.S. and Sherwood, L.M. (1987) Calcif. Tissue Int. 41, 48–51.
95. Binderman, I. and Somjen, D. (1984) Endocrinology 115, 430–432.
96. Wilhelm, F., Ross, F.P. and Norman, A.W. (1986) Arch. Biochem. Biophys. 249, 88–94.
97. Yamada, S., Ohmori, M. and Takayama, H. (1979) Tetrahedron Lett. 21, 1859–1862.
98. Miller, S.C., Halloran, B.P., DeLuca, H.F., Yamada, S., Takayama, H. and Jee, W.S. (1981) Calcif. Tissue Int. 33, 489–97.
99. Brommage, R., Jarnagin, K., DeLuca, H.F., Yamada, S. and Takayama, H. (1983) Am. J. Physiol. 244, E298-E304.
100. Jarnagin, K., Brommage, R., DeLuca, H.F., Yamada, S. and Takayama, H. (1983) Am. J. Physiol. 244, E290-E297.
101. Hebert, S.C. and Brown, E.M. (1995) Curr. Opin. Cell Biol. 7, 484–492.
102. Rogers, K.V., Dunn, C.K., Conklin, R.L., Hadfield, S., Petty, B.A., Brown, E.M., Hebert, S.C., Nemeth, E.F. and Fox, J. (1995) Endocrinology 136, 499–504.
103. Tucker, G.d., Gagnon, R.E. and Haussler, M.R. (1973) Arch. Biochem. Biophys. 155, 47–57.
104. Okuda, K.I. (1994) J. Lipid Res. 35, 361–372.
105. Ohyama, Y., Masumoto, O., Usui, E. and Okuda, K. (1991) J. Biochem. 109, 389–393.
106. Axen, E., Postlind, H., Sjoberg, H. and Wikvall, K. (1994) Proc. Natl. Acad. Sci. U.S.A. 91, 10014–10018.
107. Axen, E., Postlind, H. and Wikvall, K. (1995) Biochem. Biophys. Res. Commun. 215, 136–141.
108. Axen, E. (1995) FEBS Lett. 375, 277–279.
109. DeLuca, H.F. and Schnoes, H.K. (1984) Annu. Rep. Med. Chem. p. 179–190.
110. DeLuca, H.F. (1984) in Vitamin D (Kumar, R., ed.), pp. 1–68, Martinus Nijhoff, Boston, MA.
111. Bhattacharyya, M.H. and DeLuca, H.F. (1973) J. Biol. Chem. 248, 2969–2973.
112. Blunt, J.W., DeLuca, H.F. and Schnoes, H.K. (1968) Biochemistry 7, 3317–3322.
113. Beckman, M.J., Johnson, J.A., Goff, J.P., Reinhardt, T.A., Beitz, D.C. and Horst, R.L. (1995) Arch. Biochem. Biophys. 319, 535–539.
114. Beckman, M.J., Horst, R.L., Reinhardt, T.A. and Beitz, D.C. (1990) Biochem. Biophys. Res. Commun. 169, 910–915.
115. Brumbaugh, P.F. and Haussler, M.R. (1973) Life Sci. 13, 1737–1746.
116. Reynolds, J.J., Holick, M.F. and DeLuca, H.F. (1974) Calcif. Tissue Res. 15, 333–339.
117. Raisz, L.G., Trummel, C.L., Holick, M.F. and DeLuca, H.F. (1972) Science (Washington, D.C.) 175, 768–769.
118. Counts, S.J., Baylink, D.J., Shen, F.H., Sherrard, D.J. and Hickman, R.O. (1975) Ann. Intern. Med. 82, 196–200.
119. Vieth, R. (1990) Bone Miner. 11, 267–272.
120. Beckman, M.J., Goff, J.P., Reinhardt, T.A., Beitz, D.C. and Horst, R.L. (1994) Endocrinology 135, 1951–1955.
121. Lore, F. and Di Cairano, G. (1986) Ann. Med. Interne 137, 206–208.
122. Caniggia, A., Nuti, R., Lore, F. and Vattimo, A. (1984) J. Endocrinol. Invest. 7, 373–378.
123. Kollenkirchen, U., Walters, M.R. and Fox, J. (1991) Am. J. Physiol. 260, E447-E452.

124. Armbrecht, H.J. and Forte, L.R. (1985) Arch. Biochem. Biophys. 242, 464–469.
125. Halloran, B.P. and Castro, M.E. (1989) Am. J. Physiol. 256, E686-E691.
126. Bolt, M.J., Meredith, S.C. and Rosenberg, I.H. (1988) Calcif. Tissue Int. 42, 273–278.
127. Baran, D.T. and Milne, M.L. (1986) J. Clin. Invest. 77, 1622–1626.
128. Corlett, S.C., Chaudhary, M.S., Tomlinson, S. and Care, A.D. (1987) Cell Calcium 8, 247–258.
129. Thierry-Palmer, M., Cullins, S., Rashada, S., Gray, T.K. and Free, A. (1986) Arch. Biochem. Biophys. 250, 120–127.
130. Ghazarian, J.G. (1990) J. Bone Min. Res. 5, 897–903.
131. Arabian, A., Grover, J., Barre, M.G. and Delvin, E.E. (1993) J. Steroid Biochem. Mol. Biol. 45, 513–516.
132. Garabedian, M., Holick, M.F., Deluca, H.F. and Boyle, I.T. (1972) Proc. Natl. Acad. Sci. U.S.A. 69, 1673–1676.
133. Boyle, I.T., Gray, R.W. and DeLuca, H.F. (1971) Proc. Natl. Acad. Sci. U.S.A. 68, 2131–2134.
134. Shigematsu, T., Horiuchi, N., Ogura, Y., Miyahara, T. and Suda, T. (1986) Endocrinology 118, 1583–1589.
135. Tanaka, Y. and DeLuca, H.F. (1984) Am. J. Physiol. 246, E168-E173.
136. Horiuchi, N., Suda, T., Takahashi, H., Shimazawa, E. and Ogata, E. (1977) Endocrinology 101, 969–974.
137. Iida, K., Shinki, T., Yamaguchi, A., DeLuca, H.F., Kurokawa, K. and Suda, T. (1995) Proc. Natl. Acad. Sci. U.S.A. 92, 6112–6116.
138. Chen, M.L., Boltz, M.A. and Armbrecht, H.J. (1993) Endocrinology 132, 1782–1788.
139. Mandel, M.L., Moorthy, B. and Ghazarian, J.G. (1990) Biochem. J. 266, 385–392.
140. Koyama, H., Inaba, M., Nishizawa, Y., Ohno, S. and Morii, H. (1994) J. Cell. Biochem. 55, 230–240.
141. Kawashima, H., Torikai, S. and Kurokawa, K. (1981) Proc. Natl. Acad. Sci. U.S.A. 78, 1199–1203.
142. Costa, E.M. and Feldman, D. (1986) Biochem. Biophys. Res. Commun. 137, 742–747.
143. Wiese, R.J., Uhland-Smith, A., Ross, T.K., Prahl, J.M. and DeLuca, H.F. (1992) J. Biol. Chem. 267, 20082–20086.
144. Goff, J.P., Reinhardt, T.A., Beckman, M.J. and Horst, R.L. (1990) Endocrinology 126, 1031–1035.
145. Sandgren, M.E. and DeLuca, H.F. (1990) Proc. Natl. Acad. Sci. U.S.A. 87, 4312–4314.
146. Reinhardt, T.A. and Horst, R.L. (1990) Endocrinology 127, 942–948.
147. Uhland-Smith, A. and DeLuca, H.F. (1993) Biochim. Biophys. Acta 1176, 321–326.
148. Tanaka, Y. and Deluca, H.F. (1973) Arch. Biochem. Biophys. 154, 566–574.
149. Tanaka, Y., Frank, H. and DeLuca, H.F. (1973) Science (Washington, D.C.) 181, 564–566.
150. Galante, L., Colston, K.W., Evans, I.M., Byfield, P.G., Matthews, E.W. and MacIntyre, I. (1973) Nature (London) 244, 438–440.
151. Haussler, M., Hughes, M., Baylink, D., Littledike, E.T., Cork, D. and Pitt, M. (1977) Adv. Exp. Med. Biol. 81, 233–250.
152. Sriussadaporn, S., Wong, M.S., Pike, J.W. and Favus, M.J. (1995) J. Bone Miner. Res. 10, 271–280.
153. Wongsurawat, N., Armbrecht, H.J., Zenser, T.V., Forte, L.R. and Davis, B.B. (1984) J. Endocrinol. 101, 333–338.
154. Menaa, C., Vrtovsnik, F., Friedlander, G., Corvol, M. and Garabedian, M. (1995) J. Biol. Chem. 270, 25461–25467.
155. Tanaka, Y., Chen, T.C. and Deluca, H.F. (1972) Arch. Biochem. Biophys. 152, 291–298.
156. Tanaka, Y. and DeLuca, H.F. (1971) Proc. Natl. Acad. Sci. U.S.A. 68, 605–608.

157. Tanaka, Y. and DeLuca, H.F. (1983) Biochem. J. 214, 893–897.
158. Lorenc, R., Tanaka, Y., DeLuca, H.F. and Jones, G. (1977) Endocrinology 100, 468–472.
159. Horiuchi, N., Takahashi, H., Matsumoto, T., Takahashi, N., Shimazawa, E., Suda, T. and Ogata, E. (1979) Biochem. J. 184, 269–275.
160. Welsh, J., Weaver, V. and Simboli-Campbell, M. (1991) Biochem. Cell Biol. 69, 768–770.
161. Spanos, E., Colston, K.W., Evans, I.M., Galante, L.S., Macauley, S.J. and Macintyre, I. (1976) Mol. Cell. Endocrinol. 5, 163–167.
162. Spanos, E., Barrett, D.I., Chong, K.T. and MacIntyre, I. (1978) Biochem. J. 174, 231–236.
163. Spanos, E., Barrett, D., MacIntyre, I., Pike, J.W., Safilian, E.F. and Haussler, M.R. (1978) Nature (London) 273, 246–247.
164. Caniggia, A., Lore, F., di Cairano, G. and Nuti, R. (1987) J. Steroid Biochem. 27, 815–824.
165. Civitelli, R., Agnusdei, D., Nardi, P., Zacchei, F., Avioli, L.V. and Gennari, C. (1988) Calcif. Tissue Int. 42, 77–86.
166. Breslau, N.A. (1988) Am. J. Med. Sci. 296, 417–425.
167. Hartwell, D., Hassager, C., Overgaard, K., Riis, B.J., Podenphant, J.R. and Christiansen, C. (1990) Acta Endocrinol. 122, 715–721.
168. DeLuca, H.F. (1981) Curr. Med. Res. Opin. 7, 279–293.
169. Cunningham, J., Bikle, D.D. and Avioli, L.V. (1984) Kidney Int. 25, 47–52.
170. Portale, A.A., Halloran, B.P., Harris, S.T., Bikle, D.D. and Morris, R.C., Jr. (1992) Am. J. Physiol. 263, E1164-E1170.
171. Labuda, M., Morgan, K. and Glorieux, F.H. (1990) Am. J. Hum. Genet. 47, 28–36.
172. Labuda, M., Fujiwara, T.M., Ross, M.V., Morgan, K., Garcia-Heras, J., Ledbetter, D.H., Hughes, M.R. and Glorieux, F.H. (1992) J. Bone Min. Res. 7, 1447–1453.
173. Bronner, F. (1990) Miner. Electrolyte Metab. 16, 94–100.
174. Wasserman, R.H., Corradino, R.A. and Taylor, A.N. (1968) J. Biol. Chem. 243, 3978–3986.
175. Wasserman, R.H. and Taylor, A.N. (1968) J. Biol. Chem. 243, 3987–3993.
176. Christakos, S., Gabrielides, C. and Rhoten, W.B. (1989) Endocr. Rev. 10, 3–26.
177. Perret, C., Lomri, N. and Thomasset, M. (1989) Adv. Exp. Med. Biol. 255, 241–250.
178. Minghetti, P.P., Cancela, L., Fujisawa, Y., Theofan, G. and Norman, A.W. (1988) Mol. Endocrinol. 2, 355–367.
179. Kumar, R. (1991) Kidney Int. 40, 1177–1189.
180. Kumar, R. (1990) J. Am. Soc. Nephrol. 1, 30–42.
181. van Corven, E.J., Roche, C. and van Os, C.H. (1985) Biochim. Biophys. Acta 820, 274–282.
182. Matsumoto, T., Fontaine, O. and Rasmussen, H. (1980) Biochim. Biophys. Acta 599, 13–23.
183. Bachelet, M., Lacour, B. and Ulmann, A. (1982) Miner. Electrolyte Metab. 8, 261–266.
184. Peterlik, M. and Wasserman, R.H. (1980) Horm. Metabol. Res. 12, 216–219.
185. Garabedian, M., Tanaka, Y., Holick, M.F. and DeLuca, H.F. (1974) Endocrinology 94, 1022–1027.
186. Talmage, R.V. (1970) Am. J. Anat. 129, 467–476.
187. Frost, H.M. (1966) Bone Dynamics in Osteoporosis and Osteomalacia. Henry Ford Hospital Surgical Monograph Series. Charles A. Thomas Co., Springfield, MA.
188. Tanaka, Y. and Deluca, H.F. (1971) Arch. Biochem. Biophys. 146, 574–578.
189. Tanaka, Y., Frank, H. and DeLuca, H.F. (1972) J. Nutr. 102, 1569–1577.
190. Sugimoto, T., Fukase, M., Tsutsumi, M., Nakada, M., Hishikawa, R., Tsunenari, T., Yoshimoto, Y. and Fujita, T. (1986) Endocrinology 118, 1808–1813.

191. Holtrop, M.E., Cox, K.A., Carnes, D.L. and Holick, M.F. (1986) Am. J. Physiol. 251, E234-E240.
192. DeLuca, H.F. (1986) in Nutrition and Aging (Hutchinson, M.L. and Munroe, H.N., eds.), pp. 217–234, Academic Press, Orlando, FL.
193. Lian, J.B., Carnes, D.L. and Glimcher, M.J. (1987) Endocrinology 120, 2123–2130.
194. Prince, C.W. and Butler, W.T. (1987) Collagen Relat. Res. 7, 305–313.
195. Noda, M., Vogel, R.L., Craig, A.M., Prahl, J., DeLuca, H.F. and Denhardt, D.T. (1990) Proc. Natl. Acad. Sci. U.S.A. 87, 9995–9999.
196. Rowe, D.W. and Kream, B.W. (1982) J. Biol. Chem. 257, 8009–8015.
197. Price, P.A. (1984) in Vitamin D, Basic and Clinical Aspects (Kumar, R., ed.), pp. 297–410, Martinus-Nijhoff, Boston, MA.
198. Holtrop, M.E., Cox, K.A., Clark, M.B., Holick, M.F. and Anast, C.S. (1981) Endocrinology 108, 2293–2301.
199. Gray, T.K. and Cohen, M.S. (1985) Surv. Immunol. Res. 4, 200–212.
200. Huffer, W.E. (1988) Lab. Invest. 59, 418–442.
201. Rigby, W.F.C., Denome, S. and Fanger, M.W. (1987) J. Clin. Invest. 79, 1659–1664.
202. Yamamoto, M., Kawanobe, Y., Takahashi, H., Shimazawa, E., Kimura, S. and Ogata, E. (1984) J. Clin. Invest. 74, 507–513.
203. Suda, T., Takahashi, N. and Abe, E. (1992) J. Cell. Biochem. 49, 53–58.
204. Suda, T., Shinki, T. and Takahashi, N. (1990) Annu. Rev. Nutr. 10, 195–211.
205. Link, R. and DeLuca, H.F. (1985) in The Receptors (Conn, P.M., ed.), pp. 1–35, Academic Press, New York.
206. Rook, G.A., Steele, J., Fraher, L., Barker, S., Karmali, R., O'Riordan, J. and Stanford, J. (1986) Immunology 57, 159–163.
207. Kumar, R., Schnoes, H.K. and DeLuca, H.F. (1978) J. Biol. Chem. 253, 3804–3809.
208. Haussler, M.R., Mangelsdorf, D.J., Komm, B.S., Terpening, C.M., Yamaoka, K., Allegretto, E.A., Baker, A.R., Shine, J., McDonnell, D.P., Hughes, M. and 13 other authors (1988) Recent Prog. Horm. Res. 44, 263–305.
209. Zierold, C., Darwish, H.M. and DeLuca, H.F. (1994) Proc. Natl. Acad. Sci. U.S.A. 91, 900–902.
210. Ohyama, Y., Ozono, K., Uchida, M., Shinki, T., Kato, S., Suda, T., Yamamoto, O., Noshiro, M. and Kato, Y. (1994) J. Biol. Chem. 269, 10545–10550.
211. Okuda, K., Usui, E. and Ohyama, Y. (1995) J. Lipid Res. 36, 1641–1652.
212. Nishimura, A., Shinki, T., Jin, C.H., Ohyama, Y., Noshiro, M., Okuda, K. and Suda, T. (1994) Endocrinology 134, 1794–1799.
213. Ohyama, Y., Hayashi, S. and Okuda, K. (1989) FEBS Lett. 255, 405–408.
214. Ohyama, Y. and Okuda, K. (1991) J. Biol. Chem. 266, 8690–8695.
215. Ohyama, Y., Noshiro, M. and Okuda, K. (1991) FEBS Lett. 278, 195–198.
216. Chen, K.S., Prahl, J.M. and DeLuca, H.F. (1993) Proc. Natl. Acad. Sci. U.S.A. 90, 4543–4547.
217. Hahn, C.N., Baker, E., Laslo, P., May, B.K., Omdahl, J.L. and Sutherland, G.R. (1993) Cytogenet. Cell Genet. 62, 192–193.
218. Zierold, C., Darwish, H.M. and DeLuca, H.F. (1995) J. Biol. Chem. 270, 1675–1678.
219. Hahn, C.N., Kerry, D.M., Omdahl, J.L. and May, B.K. (1994) Nucleic Acids Res. 22, 2410–2416.
220. Chen, K.S. and DeLuca, H.F. (1995) Biochim. Biophys. Acta 1263, 1–9.
221. Inaba, M., Burgos-Trinidad, M. and DeLuca, H.F. (1991) Arch. Biochem. Biophys. 284, 257–263.
222. Bikle, D.D., Siiteri, P.K., Ryzen, E. and Haddad, J.G. (1985) J. Clin. Endocrinol. Metab. 61, 969–975.

223. Tanaka, Y., Castillo, L. and DeLuca, H.F. (1977) J. Biol. Chem. 252, 1421–1424.
224. Tanaka, Y., Wichmann, J.K., Schnoes, H.K. and DeLuca, H.F. (1981) Biochemistry 20, 3875–3879.
225. Akiyoshi-Shibata, M., Sakaki, T., Ohyama, Y., Noshiro, M., Okuda, K. and Yabusaki, Y. (1994) Eur. J. Biochem. 224, 335–343.
226. Beckman, M.J., Tadikonda, P., Werner, E., Prahl, J.M., Yamada, S. and DeLuca, H.F. (1996) Biochemistry, in press.
227. Evans, R.M. (1988) Science (Washington, D.C.) 240, 889–895.
228. Kream, B.E., Reynolds, R.D., Knutson, J.C., Eisman, J.A. and DeLuca, H.F. (1976) Arch. Biochem. Biophys. 176, 779–787.
229. Eisman, J.A., Hamstra, A.J., Kream, B.E. and DeLuca, H.F. (1976) Science (Washington, D.C.) 193, 1021–1023.
230. Simpson, R.U. and DeLuca, H.F. (1982) Proc. Natl. Acad. Sci. U.S.A. 79, 16–20.
231. Allegretto, E.A., Pike, J.W. and Haussler, M.R. (1987) Biochem. Biophys. Res. Commun. 147, 479–485.
232. Allegretto, E.A., Pike, J.W. and Haussler, M.R. (1987) J. Biol. Chem. 262, 1312–1319.
233. Pike, J.W., Marion, S.L., Donaldson, C.A. and Haussler, M.R. (1983) J. Biol. Chem. 258, 1289–1296.
234. Dame, M.C., Pierce, E.A., Prahl, J.M., Hayes, C.E. and DeLuca, H.F. (1986) Biochemistry 25, 4523–4534.
235. McDonnell, D.P., Mangelsdorf, D.J., Pike, J.W., Haussler, M.R. and O'Malley, B.W. (1987) Science (Washington, D.C.) 235, 1214–1217.
236. Baker, A.R., McDonnell, D.P., Hughes, M., Crisp, T.M., Mangelsdorf, D.J., Haussler, M.R., Pike, J.W., Shine, J. and O'Malley, B.W. (1988) Proc. Natl. Acad. Sci. U.S.A. 85, 3294–3298.
237. Burmester, J.K., Maeda, N. and DeLuca, H.F. (1988) Proc. Natl. Acad. Sci. U.S.A. 85, 1005–1009.
238. McDonnell, D.P., Scott, R.A., Kerner, S.A., O'Malley, B.W. and Pike, J.W. (1989) Mol. Endocrinol. 3, 635–644.
239. Hughes, M.R., Malloy, P.J., Kieback, D.G., Kesterson, R.A., Pike, J.W., Feldman, D. and O'Malley, B.W. (1988) Science (Washington, D.C.) 242, 1702–1705.
240. Wiese, R.J., Goto, H., Prahl, J.M., Marx, S.J., Thomas, M., Al-Aqeel, A. and DeLuca, H.F. (1993) Mol. Cell. Endocrinol. 90, 197–201.
241. Nakajima, S., Hsieh, J.C., MacDonald, P.N., Galligan, M.A., Haussler, C.A., Whitfield, G.K. and Haussler, M.R. (1994) Mol. Endocrinol. 8, 159–172.
242. Forman, B.M. and Samuels, H.H. (1990) Mol. Endocrinol. 4, 1293–1301.
243. Sone, T., Kerner, S. and Pike, J.W. (1991) J. Biol. Chem. 266, 23296–23305.
244. Lee, M.S., Kliewer, S.A., Provencal, J., Wright, P.E. and Evans, R.M. (1993) Science (Washington, D.C.) 260, 1117–1121.
245. Umesono, K. and Evans, R.M. (1989) Cell 57, 1139–1146.
246. Whitfield, G.K., Hsieh, J.C., Jurutka, P.W., Selznick, S.H., Haussler, C.A., MacDonald, P.N. and Haussler, M.R. (1995) J. Nutr. 125, 1690S–1694S.
247. Carlberg, C. (1995) Eur. J. Biochem. 231, 517–527.
248. MacDonald, P.N., Dowd, D.R. and Haussler, M.R. (1994) Semin. Nephrol. 14, 101–118.
249. Carlberg, C., Bendik, I., Wyss, A., Meier, E., Sturzenbecker, L.J., Grippo, J.F. and Hunziker, W. (1993) Nature (London) 361, 657–660.
250. Demay, M.B., Kiernan, M.S., DeLuca, H.F. and Kronenberg, H.M. (1992) Proc. Natl. Acad. Sci. U.S.A. 89, 8097–8101.
251. Liao, J., Ozono, K., Sone, T., McDonnell, D.P. and Pike, J.W. (1990) Proc. Natl. Acad. Sci. U.S.A. 87, 9751–9755.

252. MacDonald, P.N., Haussler, C.A., Terpening, C.M., Galligan, M.A., Reeder, M.C., Whitfield, G.K. and Haussler, M.R. (1991) J. Biol. Chem. 266, 18808–18813.
253. Ross, T.K., Moss, V.E., Prahl, J.M. and DeLuca, H.F. (1992) Proc. Natl. Acad. Sci. U.S.A. 89, 256–260.
254. Sone, T., Ozono, K. and Pike, J.W. (1991) Mol. Endocrinol. 5, 1578–1586.
255. Ross, T.K., Darwish, H.M., Moss, V.E. and DeLuca, H.F. (1993) Proc. Natl. Acad. Sci. U.S.A. 90, 9257–9260.
256. Yu, V.C., Delsert, C., Andersen, B., Holloway, J.M., Devary, O.V., Naar, A.M., Kim, S.Y., Boutin, J.M., Glass, C.K. and Rosenfeld, M.G. (1991) Cell 67, 1251–1266.
257. Kliewer, S.A., Umesono, K., Mangelsdorf, D.J. and Evans, R.M. (1992) Nature (London) 355, 446–449.
258. MacDonald, P.N., Dowd, D.R., Nakajima, S., Galligan, M.A., Reeder, M.C., Haussler, C.A., Ozato, K. and Haussler, M.R. (1993) Mol. Cell. Biol. 13, 5907–5917.
259. Munder, M., Herzberg, I.M., Zierold, C., Moss, V.E., Hanson, K., Clagett-Dame, M. and DeLuca, H.F. (1995) Proc. Natl. Acad. Sci. U.S.A. 92, 2795–2799.
260. Haussler, M.R., Jurutka, P.W., Hsieh, J.C., Thompson, P.D., Selznick, S.H., Haussler, C.A. and Whitfield, G.K. (1995) Bone 17, 33S-38S.
261. Walters, S.N., Reinhardt, T.A., Dominick, M.A., Horst, R.L. and Littledike, E.T. (1986) Arch. Biochem. Biophys. 246, 366–373.
262. Pike, J.W. and Haussler, M.R. (1983) J. Biol. Chem. 258, 8554–8560.
263. Barsony, J., Pike, J.W., DeLuca, H.F. and Marx, S.J. (1990) J. Cell. Biol. 111, 2385–2395.
264. Barsony, J. and McKoy, W. (1992) J. Biol. Chem. 267, 24457–24465.
265. Amizuka, N. and Ozawa, H. (1992) Arch. Histol. Cytol. 55, 77–88.
266. Barsony, J. and Marx, S.J. (1991) Proc. Natl. Acad. Sci. U.S.A. 88, 1436–1440.
267. Kamimura, S., Gallieni, M., Zhong, M., Beron, W., Slatopolsky, E. and Dusso, A. (1995) J. Biol. Chem. 270, 22160–22166.
268. Hewison, M., Rut, A.R., Kristjansson, K., Walker, R.E., Dillon, M.J., Hughes, M.R. and O'Riordan, J.L. (1993) Clin. Endocrinol. 39, 663–670.
269. Simboli-Campbell, M., Franks, D.J. and Welsh, J. (1992) Cell. Signal. 4, 99–109.
270. Simboli-Campbell, M., Gagnon, A., Franks, D.J. and Welsh, J. (1994) J. Biol. Chem. 269, 3257–64.
271. Bissonnette, M., Wali, R.K., Hartmann, S.C., Niedziela, S.M., Roy, H.K., Tien, X.Y., Sitrin, M.D. and Brasitus, T.A. (1995) J. Clin. Invest. 95, 2215–2221.
272. Brown, T.A. and DeLuca, H.F. (1990) J. Biol. Chem. 265, 10025–10029.
273. Brown, T.A. and DeLuca, H.F. (1991) Arch. Biochem. Biophys. 286, 466–472.
274. Jones, B.B., Jurutka, P.W., Haussler, C.A., Haussler, M.R. and Whitfield, G.K. (1991) Mol. Endocrinol. 5, 1137–1146.
275. Pike, J.W. and Sleator, N.M. (1985) Biochem. Biophys. Res. Commun. 131, 378–385.
276. Jurutka, P.W., Terpening, C.M. and Haussler, M.R. (1993) Biochemistry 32, 8184–81892.
277. Jurutka, P.W., Hsieh, J.C., MacDonald, P.N., Terpening, C.M., Haussler, C.A., Haussler, M.R. and Whitfield, G.K. (1993) J. Biol. Chem. 268, 6791–6799.
278. Hilliard, G.M., Cook, R.G., Weigel, N.L. and Pike, J.W. (1994) Biochemistry 33, 4300–4311.
279. Hsieh, J.C., Jurutka, P.W., Galligan, M.A., Terpening, C.M., Haussler, C.A., Samuels, D.S., Shimizu, Y., Shimizu, N. and Haussler, M.R. (1991) Proc. Natl. Acad. Sci. U.S.A. 88, 9315–9319.
280. Hsieh, J.C., Jurutka, P.W., Nakajima, S., Galligan, M.A., Haussler, C.A., Shimizu, Y., Shimizu, N., Whitfield, G.K. and Haussler, M.R. (1993) J. Biol. Chem. 268, 15118–15126.
281. Hunter, T. and Karin, M. (1992) Cell 70, 375–380.

282. Darwish, H.M., Burmester, J.K., Moss, V.E. and DeLuca, H.F. (1993) Biochim. Biophys. Acta 1167, 29–36.
283. Desai, R.K., Vanwijnen, A.J., Stein, J.L., Stein, G.S. and Lian, J.B. (1995) Endocrinology 136, 5685–5693.
284. Jurutka, P.W., Hsieh, J.C., Nakajima, S., Haussler, C.A., Whitfield, G.K. and Haussler, M.R. (1996) Proc. Natl. Acad. Sci. U.S.A. 93, 3519–3524.
285. Allen, G.F., Tsai, S.Y., Tsai, M.-J. and O'Malley, B.W. (1992) Proc. Natl. Acad. Sci. U.S.A. 89, 11750–11755.
286. Allen, G.F., X., L., Tsai, S.Y., Weigel, N.L., Edwards, D.P., Tsai, M.-J. and O'Malley, B.W. (1992) J. Biol. Chem. 267, 19513–19520.
287. Arbour, N.C., Prahl, J.M. and DeLuca, H.F. (1993) Mol. Endocrinol. 7, 1307–1312.
288. Mellon, W.S. and DeLuca, H.F. (1979) Arch. Biochem. Biophys. 197, 90–95.
289. Weigel, N.L., Carter, T.H., Schrader, W.T. and O'Malley, B.W. (1992) Mol. Endocrinol. 6, 8–14.
290. MacDonald, P.N., Sherman, D.R., Dowd, D.R., Jefcoat, S.C., Jr. and DeLisle, R.K. (1995) J. Biol. Chem. 270, 4748–4752.
291. Onate, S.A., Tsai, S.Y., Tsai, M.J. and O'Malley, B.W. (1995) Science (Washington, D.C.) 270, 1354–1357.
292. Chen, J.D. and Evans, R.M. (1995) Nature (London) 377, 454–457.
293. Kurokawa, R., Soderstrom, M., Horlein, A., Halachmi, S., Brown, M., Rosenfeld, M.G. and Glass, C.K. (1995) Nature (London) 377, 451–454.
294. Horlein, A.J., Naar, A.M., Heinzel, T., Torchia, J., Gloss, B., Kurokawa, R., Ryan, A., Kamei, Y., Soderstrom, M., Glass, C.K. and Rosenfeld, M.G. (1995) Nature (London) 377, 397–404.
295. Luban, J. and Goff, S.P. (1995) Curr. Opin. Biotechnol. 6, 59–64.
296. Naveh-Many, T. and Silver, J. (1990) J. Clin. Invest. 86, 1313–1319.
297. Naveh-Many, T., Friedlaender, M.M., Mayer, H. and Silver, J. (1989) Endocrinology 125, 275–280.
298. Slatopolsky, E., Weerts, C., Thielan, J., Horst, R., Harter, H., and Martin, K.J. (1984) J. Clin. Invest. 74, 2136–2143.
299. Gonzalez, E.A. and Martin, K.J. (1995) Nephrol. Dial. Transplant. 10, 13–21.
300. Kitagawa, T., Akatsuka, A., Owada, M. and Mano, T. (1980) Contrib. Nephrol. 22, 107–119.
301. Latta, K., Hisano, S. and Chan, J.C. (1993) Pediatr. Nephrol. 7, 744–748.
302. Davisson, M.T. (1987) Genomics 1, 213–227.
303. Hruska, K.A., Rifas, L., Cheng, S.L., Gupta, A., Halstead, L. and Avioli, L. (1995) Am. J. Physiol. 268, F357-F362.
304. Glorieux, F.H., Holick, M.F., Scriver, C.R. and DeLuca, H.F. (1973) Lancet 2, 287–289.
305. Nesbitt, T., Econs, M.J., Byun, J.K., Martel, J., Tenenhouse, H.S. and Drezner, M.K. (1995) J. Bone Miner. Res. 10, 1327–1333.
306. Collins, J.F., Scheving, L.A. and Ghishan, F.K. (1995) Am. J. Physiol. 269, F439-F448.
307. Lobaugh, B. and Drezner, M.K. (1983) J. Clin. Invest. 71, 400–403.
308. Roy, S., Martel, J., Ma, S. and Tenenhouse, H.S. (1994) Endocrinology 134, 1761–1767.
309. Harrell, R.M., Lyles, K.W., Harrelson, J.M., Friedman, N.E. and Drezner, M.K. (1985) J. Clin. Invest. 75, 1858–1868.
310. Prader, A., Illig, R. and Heierli, E. (1961) Helv. Paediatr. Acta 16, 452–468.
311. Harmeyer, J.V., Grabe, C. and Winkley, I. (1982) Exp. Biol. Med. 7, 117–125.
312. Fraser, D., Kooh, S.W., Kind, H.P., Holick, M.F., Tanaka, Y. and DeLuca, H.F. (1973) N. Engl. J. Med. 289, 817–822.
313. Glorieux, F. (1991) N. Engl. J. Med. 325, 1875–1877.

314. Brooks, M.H., Bell, N.H., Love, L., Stern, P.H., Orfei, E., Queener, S.F., Hamstra, A.J. and DeLuca, H.F. (1978) N. Engl. J. Med. 298, 996–999.
315. Malloy, P.J., Hochberg, Z., Pike, J.W. and Feldman, D. (1989) J. Clin. Endocrinol. Met. 68, 263–269.
316. Pike, J.W., Dokoh, S., Haussler, M.R., Liberman, U.A., Marx, S.J. and Eli, C. (1984) Science (Washington, D.C.) 224, 879–881.
317. Balsan, S., Garabedian, M., Leberman, U.A., Eil, C., Bourdeau, N., Guillozo, H., Grimberg, R., Le Deunff, M.J., Lieberherr, M., Guimbaud, P., Broyer, M. and Marx, S.J. (1983) J. Clin. Endocrinol. Met. 57, 803–811.
318. Castells, S., Greig, F., Fusi, M.A., Finberg, L., Yasumura, S., Liberman, U.A., Eil, C. and Marx, S.J. (1986) J. Clin. Endocrinol. Metab. 63, 252–256.
319. Eil, C., Liberman, U.A., Rosen, J.F. and Marx, S.J. (1981) New Engl. J. Med. 304, 1588–1591.
320. Rosen, J.F., Fleischman, A.R., Lennette, E.T. and Henle, G. (1979) J. Pediatr. 94, 729–735.
321. Balson, S., Garabedian, M., Larchet, M., Gorski, A.M., Cournot, G., Tau, C., Bourdeau, A., Silve, C., and Ricour, C. (1986) J. Clin. Invest. 77, 1661–1667.
322. Eisman, J.A., Kelly, P.J., Morrison, N.A., Pocock, N.A., Yeoman, R., Birmingham, J. and Sambrook, P.N. (1993) Osteoporos. Int. 3, 56–60.
323. Gallagher, J.C., Riggs, B.L., Eisman, J., Hamstra, A., Arnaud, S.B. and DeLuca, H.F. (1979) J. Clin. Invest. 64, 729–736.
324. Horst, R.L., Goff, J.P. and Reinhardt, T.A. (1990) Endocrinology 126, 1053–1057.
325. Hartwell, D., Rodbro, P., Jensen, S.B., Thomsen, K. and Christiansen, C. (1990) Scand. J. Clin. Lab. Invest. 50, 115–121.
326. Riggs, B.L., Gallagher, J.C., DeLuca, H.F. and Zinsmeister, A.R. (1982) in Vitamin D. Chemical, Biochemical and Clinical Endocrinology of Calcium Metabolism (Norman, A.W., Schaefer, K., Herrath, D.v.and Grigoleit, H.-G., eds.), pp. 903–908, Walter de Gruyter, New York.
327. Morio, K. and Koide, K. (1995) Nippon Rinsho – Jpn. J.Clin. Med. 53, 958–64.
328. Ohta, H., Sugimoto, I., Masuda, A., Komukai, S., Suda, Y., Makita, K., Takamatsu, K., Horiguchi, F. and Nozawa, S. (1996) Bone 18, 227–231.
329. Sulak, P.J. (1996) Internatl J. Fertil. Menopausal Stud. 41, 85–89.
330. Ash, S.L. and Goldin, B.R. (1988) Am. J. Clin. Nutr. 47, 694–699.
331. Kohn, B., Erben, R., Zucker, H., Weiser, H. and Rambeck, W.A. (1990) J. Anim. Physiol. Anim. Nutr. (S20), 78–83.
332. Ravn, S.H., Rosenberg, J. and Bostofte, E. (1994) Eur. J. Obstetr. Gynecol. Reprod. Biol. 53, 81–93.
333. Breslau, N.A. (1994) Rheum. Dis. Clin. N. Am. 20, 691–716.
334. Adler, R.A. and Rosen, C.J. (1994) Endocrinol. Metab. Clin. N. Am. 23, 641–654.
335. Reid, I.R., Veale, A.G. and France, J.T. (1994) J. Asthma 31, 7–18.
336. Olbricht, T. and Benker, G. (1993) J. Int. Med. 234, 237–244.
337. Howard, G., Nguyen, T., Morrison, N., Watanabe, T., Sambrook, P., Eisman, J. and Kelly, P.J. (1995) J. Clin. Endocrinol. Metab. 80, 2800–2805.
338. Peacock, M. (1995) J. Bone Miner. Res. 10, 1294–1297.
339. Carling, T., Kindmark, A., Hellman, P., Lundgren, E., Ljunghall, S., Rastad, J., Akerstrom, G. and Melhus, H. (1995) Nature Med. 1, 1309–1311.
340. Fujita, T. (1993) Nippon Rinsho - Jpn. J. Clin. Med. 51, 1004–1010.
341. Orimo, H., Shiraki, M., Hayashi, T. and Nakamura, T. (1987) Bone Miner. 3, 47–52.
342. Caniggia, A., Nuti, R., Lore, F., Martini, G., Turchetti, V. and Righi, G. (1990) Metabolism 39, 43–49.

343. Tilyard, M.W., Spears, G.F., Thomson, J. and Dovey, S. (1992) N. Engl. J. Med. 326, 357–362.
344. Gallagher, J.C. (1990) Metabolism 39, 27–29.
345. Aloia, J.F. (1990) Metabolism 39, 35–38.
346. Ott, S.M. and Chesnut, C.H. (1989) Ann. Int. Med. 110, 267–274.
347. Chesnut, C.H. (1992) N. Engl. J. Med. 326, 406–408.
348. Menne, T. and Larsen, K. (1992) Semin. Dermatol. 11, 278–283.
349. Kragballe, K. (1992) J. Am. Acad. Dermatol. 27, 1001–1008.
350. Holick, M.F., Smith, E. and Pincus, S. (1987) Arch. Dermatol. 123, 1677–1681.
351. Sorensen, H., Binderup, L., Calverley, M.J., Hoffmeyer, L. and Andersen, N.R. (1990) Biochem. Pharmacol. 39, 391–393.
352. Bhalla, A.K., Amento, E.P., Clemens, T.L., Holick, M.F. and Krane, S.M. (1983) J. Clin. Endocrinol. Metab. 57, 1308–1310.
353. Provvedini, D.M., Deftos, L.J. and Manolagas, S.C. (1984) Biochem. Biophys. Res. Commun. 121, 277–283.
354. Manolagas, S.C., Yu, X.P., Girasole, G. and Bellido, T. (1994) Semin. Nephrol. 14, 129–43.
355. Hewison, M. (1992) J. Endocrinol. 132, 173–175.
356. Casteels, K., Bouillon, R., Waer, M. and Mathieu, C. (1995) Curr. Opin. Nephrol. Hyperten. 4, 313–318.
357. Rigby, W.F.C., Stacy, T. and Fanger, M.W. (1984) J. Clin. Invest. 74, 1451–1455.
358. Muscettola, M. and Grasso, G. (1988) Immunol. Lett. 17, 121–124.
359. Provvedini, D.M. and Manolagas, S.C. (1989) J. Clin. Endocrinol. Metab. 68, 774–779.
360. Iho, S., Takahashi, T., Kura, F., Sugiyama, H. and Hoshino, T. (1986) J. Immunol. 136, 4427–4431.
361. Goldman, R. (1984) Cancer Res. 44, 11–19.
362. Amento, E.P. and Cotter, A.C. (1988) J. Bone Miner. Res. 3, S217.
363. Tobler, A., Gasson, J., Reichel, H., Norman, A.W. and Koeffler, H.P. (1987) J. Clin. Invest. 79, 1700–1705.
364. Adams, J.S., Ren, S.Y., Arbelle, J.E., Horiuchi, N., Gray, R.W., Clemens, T.L. and Shany, S. (1994) Endocrinology 134, 2567–2573.
365. Yang, S., Smith, C. and DeLuca, H.F. (1993) Biochim. Biophys. Acta 1158, 279–286.
366. Mohamed, M.I., Beckman, M.J., Meehan, J. and DeLuca, H.F. (1996) Biochim. Biophys. Acta, 275–283.
367. Banbe, M. and Viccari, E. (1967) Minerva. Pediatr. 19, 377–381.
368. Stroder, J. and Kasal, P. (1970) Klin. Wochenschr. 48, 383–384.
369. Lorente, F., Fontan, G., Jara, P., Casas, C., Garcia-Rodriguez, M.C. and Ojeda, J.A. (1976) Acta Paediatr. Scand. 65, 695–699.
370. Stroder, J. and Franzen, C. (1975) Klin. Paediatr. 187, 461–464.
371. Toss, G. and Symreng, T. (1983) Int. J. Vitam. Nutr. Res. 53, 27–31.
372. Stroder, J. (1975) in Vitamin D and Problems of Uremic Bone Disease (Norman, A.W., Schaefer, K., Grigoleit, H.G., Herrath, D.v.and Ritz, E., eds.), p. 675, de Gruyter, Berlin.
373. Haug, C., Muller, F., Aukrust, P. and Froland, S.S. (1994) J. Infect. Dis. 169, 889–893.
374. Yang, S., Smith, C., Prahl, J.M., Luo, X. and DeLuca, H.F. (1993) Arch. Biochem. Biophys. 303, 98–106.
375. Mathieu, C., Waer, M., Casteels, K., Laureys, J. and Bouillon, R. (1995) Endocrinology 136, 866–872.
376. Lemire, J.M. (1992) J. Cell. Biochem. 49, 26–31.
377. Cantorna, M., Hayes, C.E. and DeLuca, H.F. (1996), Proc. Natl. Acad. Sci. U.S.A., in press.

378. Suda, T. (1989) Proc. Soc. Exp. Biol. Med. 191, 214–220.
379. Eisman, J.A., Koga, M., Sutherland, R.L., Barkla, D.H. and Tutton, P.J.M. (1989) Proc. Soc. Exp. Biol. Med. 191, 221–226.
380. Eisman, J.A. and Martin, T.J. (1989) Lancet 334, 549–550.
381. Sandgren, M., Danforth, L., Plasse, T.F. and DeLuca, H.F. (1991) Cancer Res. 51, 2021–2024.
382. Frampton, R.J., Suva, L.J., Eisman, J.A., Findlay, D.M., Moore, G.E., Moseley, J.M. and Martin, T.J. (1982) Cancer Res. 42, 1116–1119.
383. Hanchette, C.L. and Schwartz, G.G. (1992) Cancer 70, 2861–2869.
384. Schwartz, G.G. and Hulka, B.S. (1990) Anticancer Res. 10, 1307–1311.
385. Corder, E.H., Guess, H.A., Hulka, B.S., Friedman, G.D., Sadler, M., Vollmer, R.T., Lobaugh, B., Drezner, M.K., Vogelman, J.H. and Orentreich, N. (1993) Cancer Epidemiol. Biomarkers Prev. 2, 467–472.
386. Dokoh, S., Donaldson, C.A. and Haussler, M.R. (1984) Cancer Res. 44, 2103–2109.
387. Miller, G.J., Stapleton, G.E., Houmiel, K.L. and Ferrara, J.A. (1995) Clin. Cancer Res. 1, 997–1003.
388. Hedlund, T.E., Moffatt, K.A. and Miller, G.J. (1996) Endocrinology 137, 1554–1561.
389. Kreutz, M., Andreesen, R., Krause, S.W., Szabo, A., Ritz, E. and Reichel, H. (1993) Blood 82, 1300–1307.
390. Wali, R.K., Baum, C.L., Sitrin, M.D. and Brasitus, T.A. (1990) J. Clin. Invest. 85, 1296–1303.
391. Lieberherr, M., Grosse, B., Duchambon, P. and Drueke, T. (1989) J. Biol. Chem. 264, 20403–20406.
392. Wali, R.K., Baum, C.L., Sitrin, M.D., Bolt, M.J., Dudeja, P.K. and Brasitus, T.A. (1992) Am. J. Physiol. 262, G945-G953.
393. Grosse, B., Bourdeau, A. and Lieberherr, M. (1993) J. Bone Miner. Res. 8, 1059–1069.
394. Yen, A., Chandler, S., Forbes, M.E., Fung, Y.K., T'Ang, A. and Pearson, R. (1992) Eur. J. Cell Biol. 57, 210–221.
395. Mangasarian, K. and Mellon, W.S. (1993) Biochim. Biophys. Acta 1172, 55–63.
396. Kolla, S.S. and Studzinski, G.P. (1993) J. Cell. Physiol. 156, 63–71.
397. Yen, A., Forbes, M.E., Varvayanis, S., Tykocinski, M.L., Groger, R.K. and Platko, J.D. (1993) J. Cell. Physiol. 157, 379–391.
398. Labbaye, C., Zhang, J., Casanova, J.L., Lanotte, M., Teng, J., Miller, W.H., Jr. and Cayre, Y.E. (1993) Blood 81, 475–481.
399. Meenakshi, T., Ross, F.P., Martin, J. and Teielbaum, S.L. (1993) J. Cell. Biochem. 53, 145–155.
400. Honda, A., Morita, I., Murota, S.I. and Mori, Y. (1986) Biochim. Biophys. Acta 877, 423–432.
401. Kole, K.L., Gyetko, M.R., Simpson, R.U. and Sitrin, R.G. (1991) Biochem. Pharmacol. 41, 585–591.
402. Goto, H., Chen, K.S., Prahl, J.M. and DeLuca, H.F. (1992) Biochim. Biophys. Acta 1132, 103–108.
403. Ostrem, V.K. and DeLuca, H.F. (1987) Steroids 49, 73–102.
404. Hruska, K.A., Bar-Shavit, Z., Malone, J.D. and Teitelbaum, S. (1988) J. Biol. Chem. 263, 16039–16044.
405. Collins, S.J., Ruscelli, F.W., Gallagher, R.G. and Gallo, R.C. (1978) Proc. Natl. Acad. Sci. U.S.A. 75, 2458–2462.
406. Beitman, T.R., Selonick, S.G. and Collins, S.J. (1980) Proc. Natl. Acad. Sci. U.S.A. 77, 2936–2940.

407. Morel, P.A., Manolagas, S.C., Provvedini, D.M., Wegman, D.R. and Chiller, J.M. (1986) J. Immunol. 136, 2181–2186.
408. Ostrem, V.K., Lau, W.F., Lee, S.H., Perlman, K., Prahl, J., Schnoes, H.K. and DeLuca, H.F. (1987) J. Biol. Chem. 262, 14164–14171.
409. Bouillon, R., Okamura, W.H. and Norman, A.W. (1995) Endocr. Rev. 16, 200–257.
410. Skowronski, R.J., Peehl, D.M. and Feldman, D. (1995) Endocrinology 136, 20–26.

Progress in Medicinal Chemistry – Vol. 35
Edited by G.P. Ellis, D.K. Luscombe and A.W. Oxford
© 1998 Elsevier Science B.V. All rights reserved.

2 Neurokinin Receptor Antagonists

CHRISTOPHER J. SWAIN, Ph.D.

Merck, Sharp and Dohme Research Laboratories, Neuroscience Research Centre, Terlings Park, Harlow, Essex, CM20 2QR, U.K.

INTRODUCTION

A number of extensive reviews have been published which cover much of the current literature on receptor distribution, *in vitro* and *in vivo* pharmacology, potential clinical utilities of neurokinin (NK) receptor antagonists and the patent literature [1–5]. Neurokinin receptors are currently divided into three main populations. The endogenous agonists are a family of peptides that share the common C-terminal sequence 'Phe-X-Gly-Leu-Met-NH_2'. The preferred mammalian neurokinin agonist for each of the receptors is substance P (SP) for NK_1, neurokinin A (NKA) for NK_2 and neurokinin B (NKB) for NK_3. However, each of the agonists can act as a full agonist at each of the receptors albeit with reduced affinity. It is possible that under physiological conditions the actions of each agonist may be mediated by more than one receptor type.

Mammalian Tachykinins

Substance P	Arg-Pro-Lys-Pro-Gln-Gln-*Phe*-Phe-*Gly-Leu-Met* -NH_2
Neurokinin A	His-Lys-Thr-Asp-Ser-*Phe*-Val-*Gly-Leu-Met* -NH_2
Neurokinin B	Asp-Met-His-Asp-Phe-*Phe*-Val-*Gly-Leu-Met* -NH_2

All three human receptors have been cloned and expressed in cell lines facilitating rapid screening of novel agents for receptor selectivity [6a-c]. The cloning of the human receptor was of particular importance, because there are considerable species differences with respect to antagonist affinities.

BIOLOGICAL ACTIONS OF NEUROKININS AND NEUROKININ ANTAGONISTS

SUBSTANCE P-MEDIATED ACTIONS

Substance P and the related tachykinins neurokinin A and neurokinin B are mainly found in neurons, particularly unmyelinated sensory somatic and visceral fibres, in enteric sensory neurons and in a number of pathways within the brain. The release of tachykinins from the peripheral ends of these neurons may play an important role in the neurogenic inflammatory responses to local injury and inflammation by promoting the release of histamine from mast cell degranulation, and the release of cytokines from invading white cells, as well as acting directly upon blood vessels to produce vasodilation and plasma extravasation. Neurogenic inflammation within the dura mater, a pain-producing intracranial tissue [7], has been hypothe-sized to be the source of migraine headache pain (*Figure 2.1*) since it is inhib-ited by current clinically effective anti-migraine agents [8]. Within the CNS, at the central terminals of primary sensory neurons in the brainstem, cranial nerve nuclei and spinal cord dorsal horn, substance P may function as a sen-sory neurotransmitter or neuromodulator particularly with regard to the central input to pain pathways from unmyelinated C-fibres involved in noci-ception. This central antinociceptive site of action may be critical to the speed of onset and the effectiveness of NK_1 antagonists for the acute treat-ment of migraine since they would be able to block headache pain rapidly and directly. Substance P-containing afferent nerve fibres also innervate the brainstem in the region of the nucleus tractus solitarius, a CNS area in-volved in the control of emesis (*Figure 2.2*) which is indicative of a role for central NK_1 receptor antagonists as antiemetics [9,10]. Confirmation of this central site of action has come from studies using the non-CNS pene-trating NK_1 antagonist L743310 [11]. It has also been hypothesized that tachykinins within the CNS may have a role in modulating the activity of norepinephrine and dopamine pathways [12] that are involved in affective disorders and responses to stress. The presence of SP and NKA in the res-piratory tract of various mammalian species has been shown using radioim-munoassay and immunohistochemical techniques; this suggests a possible therapeutic role for NK_1 and/or NK_2 antagonists in airways disease [12].

NKA-MEDIATED ACTIONS

NK_2 receptor antagonists have been claimed to have anxiolytic properties in preclinical assays [13]. However, until very recently it had not been possible to demonstrate the existence of NK_2 receptors in brain tissue from adult rats (although present in neonatal tissue) using radioligand-binding tech-niques. However, new autoradiographic evidence for the existence of NK_2 receptors in adult rat brain using localization of a radiolabelled NK_2 recep-tor selective antagonist, $[^3H]$-SR48968 has been disclosed [14].

NKB-MEDIATED ACTIONS

The presence of NK_3 receptors in the CNS of man has not yet been con-firmed. However, a number of CNS effects have been demonstrated in other species. NK_3 receptors have been shown to modulate central monoamine function suggesting that agonists or antagonists may be clinically useful for the treatment of schizophrenia, Parkinson's disease and depression. NK_3 re-ceptors have been shown to modulate the activity of dopaminergic cell

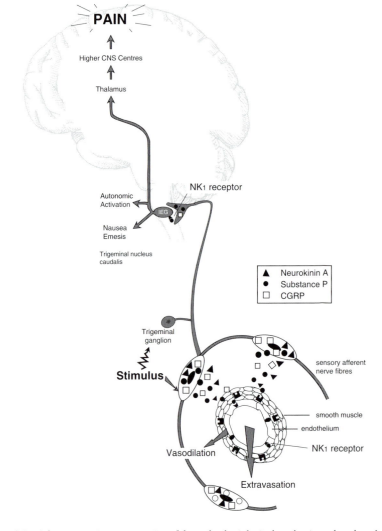

Figure 2.1. A diagrammatic representation of the pathophysiological mechanisms thought to be involved in the development of migraine headache and the potential sites of action of NK$_1$-receptor antagonists. In preclinical studies, NK$_1$-receptor antagonists antagonize the effects of substance P released from the perivascular sensory neurons (possibly via blockade of receptors located on the endothelial cells) resulting in a reduction in vasodilation and extravasation. Furthermore, NK$_1$-receptor antagonists inhibit immediate early gene (IEG) expression in the trigeminal nucleus caudalis indicating inhibition of neuronal activation which may also contribute to a central antinociceptive action of these agents.

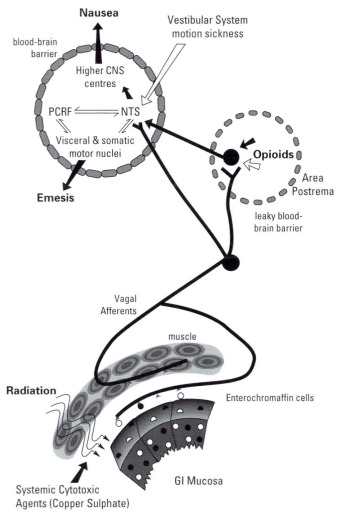

Figure 2.2. Diagrammatic representation of the emetic reflex pathways. Abdominal vagal sensory afferents containing substance P terminate within the nucleus tractus solitarius (NTS) in the brainstem. Emetogens cause activation of a number of different neuronal pathways that ultimately pass through the NTS which acts as a 'common gateway' and then onto other centres in the brain to produce the emetic reflex response and the sensation of nausea. Cytotoxic agents and radiation trigger the emetic reflex via activation of the peripheral endings of sensory neurons, opioids act in the area postrema (with efferents from the area postrema projecting to the NTS) and motion causes activation of the vestibular system which also has inputs to the NTS. NK_1-receptor antagonists may prevent emesis by an action in the NTS although other sites of action cannot be excluded.

bodies in the ventral tegmental area and substantia nigra in rodents suggesting that an NK_3 receptor *antagonist* may exhibit antipsychotic activity [15]. On the other hand, central infusion of the peptide NK_3 receptor agonist senktide in rats elicits the 5-HT behavioural syndrome [16] and increases release of norepinephrine suggesting that NK_3 *agonists* may have utility as antidepressants.

NK_1 RECEPTOR-SELECTIVE NON-PEPTIDE ANTAGONISTS

DIAMINES

Pfizer discovered the first non-peptide NK_1 receptor antagonist CP96345 (1) in 1991 [17]. An investigation of the structure-activity relationships within this series of quinuclidines highlighted the crucial importance of the *ortho*-methoxy substitution of the benzylamine side-chain [18], removal of the methoxy group resulting in a 20-fold drop in affinity. Whilst (1) is a potent antagonist of substance P *in vitro* and *in vivo,* it also has significant affinity for a number of ion channels which compromise its use *in vivo* [19].

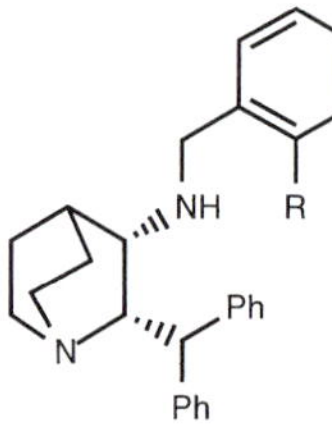

	R	IC_{50} (nM)
(1)	OMe	0.77
(2)	H	16
(3)	Cl	33
(4)	Et	17

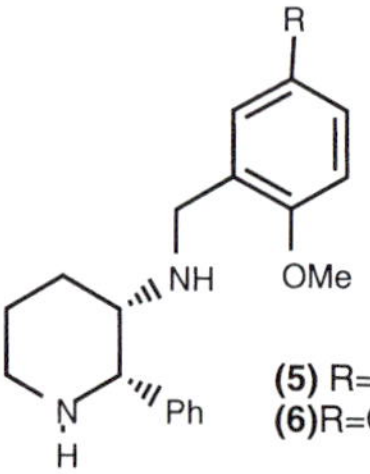

(5) R=H
(6) R=OCF_3

More recently, the piperidine analogue CP99994 (5) was reported to have excellent affinity and selectivity *in vitro* [human NK_1 receptor (hNK_1), human NK_2 receptor (hNK_2), human NK_3 receptor (hNK_3)] (hNK_1 IC_{50} 0.6 nM; hNK_2 IC_{50} > 1000 nM; hNK_3 IC_{50} > 1000 nM) and considerably reduced ion channel activity compared with (1) [20]. However, CP99994 apparently has poor oral activity in a number of species. The poor oral bioavailability of CP99994 would now appear to have been addressed with the discovery of CP122721 (6), which is a potent and selective NK_1 antagonist (hNK_1 IC_{50} 0.14 nM) but has significant affinity for the L-type calcium channel (390 nM) [21]. CP122721 is active in assays of NK_1 agonist-induced dermal extravasation (0.01–0.05 mg/kg p.o.), substance P-induced locomotor hyperactivity (0.2 mg/kg) and against cisplatin-induced emesis (0.3 mg/kg i.v.) [22].

Pfizer and others have continued to explore novel bicyclic diamines. This approach has been used successfully by a number of groups [23, 24], although the resulting highly rigid cage structures appear not to show any advantage over the simpler diamines.

Glaxo postulated that the 5-position of the aromatic ring, *para* to the methoxy, was a likely site for hydroxylation. Introduction of a variety of substituents into the 5-position in an effort to block metabolism resulted in the discovery of the tetrazole GR203040 (7) which had an improved *in vivo* profile over CP99994 [25]. In particular, there is a considerable reduction in clearance (CP99994: 94 mL/min/kg; GR203040: 24 mL/min/kg) and improvement in oral bioavailability in dogs (CP99994: <10%; GR203040: 76%). GR203040 has excellent *in vitro* affinity (hNK$_1$ IC$_{50}$ 0.06nM). In ferrets, GR203040 (0.1 mg/kg s.c.) significantly reduced emesis induced by a wide range of emetogens including X-irradiation, cisplatin and morphine. Interestingly, while the isomeric tetrazole (8) and the pyridine (9) afford excellent affinity they have reduced *in vivo* potency [25]. GR203040 was initially selected for clinical evaluation in emesis and migraine but has now been replaced by GR205171 (10) which also has excellent *in vivo* activity in the ferret against X-irradiation induced emesis (0.03 mg/kg s.c.) [26]. GR205171 is now believed to be in Phase II trials for emesis and migraine.

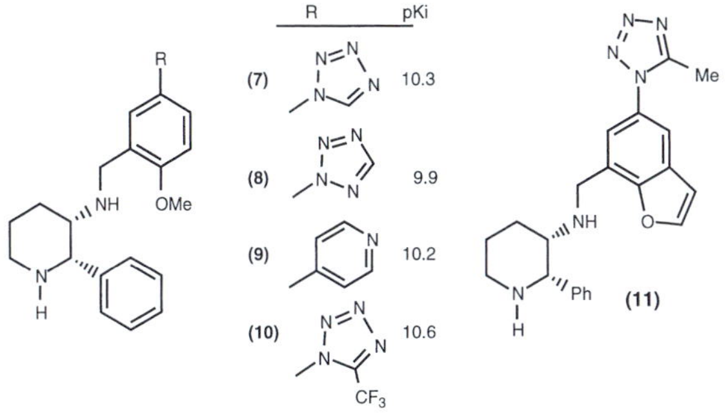

The tetrazole has also been utilized in the corresponding benzofuran analogue (11) which is reported to be active against radiation-induced emesis in ferrets (0.3 mg/kg s.c.) [27].

AMINO ETHERS

Since the first publication by the Merck group disclosing a novel series of

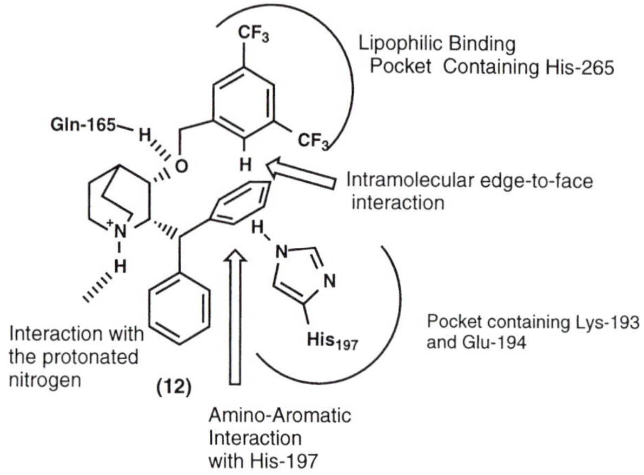

Figure 2.3

amino ether-based NK_1 antagonists (12), structure-activity relationships in this series together with a summary of the mutagenesis studies that have helped to define the pharmacophore have been reviewed [28]. On the basis of the structure-activity data for the amino ethers and the data available for the corresponding amines, four key features of the pharmacophore have been identified and are depicted in *Figure 2.3*. Firstly, an interaction with the protonated bridgehead nitrogen; whilst there are a number of acidic residues within the transmembrane region, there is no evidence to suggest that any of these form an ion-pair with the protonated nitrogen. Secondly, the benzhydryl group forms a key element of the structure of the high affinity NK_1 antagonists; however, only one of the phenyl rings is necessary for binding to the receptor, with the second ring probably acting as a conformational anchor. Thirdly, mutagenesis studies have suggested that His-197 forms an amino-aromatic interaction with one of the phenyl rings. Finally, single point mutations have also shown that Lys-193 and Glu-194 are in the binding pocket of the benzhydryl group [29]. The heteroatom of the benzyl ether probably acts as a hydrogen bond acceptor, and recent results suggest that Gln-165 may be involved in this interaction. The interaction of the third aromatic ring with the receptor appears to be different for the two classes of quinuclidine antagonists. It is possible that in the case of the benzyl ethers, an edge-to-face aromatic-aromatic interaction stabilises a conformation which accesses a lipophilic pocket containing His-265. In contrast, the benzylamines adopt a face-to-face conformation, perhaps allowing the *ortho* methoxy to engage a second hydrogen bonding interaction, while mutagenesis studies have failed to identify the residue involved, a possible interaction with the peptide backbone cannot be excluded.

The issue of calcium channel-binding was elegantly addressed [30] in the amino ether series by modulation of the pKa of the amine. Introduction of polar electron-withdrawing groups reduce affinity for the L-type calcium channel while retaining high affinity for the NK_1 receptor. The 3,5-bis(trifluoromethyl)benzyl ether is also present in the *N*-substituted 2-phenyl-3-benzyloxypiperidine series, for example, the triazole (13) and the triazolinone L741671 (14) [31]. Both compounds have excellent affinity (13), (hNK$_1$ IC$_{50}$ 0.18 nM); (14), (hNK$_1$ IC$_{50}$ 0.1 nM), and block the extravasation induced by substance P (0.5 pmol) injected into the dorsal skin in the guinea-pig [(14) ID$_{50}$ 0.007 mg/kg p.o.]. They are also active in the trigeminal dural extravasation model of migraine (10–100 μg/kg) and block cisplatin-induced emesis in ferrets (ID$_{50}$ 0.3 mg/kg). The corresponding morpholine, L742694 (15) (hNK$_1$ IC$_{50}$ 0.1 nM) has also been described and has similar affinity but an improved duration of action [32]. L742694 is a highly CNS-penetrating compound and is a potent anti-emetic (ID$_{50}$ 0.18 mg/kg i.v.) [33].

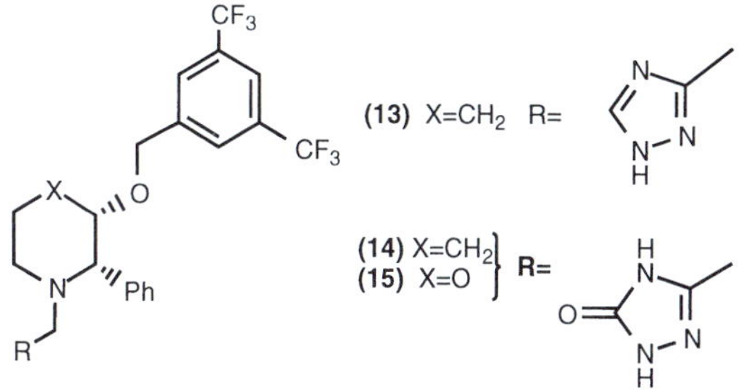

PERHYDROISOINDOLES

Since the discovery of the perhydroisoindole-based NK_1 antagonist RP67580 (16) which showed excellent affinity for the rat NK_1 receptor but 25-fold reduced affinity for the human NK_1 receptor, continued exploration of this series has identified compounds with excellent affinity at the human NK_1 receptor [34].

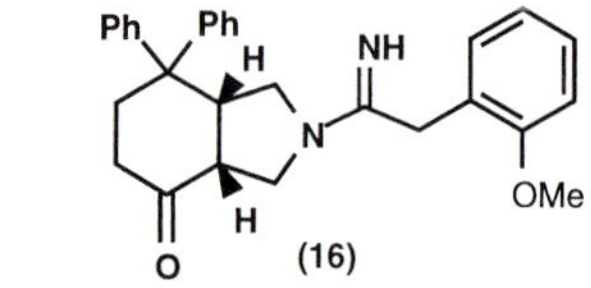

The key structural changes that confer increased affinity for the human NK_1 receptor appear to be modification of the ketone function and replacement of the amidine linking group by amide. The *ortho* methoxy has also been changed to an alkylamino group but this may be intended to improve water solubility. These compounds are claimed to be active in models of analgesia: (17) (ED_{50} 3 mg/kg p.o.), (18) (ED_{50} 1.7 mg/kg p.o.); they inhibit septide (a selective NK_1 agonist) induced extravasation (17) (ED_{50} 0.7 mg/kg s.c.), (18) (ED_{50} <0.1 mg/kg s.c.) [35]. Further modifications of the ketone functionality have now been disclosed in which a fourth aryl ring has been introduced to afford the tertiary alcohol RPR100893 (19). Interestingly, it is now the opposite enantiomer that has high affinity. The *in vitro* and *in vivo* pharmacological profiles have been extensively evaluated [36,37]. In IM9 cells, RPR100893 has an affinity for the hNK_1 receptor of 30 nM, whilst in rat brain the affinity is 1417 nM. Compound (19) was active in models of analgesia (formalin paw, ED_{50} 3.1 mg/kg s.c., ED_{50} 11 mg/kg p.o.), inflammation (septide-induced extravasation, ED_{50} 30.1 mg/kg i.v., ED_{50} 0.33 mg/kg p.o.), and migraine (dural extravasation, ED_{50} 2.5 ng/kg i.v., ED_{50} 0.5 mg/kg p.o.). Recent work has resulted in removal of the gem diphenyl moiety and the introduction of polar substituents at the 5-position of the perhydroisoindoline ring [38], in particular, 5-cyanomethyl (RPR111905, 20) (hNK_1 IC_{50} 0.3 nM). This compound shows good oral activity (dural extravasation, ED_{50} 0.09 mg/kg p.o.) but poor CNS penetration (formalin test, ED_{50} 17mg/kg p.o.).

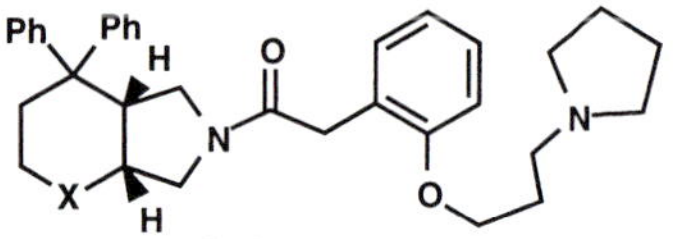

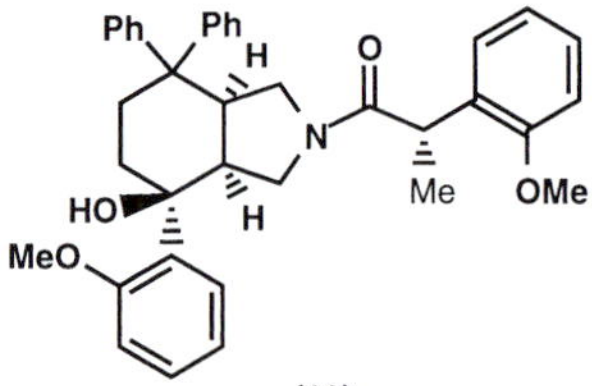

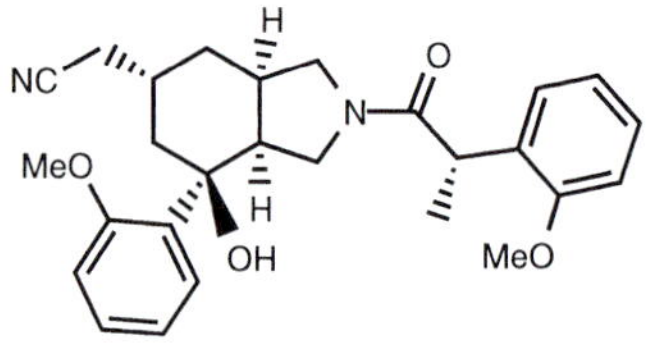

4-AMINO-2-BENZYLPIPERIDINE AMIDES

On the assumption that the two aromatic residues in truncated C-terminal analogues of substance P were essential so as to retain modest affinity, Ciba-Geigy designed a series of 4-amino-2-benzylpiperidine amides [39]. 3,5-Disubstitution on the benzamide ring was found to increase affinity and at least one small substituent on the exocyclic nitrogen is necessary for good affinity. A compound from this series, CGP49823 (21) which has relatively modest affinity (11nM) was reported to be active in the rat social interaction and forced swim tests for anxiolytic and antidepressant activity, respectively [40,41]. CGP49823 is reported to be in Phase I clinical trials as a potential treatment for anxiety disorders. Recently, analogues (22) and (23) of CGP49823 with additional N-acyl substituents have been described in patents [42, 43]. As with other recent patent filings in this series, the preferred compounds are 3,5-bis(trifluoromethyl)benzamides.

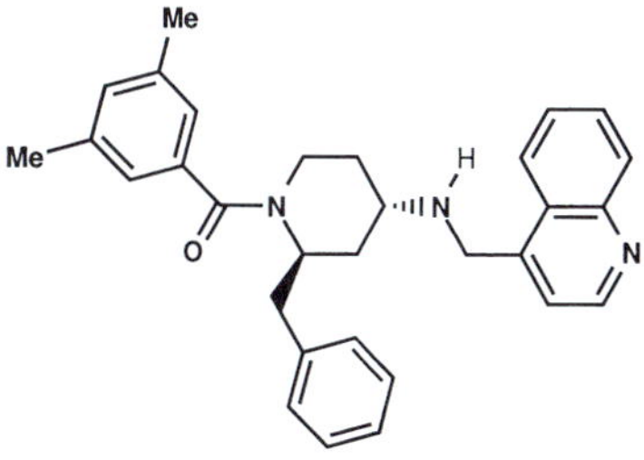

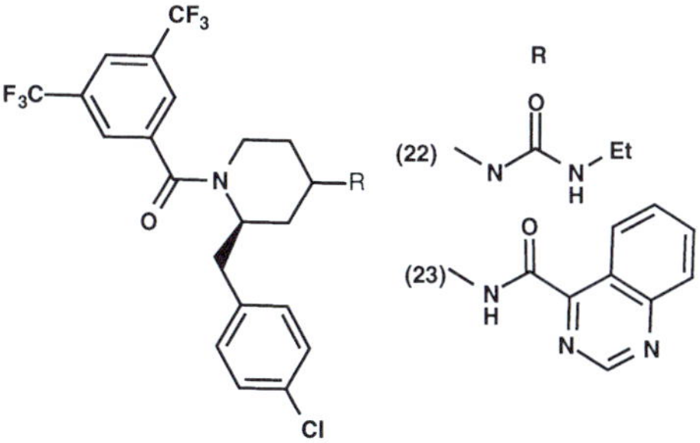

PIPERIDINE AMIDES

A novel N-acylated 3-(3,4-dichlorophenyl)piperidine, SR140333 (24) that displays high affinity both for rat and human NK_1 receptors (rat NK_1 IC_{50} 0.02 nM, human NK_1 IC_{50} 0.01 nM) has been identified by Sanofi scientists [44]. This compound is structurally related to NK_2 and NK_3 antagonists that have also been disclosed by the same group but shows excellent selectivity for the NK_1 receptor. Selectivity for the NK_1 receptor was achieved by introduction of the conformational restraint imposed by the piperidine ring and replacement of the benzoyl group by phenacyl. In the rat, (24) given intravenously strongly inhibited the plasma extravasation induced by sciatic nerve stimulation, mustard oil application and substance P [45].

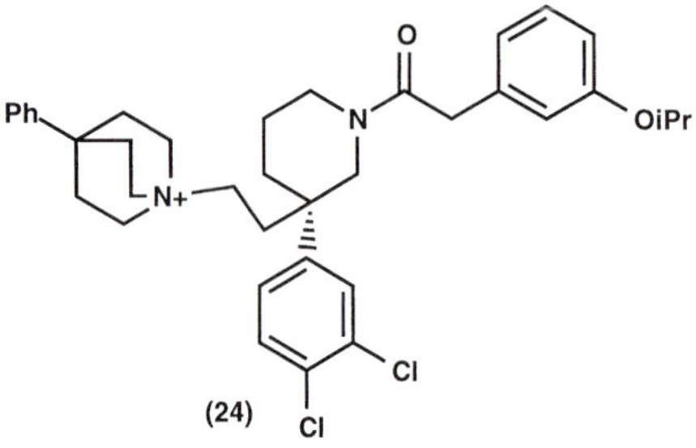

(24)

SR140333 is reported to be in Phase I trials for inflammation and migraine but would be expected to show poor CNS penetration due to its quaternary nitrogen moiety.

QUINOLINE AND 1,7-NAPHTHYRIDINE AMIDES

A new class of amide-based NK_1 antagonists (25) and (26) have been described by scientists at Takeda, again bearing the 3,5-bis(trifluoromethyl) substitution and the detailed SAR in this series has recently been published [46]. While a variety of benzyl substitutions gave excellent *in vitro* affinity (25) (hNK_1 IC_{50} 0.21 nM), only compounds containing the 3,5-bis(trifluoromethyl) group showed good oral activity in assays of capsaicin-induced extravasation (25) (0.017 mg/kg i.v., 0.068 mg/kg p.o.) compared with CP99994 (5) (0.017 mg/kg i.v., 8.7 mg/kg p.o.).

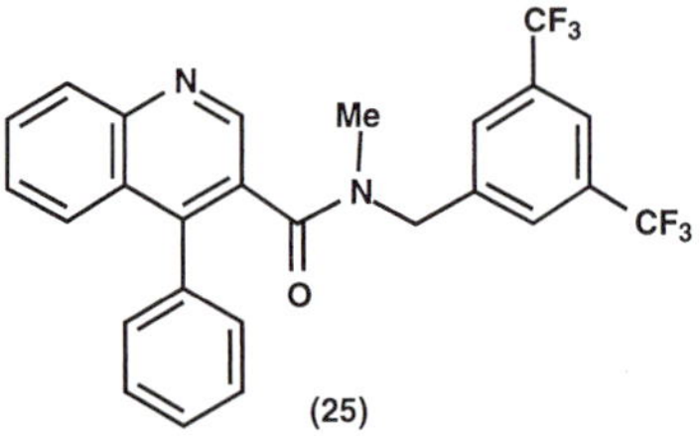

(25) 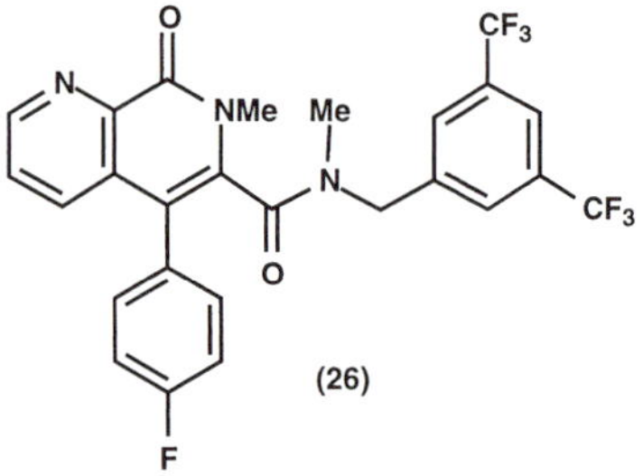   (26)

TRYPTOPHAN ANALOGUES

Tryptophan esters

Directed screening of compound libraries has identified several novel structural classes of NK_1 antagonists. Merck identified a novel series of trypto-

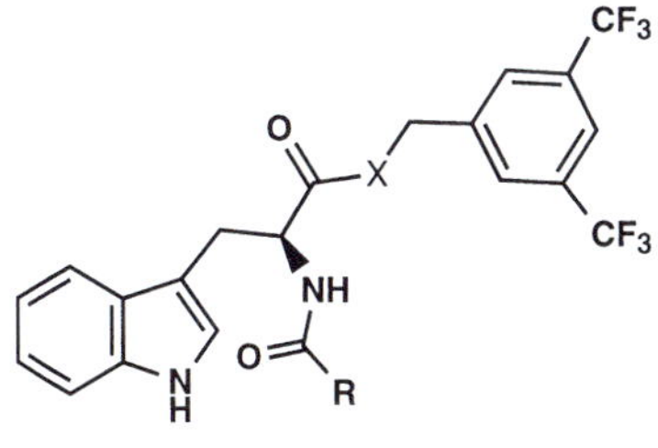

		hNK$_1$ IC$_{50}$ (nM)
(27) R=Me	X=O	1.6
(28) R= -(CH$_2$)$_3$NMe$_2$	X=CH$_2$	0.7
(29) R= 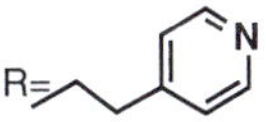	X=CH$_2$	0.47
(30) R= 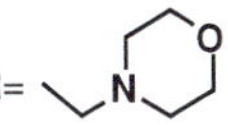	X=CH$_2$	0.17
(31) R= 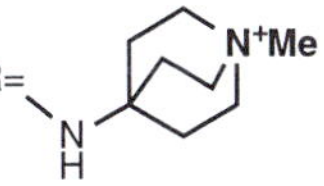	X=CH$_2$	0.37

phan esters (27) and the structure-activity relationship of this series has been described in detail [47]. Interestingly, the 3,5-bis(trifluoromethyl) substituted ring again gives the highest NK$_1$ affinity and *in vivo* activity. Subsequent work showed that it was not the benzyl ether oxygen that was required for binding to the receptor but the carbonyl oxygen of the ester. Consequently, the corresponding ketones, with a variety of groups linked to the nitrogen [*Table 2.1* (28–31)], have now been described [48]. Of particular importance to our understanding of the site of action of NK$_1$ antagonists has been the non-CNS penetrating quaternary amine L743310 (31).

Comparison of the pharmacological profile of L741671 (14) and L743310 (31) showed that they have similar receptor affinity, and have equivalent functional activity in the periphery (ID$_{50S}$ of 1.6 and 2 μg/kg i.v., respectively) as measured by inhibition of plasma extravasation evoked in the oesophagus of guinea-pigs by resiniferatoxin (7 nmol/kg i.v.). In contrast, when tested against cisplatin-induced emesis in ferrets, L741671 (14) was a potent anti-emetic (ID$_{50}$ 0.3 mg/kg i.v.) whilst L743310 was inactive at 3 and 10 mg/kg i.v. [11]. Moreover, direct injection of (14) or (31) (30 mg) into the vicinity of the nucleus tractus solitarius inhibited vomiting induced by cisplatin. The results indicate that CNS penetration is essential for the anti-emetic activity of NK_1 antagonists.

Tryptophan amides

Using a similar screening strategy, based on the assumed importance of key aromatic residues, a team at Lilly discovered a range of *N*-acylated trypto-phan amides and esters [49], including compound (32) (hNK$_1$ IC$_{50}$ 68 nM). This compound probably defines the point of divergence between the medic-inal chemistry efforts of Merck and Lilly with Merck concentrating on com-pounds with a backbone carbonyl and Lilly investigating the *N*-acetyl ana-logues. A limited study of aromatic substituents suggested that it is the 2-methoxy substituent that is preferred in this series. Interestingly, in contrast to the Merck tryptophans, it is the opposite *R*-isomer that is clearly the most active. The *in vivo* evaluation of LY303870 (33), shows it to be a potent NK_1 receptor antagonist (hNK$_1$ 0.2 nM) with good selectivity (> 1μM vs. 70 other receptors and channels, including calcium channels). Using a new technique to assess dural extravasation in guinea-pigs, (33) was equipotent with the 5-HT$_{1D}$ agonist sumatriptan when administered intravenously. LY303870 was more potent than sumatriptan when given orally (ID$_{50}$ 0.1 mg/kg vs 3 mg/kg), and as an aerosol, (33) completely inhibited dural extravasation at 10 mg /kg [50].

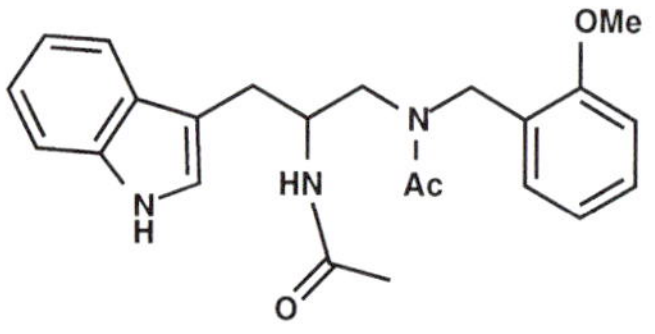

(32) 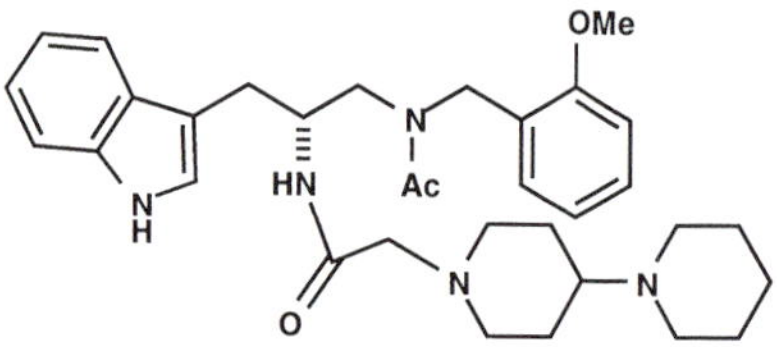   (33)

In contrast to the excellent activity of LY303870 in peripheral assays of NK_1 receptor function, very high doses were required to obtain activity in CNS assays [51]. Doses of up to 30 mg/kg were required to inhibit NK_1 ago-nist-induced hyperalgesia and formalin-induced nociception in rats, and even higher doses (up to 60 mg/kg) were needed to inhibit amphetamine-induced locomotor activity in mice [51].

PEPTIDE-DERIVED ANTAGONISTS

Scientists at Fujisawa have replaced the cyclic peptide FK224 (34), pre-viously in development for asthma, with a highly modified tri-peptide

FK888 (35) because of its superior pharmacological profile [52]. FK888 inhibited substance P-binding in guinea-pig lung with a K_i of 0.69 nM and antagonized the airway constriction induced by 10 nmol/kg substance P with an ED_{50} of 0.0032 mg/kg i.t. but was inactive p.o. [53]. This work was continued, in particular in modifying the central proline residue to afford a pyrrolidone (36) (hNK$_1$ 1.37 nM) and the terminal heterocycle (37) [54].

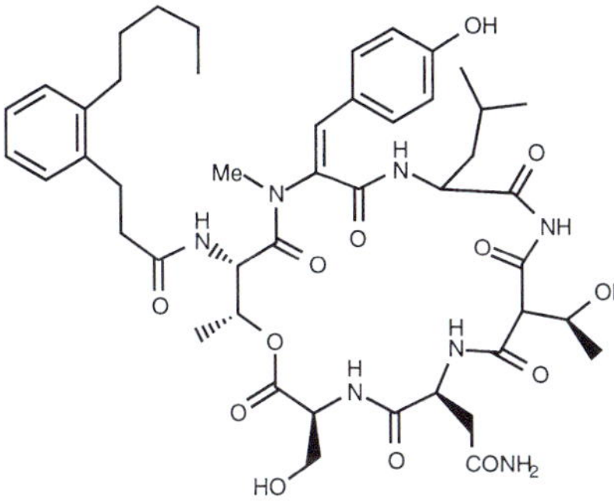

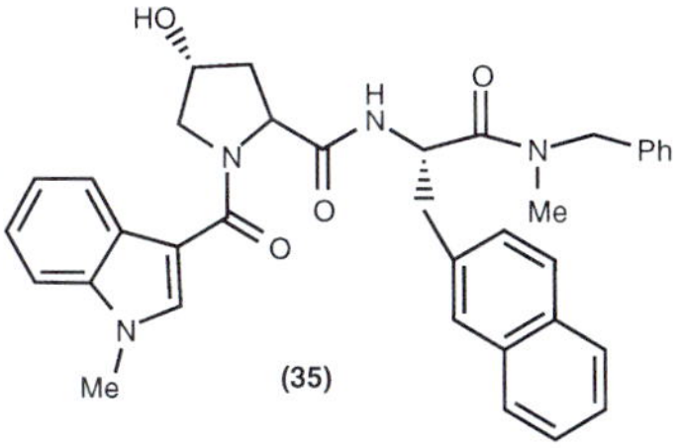

DUAL NK$_1$/NK$_2$ RECEPTOR ANTAGONISTS

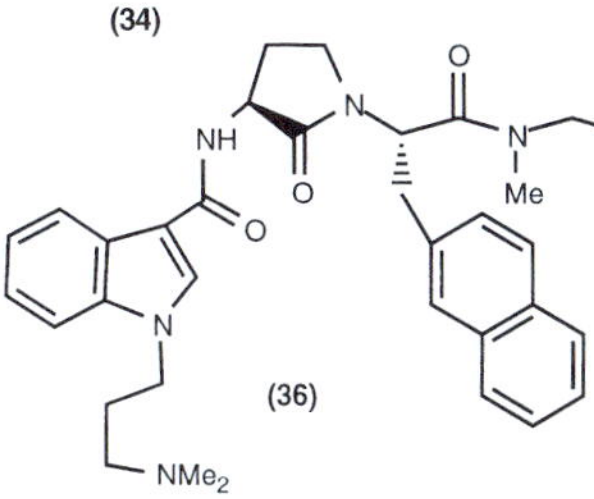

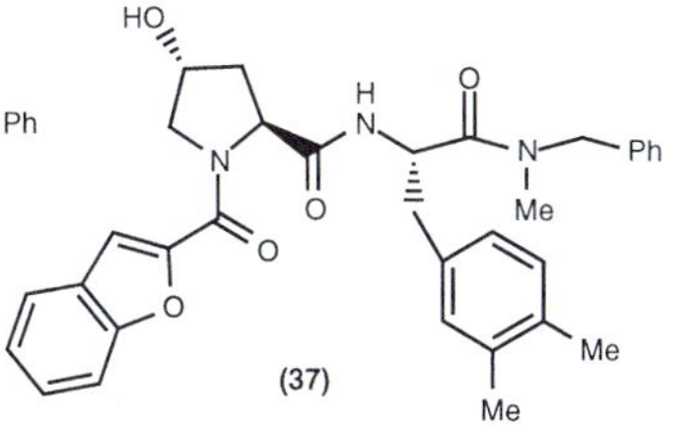

A number of groups have used the NK_2-selective antagonist SR48968 (38) as a starting point to devise dual NK_1/NK_2 antagonists for the treatment of asthma. This approach was highlighted by Merck who reported a series of compounds in which the substituted piperidine ring has been replaced by a spiro system in L743986 (39), (hNK$_1$ IC$_{50}$ 3 nM, hNK$_2$ IC$_{50}$ 12 nM, hNK$_3$ IC$_{50}$ 220 nM) [55]. This compound was active *in vivo*, in a guinea-pig model of plasma extravasation evoked by resiniferatoxin, which is known to be mediated through the NK_1 receptor; L743986 displayed 67% inhibition 20 min after a 0.3 mg/kg intravenous dose. In addition, L743986 was dosed orally in this model, and proved to be active, with an ID_{50} of 3 mg/kg 1h after dosing.

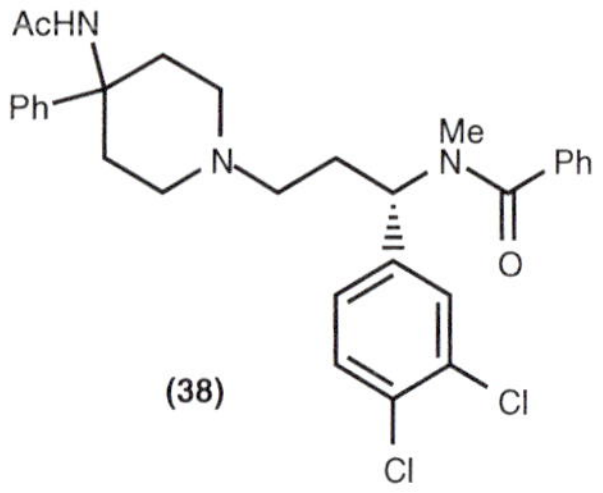

(38)

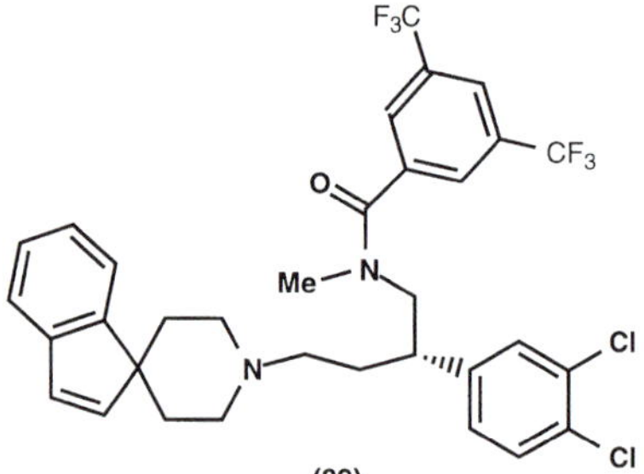

(39)

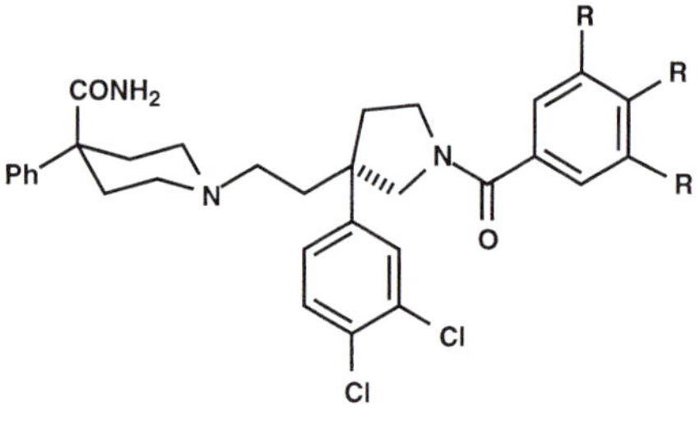

(40)) R=OMe
(41) R=H

A related series was reported by Marion Merrell Dow [56], including MDL105212 (40), a novel non-selective neurokinin receptor antagonist (hNK$_1$ IC$_{50}$ 3.1 nM, hNK$_2$ IC$_{50}$ 8.4 nM, hNK$_3$ IC$_{50}$ 21 nM). Molecular modelling studies using low-energy conformations of MDL103220 (41) (NK$_1$ 161 nM, NK$_2$ 2.2 nM) with CP96345 (1) suggested an overlap of the benzamide moiety of (41) with the 2-methoxybenzylamine of (1). Introduction of a 3-methoxy substituent into (41) afforded an increase in NK$_1$ affinity, and a subsequent study of methoxy substitution identified MDL105212 (40).

NK$_2$ RECEPTOR-SELECTIVE NON-PEPTIDE ANTAGONISTS

In 1992, workers at Sanofi reported the first NK$_2$-selective non-peptide antagonist SR48968 (38) [57]; *in vitro* SR48968 displays excellent affinity (NK$_2$ 0.5 nM) and selectivity (> 1μM NK$_1$ and NK$_2$).

More recently, a novel class of NK$_2$ antagonists have been reported by Glaxo and an example of this class [GR159897 (42)] has been evaluated in some detail [58]. This compound shows high affinity and good NK$_2$ receptor selectivity with the affinity of the *R*-enantiomer of the sulphoxide being some 10-fold higher than the *S*-enantiomer. From the assignment of NK$_2$

subtype selectivity based on the Menarini tissue assays, this compound would be a NK_{2b} selective agent. *In vitro* the compound showed excellent activity on rat colon $pA_2 = 10$, human bronchus = 8.6, human ileum = 9.5.

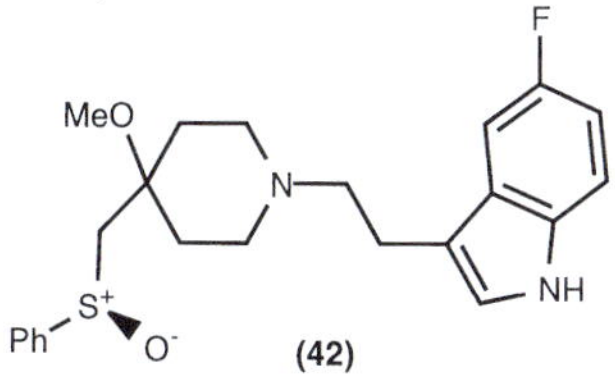

(42)

Compound (42) given *in vivo* (3 mmol/kg i.v.) blocks bronchoconstriction induced by the selective NK_2 agonist GR64349 [59]. Claims for anxiolytic activity with NK_2 receptor antagonists in preclinical assays were originally made by Glaxo [13]. Continued assessment of the role of NK_2 receptors in anxiety paradigms has shown that in the social interaction test, injection of the NK_2 agonist GR64349 (100 pmol) into the dorsal raphe nucleus of the brain gave a reduction in activity under low light conditions that was taken to be predictive of an anxiogenic action. The same work has also shown in the elevated plus maze that (42), given directly into the dorsal raphe nucleus of the brain at doses of 3–300 pmol, produced dose-related anxiolysis with an effect as large as that produced by systemic benzodiazepines. The black and white box was also used to test for anxiolysis and in this test both (42) and the Sanofi NK_2 antagonist SR48968 (38) [60] were active at systemic doses of < 1 μg/kg s.c. In the marmoset human threat test, both were anxiolytic at 50 μg/kg s.c., with a ceiling of effect 60% of that produced by a benzodiazepine. Independent support for the involvement of NK_2 receptors in anxiogenesis came recently from Calixto and colleagues [61] who showed that the NK_2 receptor antagonists SR48968 and GR100679 had anxiolytic-like effects when administered either by direct central injection into the raphe nucleus, or sytemically at very low doses and across a wide dose range (0.02 to 200 μg/kg s.c.).

NK_3 RECEPTOR-SELECTIVE NON-PEPTIDE ANTAGONISTS

Given the high affinity but structurally related NK_1 and NK_2 antagonists identified by workers at Sanofi, it is perhaps not surprising that they should report the first non-peptide NK_3 antagonist SR142801 (43) [62]. *In vitro*, (43) has excellent affinity and receptor selectivity (hNK_3 0.21 nM) and *in vivo,* it shows CNS activity blocking the turning behaviour induced by intrastriatal injection of the NK_3-selective agonist senktide in gerbils.

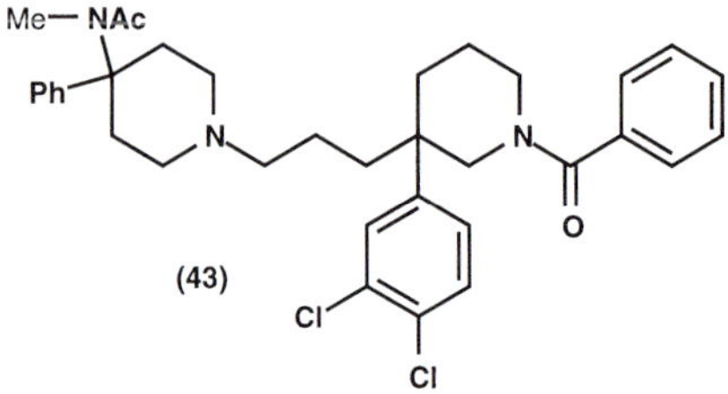

(43)

NK$_3$ receptors have been characterized in several animal species, but there is some debate as to their importance in man, and hence the possible clinical utility of NK$_3$ antagonists [63]. SmithKline Beecham has continued to show the most active interest in the identification of NK$_3$ selective compounds. A new patent extends the company's coverage of quinolines as NK$_3$ antagonists for the treatment of CNS disease, pulmonary disorders, and dermatitis [64]. The specified compound (44) is reported to have an affinity at the NK$_3$ receptor of 150 nM. Another paper [65] describes the SAR of the quinolinecarboxamide series. The methoxy derivative (45), which was originally designed to be an NK$_1$ antagonist, was found to have no affinity for the NK$_1$ receptor, but was moderately active at the NK$_3$ receptor (hNK$_3$ 533 nM). Subsequent work led to the identification of (46), which is a potent and selective NK$_3$ receptor antagonist (hNK$_1$ >100,000 nM; hNK$_2$ 144 nM; hNK$_3$ 1 nM) with no affinity for a range of other receptors and ion-channels. Compounds such as (46) should be valuable for defining the utility of NK$_3$ receptor antagonists.

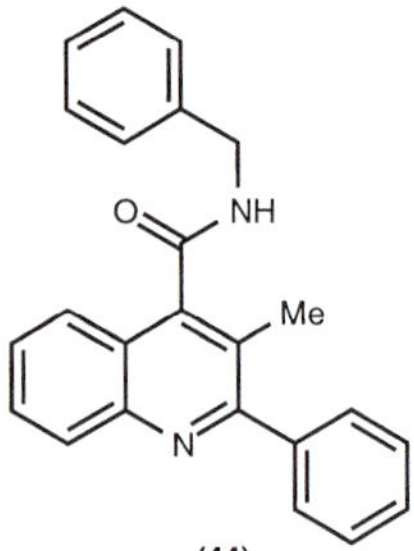

(44) 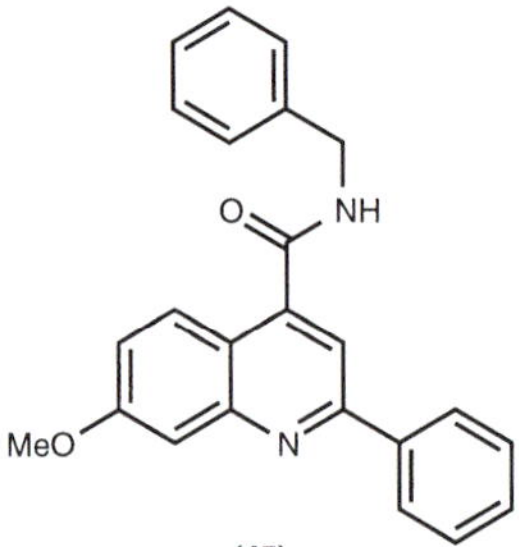   (45) 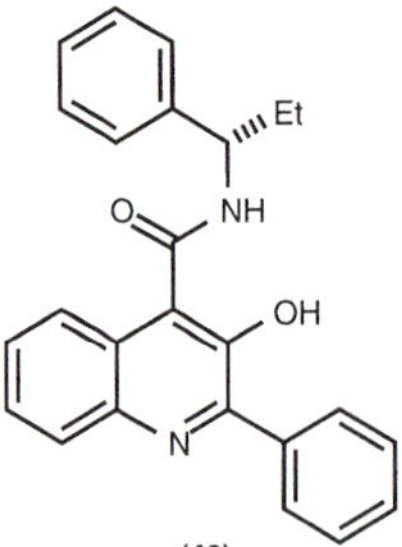   (46)

CLINICAL STUDIES

NK$_1$ receptor antagonists have long been proposed to have potential clinical utility in a wide range of pathological conditions, including pain, anxiety, arthritis, asthma, emesis, migraine, cancer, and schizophrenia. To date, reported clinical trials have focused on the use of neurokinin antagonists in asthma, pain, migraine and emesis.

Early clinical trials for the treatment of asthma with Pfizer's selective NK_1 antagonist CP99994 (5) (hNK_1 0.48 nM) gave disappointing results. In asthmatic subjects treated with hypertonic saline to induce bronchoconstriction and coughing, CP99994 (0.25 mg/kg i.v.) showed no inhibition of cough or bronchoconstriction and no significant bronchodilator response [66]. The dose resulted in plasma concentrations of 30–70 ng/mL which, based on preclinical data in the guinea-pig, should have adequately blocked human airway NK_1 receptors. A possible explanation for the lack of efficacy is that saline challenge does not activate the neurokinin system. A second clinical trial with CP99994 in painful peripheral neuropathy also showed no significant efficacy over placebo at doses of up to 0.2 mg/kg i.v. [67]. However, a recent paper describes two placebo-controlled studies of the same compound in dental pain following oral surgery [68]. The study compared CP99994 (0.75 mg/kg i.v. over 5 h starting infusion 2 h prior to surgery) with ibuprofen (600 mg/p.o. 30 min prior to surgery) and placebo in 60 patients undergoing oral surgery. Treatment with CP99994 resulted in significantly less pain than placebo at 90 min post surgery; however, the analgesic effect progressively decreased with time and pain levels became comparable with placebo from 150–240 min post surgery. In a variation of this study, 18 subjects received the same regimens of CP99994, ibuprofen or placebo 30 min prior to surgery in order to maximize drug levels during pain onset. CP99994 alleviated pain at 60, 90 and 120 min but was less effective than ibuprofen. This study demonstrated that CP99994 was analgesic immediately postsurgery but that the analgesic effect was not long lasting. This is presumably a consequence of the poor pharmacodynamic profile seen with CP99994 and emphasises the need for a good duration of action in any potential clinical candidate.

This trial provides the first clinical evidence that blockade of NK_1 receptors results in analgesia in man. The dose used in the earlier study of painful peripheral neuropathy was considerably lower than the dose used in this trial, raising the possibility that the result was a false negative and that the outcome may have been different had higher doses been administered.

Pfizer have now replaced CP99994 in the clinic with the 5-trifluoromethoxy derivative CP122721 (6). This compound has approximately three-fold higher affinity than CP99994 for the NK_1 receptor (hNK_1 0.14 nM) and is orally bioavailable. CP122721 has significant ion channel activity (390 nM, L-type calcium channel) which may compromise high dose i.v. administration in the clinic. It has recently been evaluated in a dose-ranging antiemetic clinical trial [69]: 12 patients with advanced solid tumours received CP122721 (20, 100 and 200 mg/p.o.) in conjunction with a 5-HT$_3$ antagonist and dexamethasone. An additional 5 patients received CP122721 (200 mg/p.o.) alone. Treatment was administered 30 min prior to receiving a high dose of cisplatin (100 mg/m^2). All five patients receiving CP122721 vomited (median 1 emetic event; range 1–3 events). Delayed emesis occurred in 3/17 patients - much fewer than anticipated. Control of acute vomiting was maintained or improved in those patients receiving combination therapy. No side-effects were reported and further trials are in progress.

Rhone Poulenc Rorer have developed RPR100893 (19), a selective NK$_1$ receptor antagonist (hNK$_1$ 13 nM) which is active in pre-clinical migraine assays in guinea-pigs (capsaicin-induced dural extravasation: ID$_{50}$ 0.008 mg/kg p.o.). However, RPR100893 has been reported to be ineffective at doses of up to 20 mg in Phase II clinical trials for the treatment of migraine [70]. Oral doses of 1, 5 and 20 mg RPR100893 were reported to be no better than placebo; no plasma drug levels or evidence of functional activity at these doses were disclosed. On the assumption that the doses used produced reasonable plasma drug levels, this lack of activity against migraine may be related to the poor brain penetrability of RPR100893. This result suggests that a peripheral NK$_1$ antagonist action alone may not be sufficient to give headache relief especially within the 2 h time-frame used for efficacy evaluation in clinical trials. Thus, NK$_1$ receptor antagonists may rely for their anti-migraine effects on inhibition of central nociceptive pathways in the trigeminal nuclei where headache pain could be blocked rapidly and directly. Further clinical trials with more potent, brain-penetrating development candidates, as well as trials to test the efficacy of neurokinin antagonists in the prophylactic relief of migraine should resolve this issue.

Goldstein [71] reported that the results of Lilly's NK$_1$ antagonist lanepitant LY303870 (33) in acute migraine trials were clearly negative. The doses used were 30, 80 and 240 mg. However, absorption in patients was only 7% of that seen in volunteers with only low ng/ml levels of drug seen in plasma. Since this compound is thought to be very poorly brain-penetrating, these results do not rule out the possibility that an NK$_1$ antagonist must act centrally to evoke an acute anti-migraine effect. In a second trial, lanepitant was ineffective in providing pain relief in osteoarthritis patients at 10, 30, 100 and 300 mg twice daily.

The chronic symptoms of asthma (inflammation and mucus secretion) are thought to involve NK$_1$ receptors, while the acute bronchoconstriction component is mediated *via* the NK$_2$ receptor [72]. There have been efforts by many companies to develop dual NK$_1$/NK$_2$ antagonists to treat both the acute and chronic symptoms. Fujisawa's cyclopeptide dual NK$_1$/NK$_2$ antagonist FK224 (34) (hNK$_1$ IC$_{50}$ 37 nM; hNK$_2$ IC$_{50}$ 72 nM) has been evaluated in two models of bronchoconstriction. In an early placebo-controlled study [73], FK224 (4 mg by inhalation) inhibited bradykinin-induced bronchoconstriction and coughing in subjects with stable asthma. Since FK224 has no significant affinity for the bradykinin receptor, this supports the theory that tachykinins are involved in the bronchoconstriction response. However, a recent study [74] using a similar protocol examined bronchoconstriction induced by neurokinin A, the endogenous ligand with affinity for both the NK$_1$ and NK$_2$ receptors. As before, the compound was administered by inhalation 30 min prior to the NK$_A$ challenge. In this study, FK224 had no significant effect on baseline lung function and offered no protection against neurokinin A-induced bronchoconstriction.

The authors interpreted the results as follows: FK224 is a dual NK$_1$/NK$_2$ antagonist in a variety of animal assays; however, the reported affinities are quite low (substance P-induced contraction of guinea-pig ileum, pA$_2$ 6.88; NK$_A$-induced contraction of rat vas deferens, pA$_2$ 7.50). These data suggest that the dose of FK224 used in the clinic may have been too low to effectively block neurokinin receptors in the airways and casts doubt on the interpretation of the earlier bradykinin study.

Interestingly, FK224 has been superseded in development by the NK$_1$ selective tripeptide FK888 (35) (hNK$_1$ IC$_{50}$ 0.7 nM), which is also targeted at asthma and pulmonary disease [1]. FK888 is currently in Phase II trials and has recently been evaluated in asthmatic patients. The double-blind, placebo-controlled crossover trial showed that FK888 (2.5 mg by inhalation) had no effect on acute exercise-induced airway narrowing, but caused a significant reduction in recovery times [75]. Plasma levels of 6–18 ng/mL were achieved 20 to 70 min after administration, which should be adequate to block NK$_1$ receptors in the airways. The authors suggested that the mechanism of action of FK888 may be *via* blockade of NK$_1$ receptors leading to inhibition of vascular engorgement and airway wall oedema rather than any bronchodilator response.

CONCLUSIONS

The search for selective neurokinin antagonists has uncovered an unparalleled diversity of structural classes affording high affinity receptor antagonists. This, coupled with knowledge provided by molecular biology, has provided an exciting insight into the molecular interactions involved in binding. The new selective, high affinity antagonists have improved our understanding of neurokinin pharmacology which has provided the rationale for the ongoing clinical trials.

REFERENCES

1 Maggi, C.A., Patacchini, R., Rovero, P. and Giachetti, A. (1993) J. Auton. Pharmacol. 13, 23–93.
2 Regoli, D., Boudon, A. and Fauchere, J.-L. (1994) Pharmacol. Rev. 46, 551–599.
3 Longmore, J., Swain, C.J. and Hill, R. (1995) Drug News Perspect. 8, 5–23.
4 McLean, S. (1996) Med. Res. Rev. 16, 297–317.
5 Swain, C.J. (1996) Exp. Opin. Ther. Patents 6, 367–378.
6a hNK$_1$: Takeda, Y., Chou, K.B., Takeda, J., Sachais, B.S. and Krause, J.E. (1991) Biochem. Biophys. Res. Commun.179, 1232–1240.
6b hNK$_2$: Gerard, N.P., Eddy, R.L., Shows, T.B. and Gerard, C. (1990) J. Biol. Chem. 265, 20455–20462.
6c hNK$_3$: Takahashi, K., Tanaka, A., Hara, M. and Nakanishi, S. (1992) Eur. J. Biochem. 204, 1025–1033.
7 Blau, J.N. and Dexter, S.L. (1981) Cephalagia 1, 143–147.
8 Moskowitz, M.A. (1992) Trends Pharmacol. Sci. 13, 307–317.
9 Amin, A.H., Crawford, T.B.B. and Gaddum, J.H. (1954) J. Physiol. 126, 596–618
10 Andrews, P.L.R. and Bhandari, P. (1993) Neuropharmacology 32, 799–806.
11 Tattersall, F.D., Rycroft, W., Francis, B., Pearce, D., Merchant, K., MacLeod, A.M., Ladduwahetty, T., Keown, L., Swain, C., Baker, R., Cascieri, M., Ber, E., Metzger, J., MacIntyre, D.E., Hill, R.G. and Hargreaves, R.J. (1996) Neuropharmacology 35,1121–1129.
12 Otsuka, M. and Yoshioka, K. (1993) Physiol. Rev. 73, 229–308.
13 Walsh, D.M., Stratton, S.C., Harvey, F.J., Beresford, I.J.M. and Hagan, R.M. (1995) Psychopharmacology 121, 186–191.
14 Stratton, S.C., Beresford, I.J.M. and Hagan, R.M. (1996) Br. J. Pharmacol. 117, 295P.
15 Dietl, M.M. and Palacios, J.M. (1991) Brain Res. 539, 211–222.
16 Elliott, P.J., Mason, G.S., Stephens-Smith, M. and Hagan, R.M. (1991) Neuropeptides 19, 119–126.
17 Snider, R.M., Constantine, J.W., Lowe III, J.A., Longo, K.P., Lebel, W.S., Woody, H.A., Drozda, E., Desai, M.C., Vinick, F.J., Spencer, R.W. and Hess, H-J. (1991) Science (Washington, D.C.) 251, 435–437.
18 Lowe III J.A., Drozda, S.E., Snider, R.M., Longo, K.P., Zorn, J. Morrone, S.H., Jackson, E.R., McLean, S., Bryce, D.K., Bordner, J., Nagahisa, A., Kanai, Y., Suga, O. and Tsuchiya, M. (1992) J. Med.Chem. 35, 2591–2600.
19 Caeser, M., Seabrook, G.R. and Kemp, J.A. (1993) Br. J. Pharmacol. 109, 918–924.
20 Rosen, T., Seegar, F., McLean, S., Desai, M.C., Guarino, K.J., Bryce, D., Pratt, K. and Heym, J. (1993) J. Med. Chem. 36, 3197–3201.
21 McLean, S., Fossa, A., Zorn, S. and Rosen, T. (1995) Presented at Tachykinins '95, Florence October 16–18.
22 Fahy, J.V., Wong, H.F., Geppetti, P., Reis, J.M., Harris, S.C., Maclean, D.B., Nadel, J.A. and Boushey, H.A. (1995) Am. J. Resp. Crit. Care Med. 152, 879–884.
23 Howard, H.R., Shenk, K.D., Coffman, K.C., Bryce, D.K., Crawford, R.T. and McLean, S.A. (1995) Bioorg. Med. Chem.Lett. 5, 111–114.

24 Lowe, III, J.A., Drozda, S.E., Snider, M., Longo, K.P., and Rizzi, J.P. (1993) Bioorg. Med. Chem. Lett. 3, 921–924.

25 Ward, P., Armour, D.R., Bays, D.E., Evans, B., Giblin, G.M.P., Heron, N., Hubbard, T., Liang, K., Middlemiss, D., Mordaunt, J., Naylor, A., Pegg, N.A., Vinader, V., Watson, S.P., Bountra, C. and Evans, D.C. (1995) J. Med. Chem. 38, 4985–4992.

26 Armour, D.R., Chung, K.M.L., Congreve, M., Evans, B., Guntrip, S., Hubbard, T., Kay, C., Middlemiss, D., Mordaunt, J.E., Pegg, N.A., Vinader, M.V., Ward, P. and Watson, S.P. (1996) Bioorg. Med. Chem. Lett. 6, 1015–1020.

27 Matsuo, M., Hagiwara, D., Miyake, H., Igari, N., Murano, K. W.O. Patent 9500536 A1; (1995) Chem. Abstr. 123, 144654.

28 Swain, C.J., Seward, E.M., Cascieri, M.A., Fong, T., Herbert, R., MacIntyre, D.E., Merchant, K., Owen, S.N., Owens, A.P., Sabin, V., Teall, M., VanNiel, M.B., Williams, B.J., Sadowski, S., Strader, C., Ball, R. and Baker, R. (1995) J. Med. Chem. 38, 4793–4805.

29 Gether, U., Nilsson, L., Lowe III, J.A. and Schwartz, T.W. (1994) J. Biol. Chem. 269, 23959–23964.

30 Williams, B.J., Teall, M., McKenna, J., Harrison, T., Swain, C.J., Cascieri, M.A., Sadowski, S., Strader, C. and Baker, R. (1994) Bioorg. Med. Chem. Lett. 4, 1903–1908.

31 Ladduwahetty, T., Baker, R., Cascieri, M.A., Chambers, M.S., Haworth, K., Keown, L.E., MacIntyre, D.E., Metzger, J.M., Owen, S., Rycroft, W., Sadowski, S., Seward, E.M., Shepheard, S.L., Swain, C.J., Tattersall, F.D., Watt, A.P., Williamson, D.W. and Hargreaves, R.J. (1996) J. Med. Chem. 39, 2907–2914.

32 Hale, J.J., Mills, S.G., MacCoss, M., Shah, S.K., Qi, H., Mathre, D.J., Cascieri, M.A., Sadowski, S., Strader, C.D., MacIntyre, D.E. and Metzger, J.M. (1996) J. Med. Chem 39, 1760–1762.

33 Rupniak, N.M.J., Tattersall, F.D., Williams, A.R., Rycroft, W., Carlson, E.J., Cascieri, M.A., Sadowski, S., Ber, E.,. Hale, J.J., Mills, S.G., MacCoss, M., Seward, E., Huscroft, I., Owen, S., Swain, C.J., Hill, R. G. and Hargreaves, R. J. (1997) Eur. J. Pharmacol. In press.

34 Garret, C., Carruette, A., Fardin, V., Moussaoui, S., Peyronel, J-F., Blanchard, J-C. and Laduron, P.M. (1991) Proc. Natl. Acad. Sci. U.S.A. 88, 10208–10212.

35 Achard, D., Truchon, A. and Peyronel, J.-F. (1994) Bioorg. Med. Chem. Lett. 4, 669–672.

36 Tabart, M. and Peyronel, J.-F. (1994) Bioorg. Med. Chem. Lett. 4, 673–676.

37 Moussaoui, S. M., Monttier, F., Carruette, M.A., Fardin, V., Floch, C. and Garret, C. (1994) Neuropeptides Supplement 1 26, 35.

38 Achard, D., Peyronel J.F., Carruette, A., Collemine, P., Fardin, V., Garret, C. (1996) 25th National Medicinal Chemistry Symposium, Ann Arbor, Michigan, USA, Poster 84.

39 Ofner, S., Hausser, K., Schilling, W., Vassout, A. and Veenstra, S.J. (1996) Bioorg. Med. Chem. Lett. 6, 1623–1628.

40 Subramanian, N., Ruesch, C., Anderson, G.P. and Schilling, W. (1994) J. Physiol. Pharmacol. 72 Supp 1P.

41 Vassout, A., Schaub, M., Getsch, C., Ofner, S., Schilling, W. and Veenstra, S. (1994) Neuro-peptides 26, S38.

42 Ofner, S. and Veenstra, S. J. (1995) W.O. Patent 9610562; (1996) Chem. Abstr. 125,114485.

43 Ofner, S., Veenstra, S.J. and Schilling, W. CIBA-GEIGY (1996) EP-707006-A1; (1996) Chem. Abstr. 125, 33665.

44 Emonds-Alt, X., Doutremepuich, J.D., Healulme, M., Neliat, G., Santucci, V., Steinberg, R., Vilain, P., Bichon, D., Ducoux, J.P., Proietto, V., van Brock, D., Soubrie, P., Le Fur, G. and Breliere, J.C. (1993) Eur. J. Pharmacol. 250, 403–413.

45 Juranek, I. and Lembeck, F. (1994) Proc. Br. Pharmacol. Soc., 13–16 December, P189.

46 Natsugari, H., Ikeura, Y., Kiyota, Y., Ishichi, Y., Ishimaru, T., Saga, O., Shirafuji, H., Tanaka, T., Kamo, I., Doi, T. and Otsuka, M. (1995) J. Med. Chem. 38, 3106–3120.
47 Macleod, A.M., Merchant, K.J., Brookfield, F., Lewis, R., Kelleher, F., Stevenson, G., Owens, A., Swain, C.J., Baker, R., Cascieri, M.A., Sadowski, S., Ber, E., MacIntyre, D.E., Metzger, J. and Ball, R. (1993) J. Med. Chem. 36, 2044–2045.
48 Macleod, A.M., Cascieri, M.A., Merchant, K.J., Sadowski, S., Hardwicke, S., Lewis, R.T., MacIntyre, D.E., Metzger, J.M., Fong, T.M., Shepheard, S., Tattersall, F.D., Hargreaves, R. and Baker, R. (1995) J. Med. Chem. 38, 934–941.
49 Hipskind, P.A., Howbert, J.J., Bruns, R.F., Cho, S.Y., Crowell, T.A., Foreman, M.M., Gehlert, D.R., Iyengar, S., Johnson, K.W., Krushinski, J.H., Li, D.L., Lobb, K.L., Mason, N.R., Muehl, B.S., Nixon, J.A., Phebus, L.A., Regoli, D., Simmons, R.M., Threkeld, P.G., Waters, D.C. and Gitter, B.D. (1996) J. Med. Chem. 39, 736–748.
50 Gitter, B.D., Burns, R.F., Howbert, J., Waters, D.C., Threkeld, P.G., Cox, L.M., Nixon, J.A., Lobb, K.L., Mason, N.R., Stengel, P.W., Cockerham, S.L., Silaugh, S.A., Gehlert, D.R., Schober, D.A., Iyengar, S., Calligaro, D.O., Regoli, D. and Hipskind, P.A. (1995) J. Pharmacol. Exp. Ther. 275, 737–744.
51 Gitter, B.D., Bruns, R.F., Howbert, J.J., Stengel, P.W., Gehlert, D.R., Iyengar, S., Calligaro, D.O., Regoli, D. and Hipskind, P.A. (1995) Presented at Tachykinins '95, Florence October 16–18.
52 Ichinose, M., Nakajima, N., Takahashi, T., Yamauchi, H., Inoue, H.and Takishima, T. (1992) Lancet, 340, 1248–1251.
53 Murai, M., Maeda, Y., Yamaoka, M., Hagiwara, D., Miyake, H., Matsuo, M. and Fujii, T. (1993) Regul. Pept. 46, 335–337.
54 Matsuo, M., Hagiwara, D., Miyake, H., Igari, N., Murano, K. W.O. Patent 9314113 A1; (1993) Chem. Abstr. 120, 135143. Matsuo, M., Hagiwara, D., Miyake, H. (1993) W.O. Patent 9321215 A1; (1994) Chem. Abstr. 121, 281227.
55 Mills, S.G., Hale, J.J., MacCoss, M., Shah, S.K., Qi, H., Cascieri, M.A., Sadowski, S., Metzger, J.M., Eiermann, G.J., Forrest, M.J., MacIntyre, D.E. and Strader, C.D. (1996) Presented at Tachykinins and Their Antagonists" meeting in London (October 10–11th).
56 Burkholder, T.P., Kudalacz, E.M., Le, T.-B., Knippenberg, R.W., Maynard, G.D., Shatzer, S.A., Webster, M.E. and Hogan, S.W. (1996) Bioorg. Med. Chem. Lett. 6, 951–956.
57 Emonds-Alt, X., Vilain, P., Goulaouic, P., Proietto, V., Van Broeck, D., Advenier, C., Naline, E., Neliat, G., Le Fur, G. and Breliere, J.C. (1992) Life Sci. 50, PL101.
58 Cooper, W.J., Adams, H.S., Bell, R., Gore, P.M., McElroy, A.B., Pritchard, J.M., Smith, P.W. and Ward, P. (1994) Bioorg. Med. Chem. Lett. 4, 1951–1956.
59 Ball, D.I., Wren, G.P.A., Pendry, Y.D., Smith, J.R., Piggott, L. and Sheldrick, R.L.G. (1993) Neuropeptides 24, 190.
60 Emonds-Alt, X., Doutremepuich, J.D., Jung, M., Proietto, E., Santucci, V., Van Broeck, D., Vilain, P., Soubrie, P., Le Fur, G. and Breliere, J.C. (1993) Neuropeptides 24, 231.
61 Teixeira, R.M., Santos, A.R.S., Rae, G.A., Calixto, J.B. and de Lima, T.C.M. (1995) Presented at Tachykinins '95 Florence October 16–18.
62 Emonds-Alt, X., Bichon, D., Ducoux, J.P., Heaulme, M., Miloux B., Poncelet, M., Proietto, V., Van Broeck, D., Vilain, P., Neliat, G., Soubrie, P., Le Fur, G. and Breliere, J.C. (1995) Life Sci. 56, PL27–32.
63 Dietl, M.M. and Palacios, J.M. (1991) Brain Res. 539, 211–222.
64 Farina, C., Giardina, G.A.M., Grugni, M. and Raveglia, L.F. (1996) W.O.Pat. 96/02509-A1; Chem. Abstr. 125, 33490.
65 Giardina, G.A.M., Sarau, H.M., Farina, C., Medhurst, A.D., Grugni, M., Foley, J.J., Raveglia, I.F., Schmidt, D.B., Rigolio, R., Vassallo, M., Vecchietti, V. and Hay, D.W.P. (1996) J. Med. Chem. 39, 2281–2284.

66 Fahy, J.V., Wong, H.H., Geppetti, P., Reis, J.M., Harris, S.C., Maclean, D.B., Nadel, J.A. and Boushey, H.A., (1995) Am. J. Respir. Crit. Care Med. 152, 879–884.

67 Suarez, G.A., Opfer-Gehrking, T.l., Maclean, D.B., and Low, P.A. (1994) Neurology 44 (Suppl 2), 373P.

68 Dionne, R.A., Max, M.B., Parada, S., Gordon, S.M. and Maclean, D.B. (1996) Clin. Pharmacol. Ther. 59, 216.

69 Kris, M.G., Radford, J., Pizzo, B., Inabinat, R., Lovelace, J., Casey, M. and Hesketh, P. (1996) Proc. Am. Soc. Clin. Oncol. 15, A1780.

70 Diener, H.C. (1995), for the RPR100893–201 Migraine Study Group: 6th International Headache Research Seminar, Copenhagen, Denmark, Poster 1.

71 Goldstein, D. (1996) Presented at 'Tachykinins and Their Antagonists' meeting in London (October 10–11th).

72 Mclean, S. (1996) Med. Res. Rev. 16, 297–317.

73 Ichinose, M., Nakajima, N.,Takahashi, T., Yamauchi, H., Inoue, H. and Takishima, T. (1992) Lancet 340, 1248–1251.

74 Joos, G.F., Van Schoor, J., Kips, J.C. and Pauwels, R.A. (1996) Am. J. Respir. Crit. Care Med. 153, 1781–1784.

75 Ichinose, M., Motohiko, M., Yamauchi, H., Kageyama, N., Tomaki, M., Oyake, T., Ouchi, Y., Hida, W., Miki, H., Tamura, G. and Shirato, K. (1996) Am. J. Respir. Crit. Care Med. 153, 936–941.

Progress in Medicinal Chemistry – Vol. 35
Edited by G.P. Ellis, D.K. Luscombe and A.W. Oxford

3 Opioid Receptor Antagonists[*]

[*]This review is dedicated to the memory of the late Drs. Sidney Archer and Hans W. Kosterlitz.

HELMUT SCHMIDHAMMER, PH.D.

Institute of Pharmaceutical Chemistry, University of Innsbruck, Innrain 52a, A-6020 Innsbruck, Austria

INTRODUCTION

The pharmacological concept of receptors, based upon the observation of rigid structure-activity relationships, stereospecificity, and the observation of maximal pharmacological responses goes back to the turn of the century. More than a hundred years ago, Fischer [1] proposed the lock-and-key model for the enzyme-glycoside system. Later, Langley [2] and Ehrlich [3] further developed this model which can be applied to receptors as well.

In the early 1950s, the determination of structural requirements of semisynthetic opioids led to the hypothesis of Beckett and Casy [4, 5] that they interacted with specific binding sites. Opioid structure-activity relationships established for literally thousands of compounds [6–9] revealed very rigid requirements for activity, including strict stereospecificity [10, 11]. Furthermore, both *in vivo* testing and bioassays clearly fulfilled the other criteria expected of a receptor-mediated action, including cross tolerance and dependence. The synthesis of the pure opioid antagonist naloxone as well as the mixed agonist-antagonist nalorphine, and their ability to reverse opioid analgesia and respiratory depression, also provided strong evidence for a receptor-mediated interaction. With such strong pharmacological evidence in favour of a receptor mechanism of action, it was not surprising that many groups of workers attempted to label it biochemically. In 1973, three laboratories reported the biochemical demonstration of opioid binding sites using tritium-labelled naloxone (Pert and Snyder [12]), tritium-labelled dihydromorphine (Terenius [13]), and tritium-labelled etorphine (Simon *et al.* [14]).

Martin [15] first proposed the existence of multiple types of opioid receptors on the basis of interactions of morphine and nalorphine, and of detailed structure-activity relationship studies which had been carried out by Portoghese [10, 16]. The demonstration of specific opioid binding sites in many species suggeseted the existence of endogenous substances for these binding sites. Hughes, Kosterlitz and their collaborators [17] first described such endogenous substances, which are peptides containing five amino acids, and called them enkephalins ('in the head').

Subsequently, numerous peptides with opioid-like effects have been found in the central nervous system and in peripheral tissues. These endogenous opioid peptides vary in size, but their amino terminals mostly share a similar enkephalin sequence of amino acids. Currently, four separate, individually gene-derived families of endogenous opioid peptides are recognized: the endorphins, the enkephalins, the dynorphins and the endomorphins [17a]. β-Endorphin interacts predominantly with μ and δ receptors, Leu-enkephalin and Met-enkephalin interact predominantly with δ receptors, dynorphin shows preference for κ receptors [17b], while endomorphins 1 and 2 exhibit selectivity for μ receptors. Endomorphins 1 and 2 differ from the conventional endogenous opioid ligands in their N-terminal sequence (Tyr-Pro vs. Tyr-Gly).

With the recent cloning and sequencing of opioid receptors, it is now well established that there are at least three types of G protein-coupled opioid receptors (μ, κ and δ) which show high sequence homology [18]. Opioid receptors are involved in the modulation of a variety of physiological effects *via* interaction with opioid peptides [19]. Therefore, it has become increasingly evident that receptor-selective opioid antagonists are valuable tools for identifying receptor types (and also subtypes) involved in the interaction with opioid agonists. A major advantage of selective opioid antagonists over selective agonists is their utility in probing the interaction of endogenous opioid peptides and new opioid agonists with opioid receptor types. In addition to their use as tools, selective opioid antagonists have potential clinical applications in the treatment of a variety of disorders where opioid peptides play a modulatory role. These include, for instance, food intake, shock, constipation, immune function, behaviour, CNS injury, alcoholism and mental disorders [20].

DETERMINATION OF RECEPTOR SELECTIVITY AND ANTAGONISM

Receptor selectivity can be determined using radioligand binding assays. Receptor binding selectivity can be determined by displacement of relatively selective radioligands from receptor sites in membrane suspensions prepared mostly from either rat or guinea-pig brain. Nowadays, cloned μ, κ and δ receptors can be used instead of the brain membrane preparations [21–24]. Recently the GTP-ase assay [24a] and the [^{35}S]GTPγS binding test [24b, c] in cell membranes or cloned opioid receptors have been introduced as functional assays for the determination of agonism and antagonism. Bioassays are being used to assess both receptor selectivity and antagonist (agonist) potency. These bioassays are based on inhibition of electrically evoked contractions of the guinea-pig ileum (GPI), mouse vas deferens (MVD), hamster vas deferens (HVD), rabbit vas deferens (LVD) and rat vas deferens (RVD). Whereas the HVD and LVD contain homogenous populations of κ and δ receptors, respectively, the opioid receptor populations in the other three tissues are heterogenous. In the GPI, μ and κ receptors exist and the MVD contains both μ and κ receptors in addition to the predominant δ receptors. In the RVD, μ receptors and, possibly, epsilon receptors mediate opioid effects [25]. *In vivo* tests in mice or rats (flick tail, hot plate,

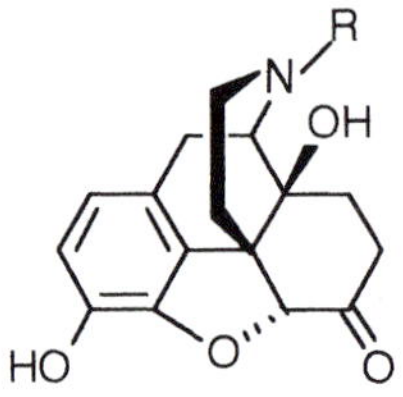

(1) R = allyl, naloxone
(2) R = cyclopropylmethyl, naltrexone

writhing) can also be used to assess antagonist potency and selectivity (measured against selective agonists).

UNIVERSAL OPIOID RECEPTOR ANTAGONISTS

Naloxone (1) was the first pure opioid antagonist to be detected and it has become an indispensable tool in opioid research. Both naloxone and its N-cyclopropylmethyl analogue naltrexone (2) are competitive antagonists at μ, κ and δ opioid receptors with some preference for μ receptors. The major criterion for the classification of an agonist effect as being opioid receptor-mediated is the ability of these antagonists to competitively antagonize this effect [26]. The unnatural (+)-isomer of naloxone exhibits 10,000-fold less affinity for opioid receptors than does the (−)-isomer [27] and is inactive as an opioid antagonist [28].

Naloxone is being used to reverse the potentially lethal respiratory depression caused by neurolept analgesia or opioid overdose. Among other pharmacological effects, naloxone antagonizes the blood pressure drop in various forms of shock [29–32], reverses neonatal hypoxic apnoea [26], counteracts chronic idiopathic constipation [34], reduces the food intake in humans [35, 36] and shows beneficial effects in CNS injuries [37].

Naltrexone appears also to be a relatively pure opioid antagonist, but with higher oral efficacy and a longer duration of action than naloxone [35]. This is probably due to biotransformation to the active metabolite 6β-naltrexol (*Scheme 3.1*), which can cross the blood-brain barrier, contributing to central opioid receptor blockade. These properties make naltrexone suitable for the management of opioid dependence and provide a new and effective modality for the physician treating addicts [36, 38]. Alcoholism is another addiction which can possibly be treated with naltrexone [38].

The main metabolite (> 70%) of naltrexone biotransformation in man is 6β-naltrexol [39] which is pharmacologically active as an opioid antagonist.

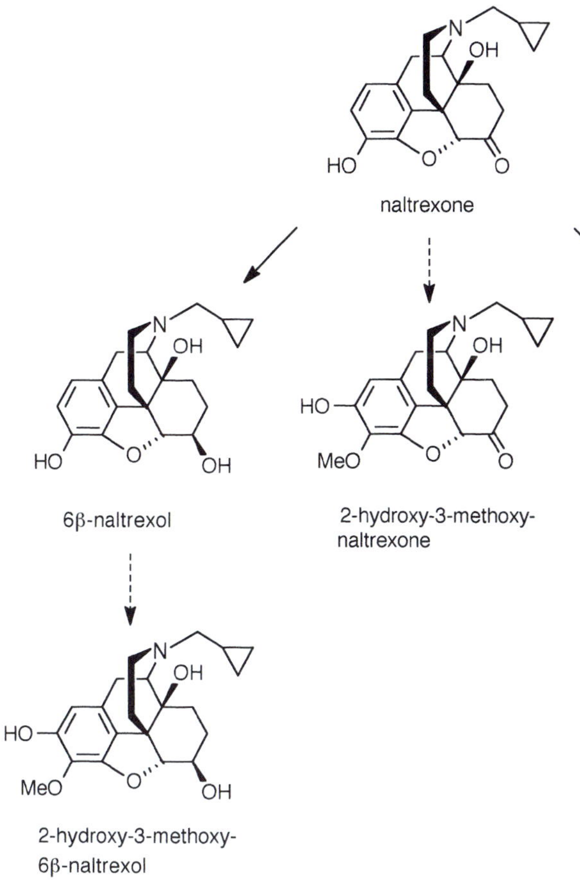

naltrexone

6β-naltrexol

2-hydroxy-3-methoxy-
naltrexone

N-noroxymorphone

2-hydroxy-3-methoxy-
6β-naltrexol

Scheme 3.1

Minor metabolites are 2-hydroxy-3-methoxy-6β-naltrexol, 2-hydroxy-3-methoxynaltrexone and *N*-noroxymorphone. The latter has agonist activity and was found in very low concentrations [40]. One major metabolite of naloxone is 6α-naloxol, which shows considerable agonist activity [41] and consequently may reduce the oral activity of naloxone.

Naloxone and naltrexone are prepared from thebaine in several steps. The first of these is the introduction of the 14-hydroxy group which can be accomplished by peroxy acid treatment of thebaine to yield 14β-hydroxycodeinone (3) [42, 43]. Catalytic hydrogenation affords 14β-hydroxydihydroco-

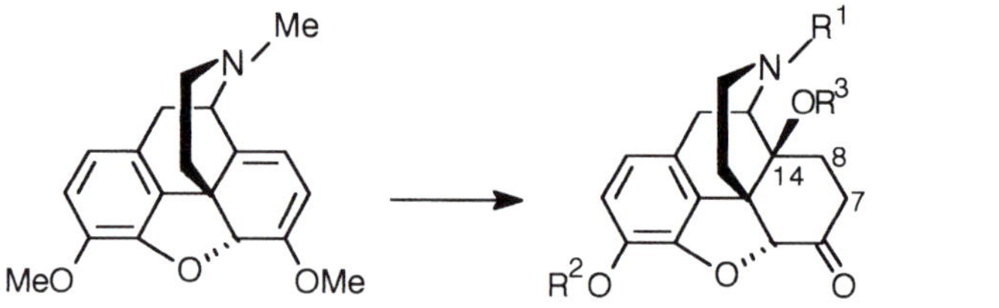

Thebaine

(3) $R^1 = R^2 = Me, R^3 = H, \Delta^{7,8}$
(4) $R^1 = R^2 = Me, R^3 = H$
(5) $R^1 = Me, R^2 = R^3 = H$
(6) $R^1 = Me, R^2 = R^3 = COMe$
(7) $R^1 = CN, R^2 = R^3 = COMe$
(8) $R^1 = R^2 = R^3 = H$

Scheme 3.2

deinone (oxycodone, 4). Recently, an improved synthesis of oxycodone has been reported. Since the performic acid treatment of thebaine also affords the N-oxide of 14β-hydroxycodeinone as a by-product, which can be easily reduced by catalytic hydrogention to oxycodone, the crude reaction product of the performic acid treatment of thebaine was hydrogenated catalytically

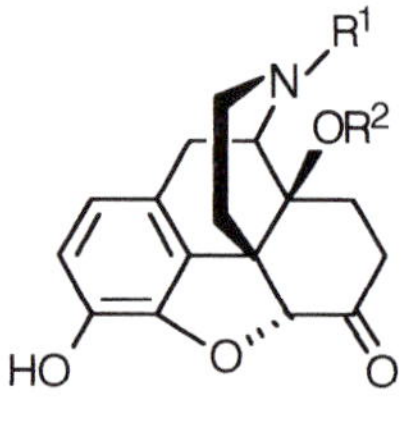

(9a) R^1 = allyl, R^2 = Me
(9b) R^1 = allyl, R^2 = Et
(9c) R^1 = cyclopropylmethyl, R^2 = Me
(9d) R^1 = cyclopropylmethyl, R^2 = Et

to afford oxycodone in *ca.* 10% higher yield [44]. Ether cleavage of oxycodone with 48% HBr yields 14β-hydroxydihydromorphinone (oxymorphone, 5) [45], which is 3,14-di-O-acetylated to give (6) prior to a von Braun N-de-

(3) $\longrightarrow$

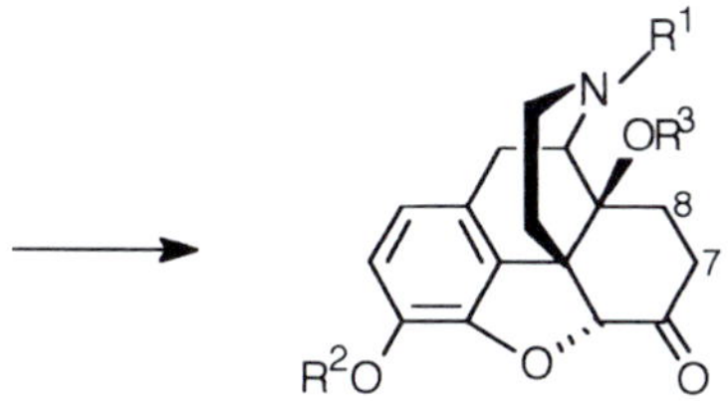

(10) $R^1 = R^2 = Me$, $R^3 = alkyl$, $\Delta^{7,8}$
(11) $R^1 = R^2 = Me$, $R^3 = alkyl$
(12) $R^1 = H$, $R^2 = Me$, $R^3 = alkyl$
(13a) $R^1 = allyl$, $R^2 = R^3 = Me$
(13b) $R^1 = allyl$, $R^2 = Me$, $R^3 = Et$
(13c) $R^1 = cyclopropylmethyl$, $R^2 = R^3 = Me$
(13d) $R^1 = cyclopropylmethyl$, $R^2 = Me$, $R^3 = Et$

Scheme 3.3

methylation with cyanogen bromide to give the *N*-cyano compound (7). Acid hydrolysis of the *N*-cyano derivative (7) affords *N*-noroxymorphone (8), which is converted to naloxone (1) on treatment with allyl bromide [46] and to naltrexone (2) on treatment with cyclopropylmethyl bromide [47] (*Scheme 3.2*).

14-*O*-Methylation and 14-*O*-ethylation of naloxone or naltrexone did not significantly alter binding affinities, antagonist potency or oral efficacy. The four naloxone and naltrexone analogues (9a-9d) appear also to be essentially devoid of any agonist activity in the MVD agonist assay. *In vivo* (hot-water tail flick test determined in rats) they were pure opioid antagonists [48].

The key step in the preparation of (9a)-(9d) was the alkylation of 14β-hydroxycodeinone (3), achieved by treating the sodium salt of (3) in dimethylformamide with a dialkyl sulphate to afford (10a) and (10b). Catalytic hydrogenation [to (11a) and (11b)] followed by *N*-demethylation using the vinyl chloroformate method [49] gave the *N*-nor derivatives (12a) and (12b), which were *N*-alkylated with allyl bromide and cyclopropylmethyl bromide respectively, to afford compounds (13a)-(13d). 3-*O*-Dealkylation using boron tribromide [50] yielded the 14β-*O*-alkyl naloxones and naltrexones (9a)-(9d) (*Scheme 3.3*).

Another universal antagonist is the naltrexone-derived nalmefene (14). This 6-methylene derivative of naltrexone shows higher κ opioid receptor affinity [51, 52]. Nalmefene is a potent, orally active opioid antagonist with a

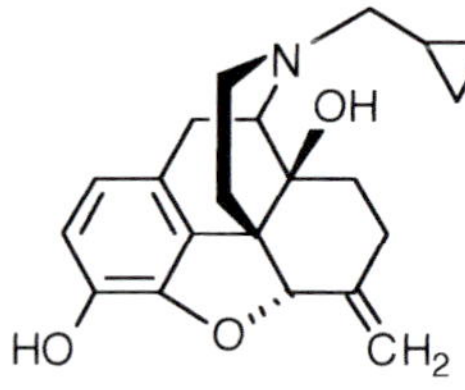

(14) Nalmefene

long duration of action, as proved in rats using the tail flick test, morphine-induced catalepsy and loss of righting reflex and the GPI. In the tail flick test, i.v. nalmefene was about twice as potent as i.v. naltrexone, whereas both exhibited similar potency after oral administration [53, 54]. It has been shown to significantly improve functional neurological recovery after spinal cord trauma [55]. It was further found that nalmefene is able to improve cellular bioenergetics after traumatic brain injury, which may in part account for the neuroprotective effects of this compound [56].

Nalmefene can be synthesized by a Wittig reaction of naltrexone with triphenylphosphonium bromide in dimethyl sulphoxide with sodium hydride as base [57].

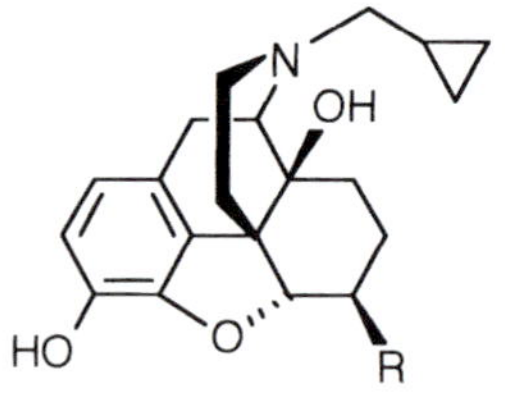

(15) β-CNA, R = N(CH$_2$CH$_2$Cl)$_2$
(16) R = N(CH$_2$CH$_2$OH)$_2$

β-Chlornaltrexamine (β-CNA, 15) another naltrexone derivative modified at C-6, is a nonequilibrium antagonist which blocks irreversibly the three major opioid receptor types (μ, κ and δ). Portoghese and his collaborators have developed this compound as the first affinity labelling agent of its class [58–61]. Compound (15) has an alkylating function at C-6 (classic nitrogen mustards) able to bind covalently to opioid receptors. In the tail flick assay in mice, β-CNA inhibited morphine-induced antinociception for 3–6 days after a single i.c.v. injection. At higher doses, β-CNA showed little agonist effect of its own.

β-CNA can be prepared in two steps from naltrexone. Reductive amination using diethanolamine and sodium cyanoborohydride gave the 6β-amino derivative (16) (the 6α-amino isomer was not formed). The conversion of (16) into (15) was effected by a modification of the triphenylphosphine-CCl_4 procedure [62] in dimethylformamide.

μ OPIOID RECEPTOR SELECTIVE ANTAGONISTS

PEPTIDES

Competitive antagonists

A series of cyclic conformationally constrained peptides related to somatostatin were designed, synthesized and tested for opioid receptor interaction by Hruby and his collaborators. Compounds (17)-(22) were found to be pure opioid antagonists (GPI) with high affinity (IC_{50} = 1.2 to 4.3 nM) and exceptional selectivity for μ over δ opioid receptors (*Table 3.1*) and with minimal or no somatostatin-like activity (ligand binding assays)[63–65].

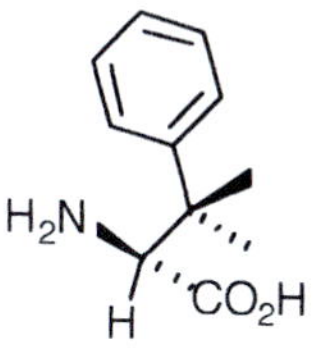

(22a) D-Phe 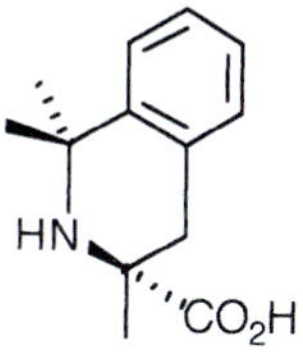   (22b) D-Tic

Table 3.1.   BINDING AFFINITIES (AS IC_{50}, NM) AND SELECTIVITIES OF COMPOUNDS (17)-(22) IN COMPETITION WITH [^{3}H]CTOP AND [^{3}H]DPDPE USING RAT BRAIN HOMOGENATES MINUS CEREBELLUM [65]

Antagonists	[^{3}H]CTOP(μ)	[^{3}H]DPDPE[δ]	μ/δ selectivity
CTP(17)	3.7	1153	312
CTOP(18)	4.3	5598	1301
CTAP(19)	2.1	5314	2530
TCTP(20)	1.2	9324	7770
TCTOP(21)	1.4	15954	11396
TCTAP(22)	1.2	1274	1060

The peptides (17)-(22) were prepared by the usual solid-phase method of peptide synthesis utilizing different resins. The structures of D-Phe (22a) and its conformationally restricted phenylalanine analogue D-Tic (tetrahydroisoquinoline-3-carboxylic acid, 22b) positioned at the N-terminal position of the somatostatin analogues are shown.

H-D-Phe-Cys-Tyr-D-Trp-Lys-Thr-Pen-Thr-NH$_2$ (CTP, 17)

H-D-Phe-Cys-Tyr-D-Trp-Orn-Thr-Pen-Thr-NH$_2$ (CTOP, 18)

H-D-Phe-Cys-Tyr-D-Trp-Arg-Thr-Pen-Thr-NH$_2$ (CTAP, 19)

H-D-Tic-Cys-Tyr-D-Trp-Lys-Thr-Pen-Thr-NH$_2$ (TCTP, 20)

H-D-Tic-Cys-Tyr-D-Trp-Orn-Thr-Pen-Thr-NH$_2$ (TCTOP, 21)

H-D-Tic-Cys-Tyr-D-Trp-Arg-Thr-Pen-Thr-NH$_2$ (TCTAP, 22)

Irreversible antagonists

The chloromethyl ketone of [D-Ala^2,Leu^5]enkephalin (DALECK) [66, 67] showed moderate preference for μ receptors over δ receptors in rat brain and was used for covalent labelling of opioid receptors [67, 68]. Somewhat higher μ selectivity was displayed by the tetrapeptide derivative H-Tyr-D-Ala-Gly-MePhe-CH$_2$Cl (DAMK), which labelled irreversibly and selectively high-affinity μ binding sites [69].

NON-PEPTIDIC LIGANDS

Competitive antagonists

Cyprodime [(-)-N-cyclopropylmethyl-4,14β-dimethoxymorphinan-6-one, 23] was reported by the author and his collaborators as the first non-peptidic, competitive pure opioid antagonist with high selectivity for μ receptors [70,

Table 3.2. BINDING AFFINITIES (IC$_{50}$, NM) AND SELECTIVITIES OF CYPRODIME AND REFERENCE DRUGS IN COMPETITION WITH [^{3}H]NALOXONE (μ), [^{3}H]TIFLUADOM (κ) AND [^{3}H]DADLE (δ) USING RAT BRAIN HOGOMENATES MINUS CEREBELLUM [71]

Antagonists	[^{3}H]Naloxone(μ)	[^{3}H]Tifluadom(κ)	[^{3}H]DADLE(δ)
Cyprodime	4.5	170	628
Naloxone	2.0	9.0	127
Naltrexone	0.5	2.6	12

71]. The pure and selective μ opioid antagonism of cyprodime was verified *in vitro* in opioid receptor binding assays (*Table 3.2*), in bioassays (GPI, MVD and RVD; *Table 3.3*) and *in vivo* (e.g., acetic acid writhing test in mice, opioid-type withdrawal jumping precipitation test in morphine-dependent mice). Cyprodime showed an antagonist potency of about one-tenth of that of naloxone *in vivo*. In receptor binding, cyprodime exhibited about one half the affinity of naloxone, and in MVD, about one-fourtieth of the μ potency of naloxone, but in contrast to naloxone, it showed good κ/μ and especially good δ/μ selectivity ratios. Cyprodime was able to antagonize sufentanil-induced respiratory depression in the dog. It was found that the onset of cyprodime, in contrast to that of naloxone, is less abrupt and that cyprodime does not exhibit sympathicotonic stimulation as does naloxone (E. Freye, personal communication). Although cyprodime shows less μ affinity than naloxone, its high μ selectivity makes it a very valuable tool in opioid research [71a–e]. Cyprodime has recently been radiolabelled with tritium [71f].

Table 3.3. POTENCIES OF CYPRODIME AND NALOXONE (K$_e$,nM) IN THE GPI, MVD AND RVD [71]

Antagonists	NM^a(μ)	EKC^b(κ)	DADLE^c[δ]	selectivity ratio κ/μ	selectivity ratio δ/μ
Cyprodime					
GPI	31	1157		37	
MVD	55.4	1551	6108	28	110
RVD	61.6		4556		74
Naloxone					
MVD	1.4	15.9	9.6	11	7

^{a}NM = normorphine ^{b}EKC = ethylketocyclazocine ^{c}DADLE = [D-Ala^2, D-Leu^5]-enkephalin.

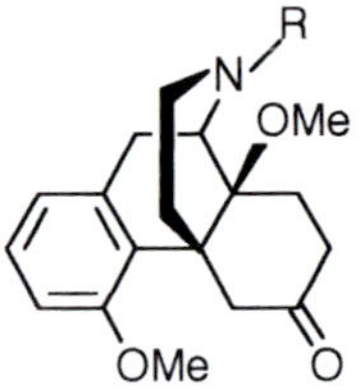

(23) cyprodime, R = cyclopropylmethyl
(24) R = Me
(25) R = CO$_2$CH$_2$CCl$_3$
(26) R = H

The first reported synthesis of cyprodime utilized as starting material the highly potent opioid agonist 4,14β-dimethoxy-N-methylmorphinan-6-one (24) which is available from oxymorphone in six steps [72]. N-Demethylation was achieved with the 2,2,2-trichloroethyl carbamate to give (25) which was cleaved reductively with activated zinc and ammonium chloride to afford N-normorphinan (26). Alkylation with cyclopropylmethyl chloride provided cyprodime (23).

Recently, a new and efficient synthesis of cyprodime from naltrexone (2) has been reported [73]. Firstly, the tetrazolyl ether (27) was formed by reaction of naltrexone with 5-chloro-1-phenyl-1H-tetrazole [74]. Catalytic hydrogenation afforded 3-deoxynaltrexone (28) which was methylated with dimethyl sulphate to give the enol ether (29). Acid hydrolysis gave the known morphinanone (30) [75] which was treated with activated zinc and ammonium chloride to yield the phenol (31). 4-O-Methylation with phenyltrimethylammonium chloride afforded cyprodime (23, *Scheme 3.4*) [75].

In an attempt to enhance the μ affinity and/or μ selectivity of cyprodime while retaining its antagonist purity and in order to further elaborate on

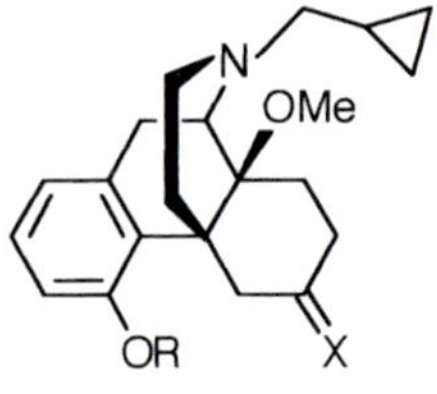

(32) R = n-Bu, X = O
(33) R = Me, X = H$_2$

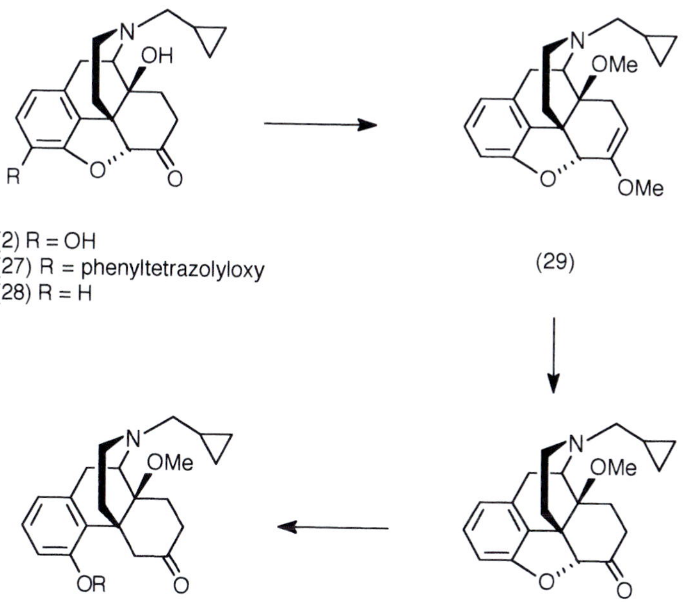

(2) R = OH
(27) R = phenyltetrazolyloxy
(28) R = H

(29)

(31) R = H

(30)

Scheme 3.4

structure-activity relationships of 14-alkoxymorphinans, an extensive study of the synthesis and biological evaluation on cyprodime-related compounds was accomplished. The 4-n-butoxy analogue (32) showed higher μ affinity than cyprodime (K_e 21 nM against normorphine) while the good κ/μ and δ/μ selectivity ratios and antagonist purity were retained. The synthesis of compound (32) was similar to the first synthesis of cyprodime [75].

Removing the 6-carbonyl function in cyprodime to form compound (33) produced only a small decrease in μ antagonist potency in the MVD, but was accompanied by an increase in κ and δ antagonist potency, resulting in a much less μ-selective compound. There was no measurable change in agonist activity, compound (33) behaving as a pure antagonist under the test conditions used. Compound (33) was prepared from cyprodime by Wolff-Kishner reduction [76].

A series of 3-hydroxy-substituted analogues [compounds (34)-(38)] of cyprodime has been synthesized in order to evaluate the role of a hydroxy group at C-3 concerning μ opioid antagonist selectivity [77]. Antagonism of μ receptor-mediated responses induced by the μ selective agonist DAMGO afforded equilibrium dissociation constants in the MVD (K_e values; *Table 3.4*) for compounds (34)-(38) which agreed closely with their affinities as determined by opioid receptor binding assays (K_i values; *Table 3.5*).

Table 3.4.   ANTAGONIST K_E VALUES ($K_e{}^a$, NM) OF COMPOUNDS (34)-(38) DETERMINED IN THE MVD [77]

Antagonists	DAMGO(μ)	CI977(κ)	DPDPE(δ)	selectivity ratio κ/μ	selectivity ratio δ/μ
(34)	5.62	368	316	65	56
(35)	24.9	174	>10000	7	>400
(36)	4.60	4.67	2272	1	494
(37)	2.92	233	106	80	36
(38)	93.8	243	8922	2.6	95
Cyprodime[b]	55.4	1551	6108	28	110
Naloxone[b]	1.4	15.9	9.6	11	7

[a]K_e = [antagonist]DR-1, where DR is the dose ratio (i.e., ratio of equiactive concentrations of the test agonist in the presence and absence of the antagonist). [b]Taken from [71].

Table 3.5.   BINDING AFFINITIES [K_i (nM)] OF COMPOUNDS (34)-(38) IN COMPETITION WITH [^{3}H]DAMGO, [^{3}H]U69593 AND [^{3}H]DPDPE USING GUINEA-PIG BRAIN HOMOGENATES [77]

Antagonists	[^{3}H]DAMGO(μ)	[^{3}H]U69593(κ)	[^{3}H]DPDPE(δ)
(34)	6.15	4.18	13.8
(35)	10.9	13.0	42
(36)	7.29	25.4	25.6
(37)	1.42	5.46	21.4
(38)	54.4	23.0	334
Cyprodime	23.7	105	61.1

Table 3.6.   ANTAGONIST $K_e{}^a$ VALUES (NM) FOR 3-HYDROXYCYPRODIME (34) IN THE MVD AND GPI AGAINST κ AGONISTS [77]

κ Agonists	MVD	GPI
CI977	368	284
U69593	354	97.5
Ethylketocyclazocine	910	3.95
Dynorphine 1–13	450	ND[b]
Tifluadom	ND[b]	1009

[a]K_e = [antagonist]DR-1, where DR is dose ratio (i.e. ratio of equiactive concentrations of the test agonist in the presence and absence of the antagonist). [b]Not determined

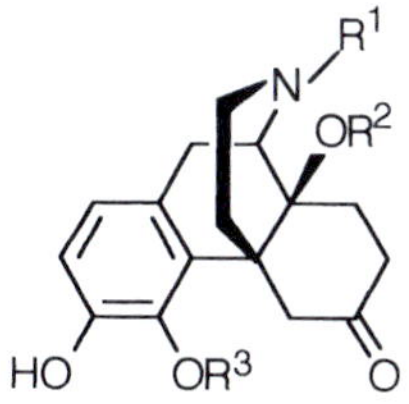

(34) R^1 = CPM, R^2 = R^3 = Me
(35) R^1 = CPM, R^2 = Me, R^3 = n-Bu
(36) R^1 = allyl, R^2 = R^3 = Me
(37) R^1 = CPM, R^2 = Et, R^3 = Me
(38) R^1 = CPM, R^2 = H, R^3 = Me
CPM = cyclopropylmethyl

Differences were apparent at κ and δ receptors. Although the compounds had high affinity for both κ and δ receptors in opioid receptor binding, they were very poor at antagonizing agonist responses mediated by κ and particularly δ agonists in the MVD. A 14-hydroxy group instead of a 14-alkoxy group leads to a reduction in affinity at all three binding sites. None of the compounds tested showed agonist potency in the MVD or the GPI. Using 3-hydroxycyprodime (34), and employing both MVD and GPI, the low affinity of this compound at κ receptors was confirmed by determination of the K_e values against a variety of κ agonists (*Table 3.6*).

The results suggest that the affinities of cyprodime and its 3-hydroxy analogues (34)-(38), measured in binding assays and by bioassay, agree for the μ receptor site but not for κ and δ receptor sites. A possible explanation is that the receptors in guinea-pig brain and the isolated tissues may be different and are distinguishable by this series of compounds.

The synthesis of 3-hydroxycyprodime (34) started from the oxymorphone derivative (39) [72]. Reductive cleavage of the 4,5-oxygen bridge was achieved with activated zinc and ammonium chloride to give phenol (40), which was *O*-methylated with phenyltrimethylammonium chloride to yield compound (41). *N*-Demethylation was accomplished with 1-chloroethyl chloroformate [78], and subsequent cleavage of the carbamate (42) in refluxing methanol afforded *N*-normorphinan (43). Alkylation with cyclopropylmethyl chloride gave (44) from which the benzyl protecting group was removed by catalytic hydrogenation to yield 3-hydroxycyprodime (34, *Scheme 3.5*).

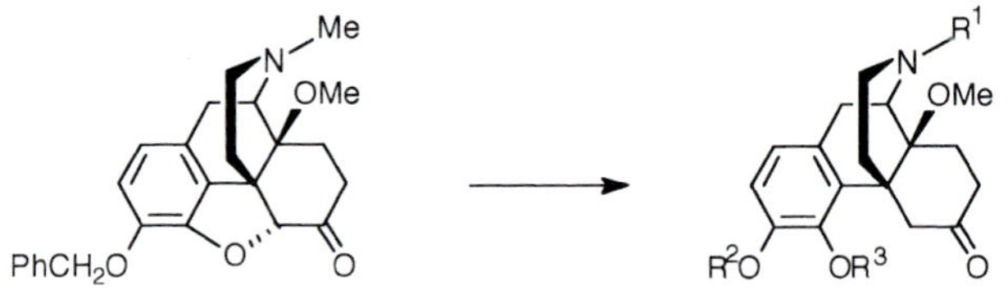

(39)

(40) R^1 = Me, R^2 = CH$_2$Ph, R^3 = H
(41) R^1 = R^3 = Me, R^2 = CH$_2$Ph
(42) R^1 = CO$_2$CHClCH$_3$, R^2 = CH$_2$Ph, R^3 = Me
(43) R^1 = H, R^2 = CH$_2$Ph, R^3 = Me
(44) R^1 = CPM, R^2 = CH$_2$Ph, R^3 = Me
CPM = cyclopropylmethyl

Scheme 3.5

Long-acting and irreversible ligands

β-Funaltrexamine (β-FNA, 45) synthesized and biologically characterized by Portoghese and his collaborators, is an irreversible μ opioid receptor antagonist [79–81]. The design rationale for β-FNA, a C-6-methylfumaramido derivative of naltrexone, was based on the corresponding nitrogen mustard analogue β-CNA (15)[58–61], which possesses potent nonequilibrium antagonist properties at all three receptor types. As the aziridinium ion derived from the nitrogen mustard is highly reactive, the authors reasoned that a Michael acceptor, such as the fumarate group, attached at an identical position would (because of its lower reactivity) confer greater selectivity in covalent bond formation with opioid receptors.

In the GPI, β-FNA produced a potent reversible agonist response, but upon incubation of the ileum preparation, it gave rise to an irreversible an-

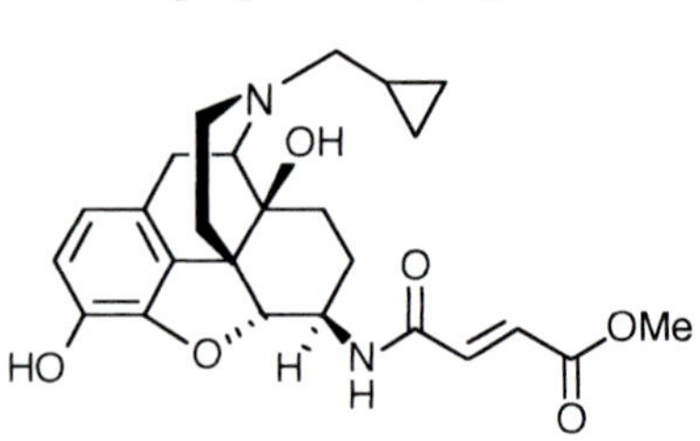

(45) β-FNA

tagonist action. β-FNA exhibited an irreversible antagonism against the μ-mediated effects of morphine, without affecting κ agonist actions. Its MVD responses suggested also μ opioid antagonism [81, 82]. The antinociceptive actions in the mouse writhing and tail flick tests were of short duration and appeared to be κ receptor mediated [80, 81].

β-FNA was synthesized by reaction of β-naltrexamine (48) with the monomethyl ester of fumaroyl chloride [79]. Amine (48) was prepared first from naltrexone by reductive amination with sodium cyanoborohydride in the presence of ammonium acetate to give 6α- and 6β-epimers (ratio *ca.* 2:1). Separation was achieved by fractional crystallization [83]. An improved synthesis of (48) was reported *via* the dibenzyliminium salt of naltrexone (46; easily accessible from naltrexone and dibenzylamine) which was reduced with sodium cyanoborohydride to give exclusively the 6β-epimer (47). Catalytic hydrogenolysis afforded β-naltrexamine (*Scheme 3.6*) [84].

Hahn, Pasternak and their collaborators found naloxazone (49) and naloxonazine (50) (*Scheme 3.7*) to be irreversible antagonists and to interact with the high affinity, μ_1 site (a putative μ receptor subtype) [85, 86]. Treatment of either rats or mice with compounds (49) or (50) eliminated the high affinity binding of a series of radiolabelled opioids. This loss of binding was associated with a dramatic shift of the analgesic dose-response curve to the right, implying that μ_1 sites mediated analgesia [87]. The data suggest that μ_1 sites are responsible for supraspinal analgesia [88]. There is a good correlation between binding and morphine's analgesic potency, both of which returned to control values after 3 days. On the other hand, μ_1 blockade did not alter the respiratory depression of morphine [89, 90] or most of the signs associated with morphine dependence [91]. The mechanism by which the long-lasting effects are produced remains to be clarified [26]. There are con-

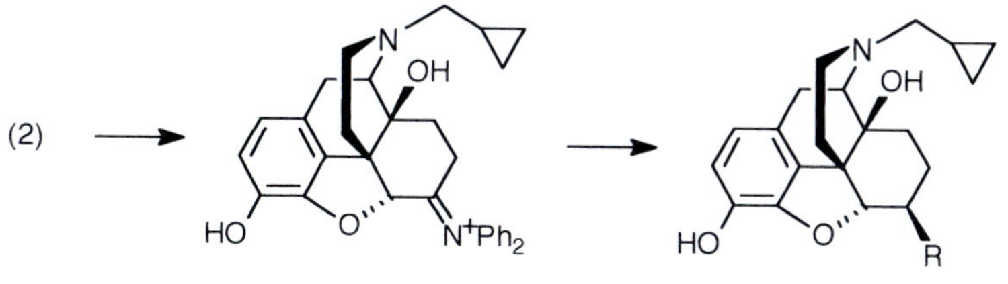

(2) ⟶   (46)   ⟶   (47) R = NPh$_2$
                                                (48) R = NH$_2$

Scheme 3.6

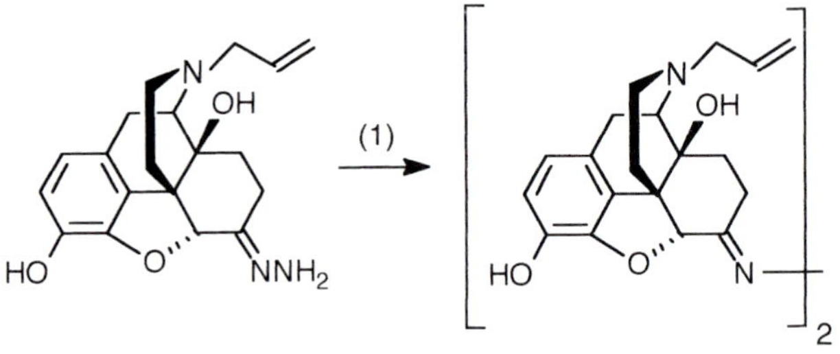

(49) naloxazone   (50) naloxonazine

Scheme 3.7

flicting reports as to the importance of covalent binding [92, 93]. Another report questions the metabolic stability of naloxonazine and suggests the involvement of a hydrolytic cleavage product and membrane phosphatides in its persistent effects [94].

The hydrazone (49) was prepared by reacting naloxone with an excess of anhydrous hydrazine [85]. Naloxazone (49) was treated with naloxone to yield the azine (50).

An isothiocyanate analogue of the μ receptor-selective strong analgesic etonitazene (51), 1-[2-(diethylamino)ethyl]-2-(4-ethoxybenzyl)-5-isothiocyanatobenzimidazole (BIT, 52) was designed, synthesized and biologically characterized by Rice and his collaborators [95, 96]. BIT is a selective acylator of μ receptors and is able to deplete membranes of μ binding sites, and as such, it has been shown to be useful in the further biochemical characterization of opioid receptors [97–100]. Interestingly, BIT exhibits morphine-like antinociceptive potency in the mouse hot-plate test [96].

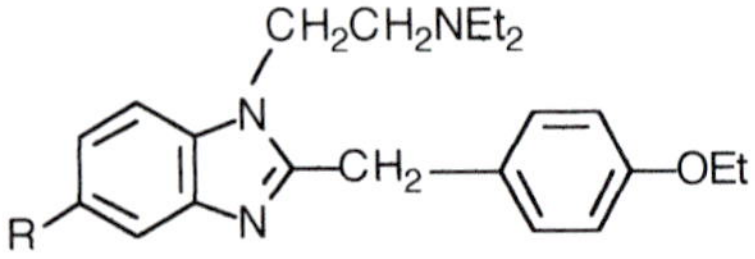

(51) Etonitazene, R = NO$_2$
(52) BIT, R = NCS
(53) R = NH$_2$

The synthesis of BIT was accomplished from etonitazene (51) in two steps. Catalytic hydrogenation afforded the amino derivative (53) [101], which was converted into the isothiocyanate (52) by reaction with thiophosgene [96].

14β-Cinnamoylaminomorphinones represent an important series of irreversible opioid ligands. Clocinnamox [14β-(4-chlorocinnamoylamino)-7,8-dihydro-N-cyclopropylmethylnormorphinone, C-CAM, 54] was developed by Lewis and his collaborators [102]. C-CAM inhibits [^{3}H]etorphine binding to rat brain membranes and antagonizes the agonistic effects of μ, κ and δ opioids in the MVD [103]. It also displayed a very long duration of antagonist action without revealing any agonist properties after peripheral administration in analgesic studies in the monkey and in the mouse [103, 104]. In the mouse warm water tail-withdrawal assay, C-CAM appeared to be a systemically active irreversible antagonist at μ receptors with no agonist activity [104]. C-CAM is also capable of eliminating binding of μ selective radioligands when administered systemically [105]. It is as yet unclear whether the irreversible effects of C-CAM are really μ selective. The mechanism of the long duration of action is uncertain. A possible explanation is the ability of the cinnamoylamino substituent of C-CAM to serve as a Michael acceptor suitable for nucleophilic addition in a similar manner to the fumaramido side-chain of β-FNA. The synthesis of C-CAM is analogous to that described below for MET-CAMO.

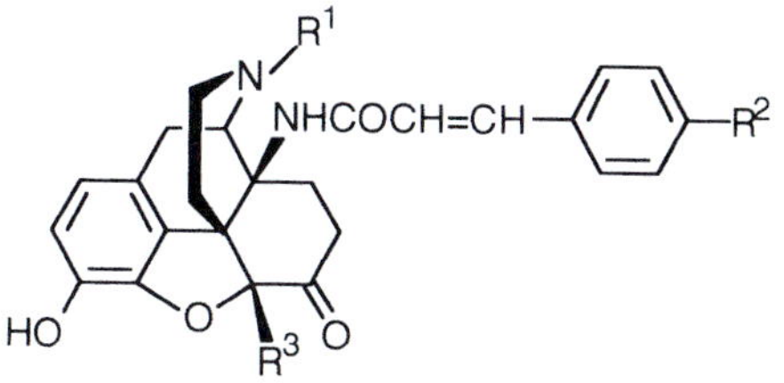

(54) C-CAM, R^1 = CH$_2$CH(CH$_2$)$_2$, R^2 = Cl, R^3 = H

(55) MET-CAMO, R^1 = R^3 = Me, R^2 = NO$_2$

(56) N-CPM-MET-CAMO, R^1 = CH$_2$CH(CH$_2$)$_2$, R^2 = NO$_2$, R^3 = Me

Since a nitro substituent is more electron-withdrawing than a chloro substituent, it was anticipated by Archer and his colleagues that replacing the p-chloro atom on the cinnamoylamino moiety with a p-nitro group would make the cinnamoylamino function a better Michael acceptor [106]. 5β-Methyl-14β-(4-nitrocinnamoylamino)-7,8-dihydromorphinone (MET-CAMO, 55) was found to be an irreversible μ selective antagonist, MET-

CAMO behaved as a nonequilibrium and μ selective ligand in bovine striatal membranes. No antinociception was produced in the mouse tail flick test when the compound was administered i.c.v. [106]. Pretreating mice for 24 h with MET-CAMO completely blocked morphine-induced μ mediated antinociception, but not antinociception mediated by either κ or δ opioid receptors [107]. Its corresponding N-cyclopropylmethyl analogue (N-CPM-MET-CAMO, 56) is also an irreversible antagonist at μ receptors but shows less selectivity than MET-CAMO [106, 108].

MET-CAMO was prepared from 5β-methylthebaine (57) [109,110] via 14β-amino-7,8-dihydro-5β-methylcodeinone (59) which was obtained by the Kirby-McLean procedure [111]. Thus, oxidation of 2,2,2-trichloroethyl N-hydroxycarbamate with sodium periodate in the presence of 5β-methylthebaine gave the adduct (58). Catalytic hydrogenation using a Pd/C catalyst in methanol in the presence of sodium acetate-acetic acid buffer yielded amine (59). Reaction with 4-nitrocinnamoyl chloride furnished amide (60) and ether cleavage using boron tribromide yielded MET-

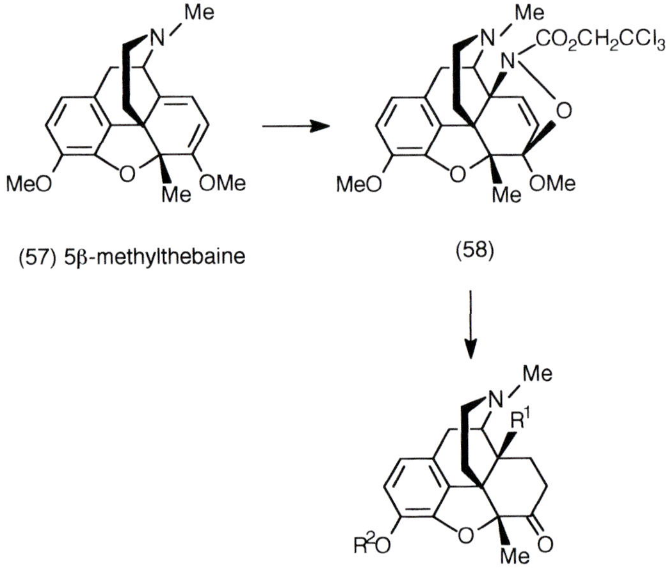

(57) 5β-methylthebaine   (58)

(59) R^1 = NH$_2$, R^2 = Me
(60) R^1 = NHCOCH=CHPh-4-NO$_2$, R^2 = Me

Scheme 3.8

CAMO (55, *Scheme 3.8*). *N*-CPM-MET-CAMO (56) was prepared from *N*-cyclopropylmethyl-5β-methylnorthebaine essentially as described for (55) [106].

Archer *et al.* prepared and evaluated a series of 14β-bromoacetamido derivatives of morphine, dihydromorphine, morphinone and dihydromorphinone [112–114]. After incubation with the disulphide bond-reducing agent dithiothreitol (*threo*-1,4-dimercapto-2,3-butanediol), followed by the addition of BAM (61), H$_2$BAM (62), BAMO (63) or H$_2$BAMO (64) and extensive washing of the membranes, greater than 90% of μ opioid binding to membranes was inhibited, while κ and δ opioid binding was not altered by alkylation of membranes with the affinity ligands [113, 114]. In the absence of dithiothreitol, these affinity ligands did not inhibit opioid binding to brain membranes in a wash-resistant manner. These studies initially suggested that there was a disulphide bond at the μ opioid binding site. However, dithiothreitol enhanced the alkylation of the opioid receptor by reducing the binding of the affinity ligands to nonspecific sulphydryl groups, not by breaking a disulphide bond at the μ opioid binding site [115]. H$_2$BAMO produced short-term antinociception at μ receptors in the mouse tail flick assay [116].

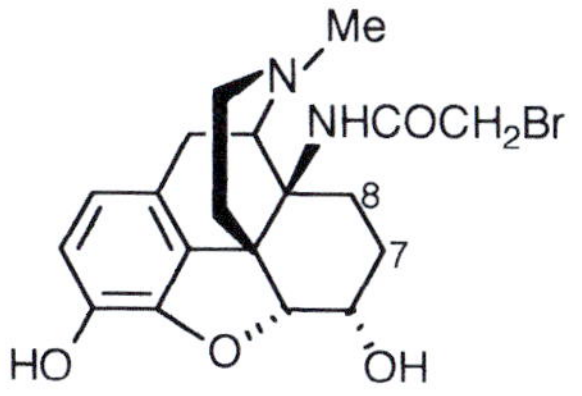

(61) BAM, $\Delta^{7,8}$
(62) H$_2$BAM

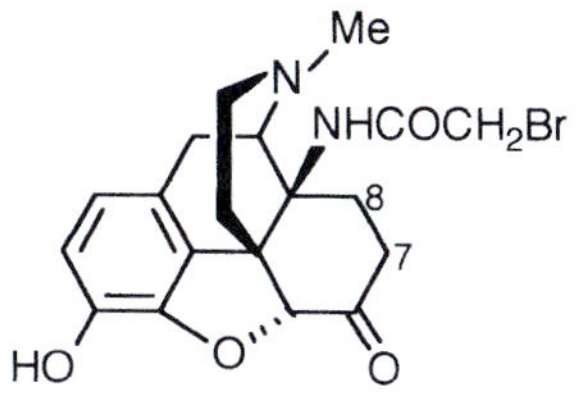

(63) BAMO, $\Delta^{7,8}$
(64) H$_2$BAMO

For the synthesis of BAM, thebaine was treated with 1-chloro-1-nitrosocyclohexane and dry hydrogen chloride to give an adduct, which was reduced with zinc and then partially hydrolysed to a mixture of 14β-aminocodeinone (65) and ketal (66). Reduction of (65) with sodium borohydride gave 14β-aminocodeine (67) [117], which on 3-*O*-demethylation with boron tribromide afforded 14β-aminomorphine (68). Bromoacetylation, followed by treatment with 0.5N hydrochloric acid yielded BAM (61)[112]. Catalytic hydrogenation of (68) followed by bromoacetylation and treatment with 1N hydrochloric acid yielded H$_2$BAM (62)[114]. Bromoacetylation of ketal (66) gave compound (69) which was 3-*O*-demethylated using boron tribro-

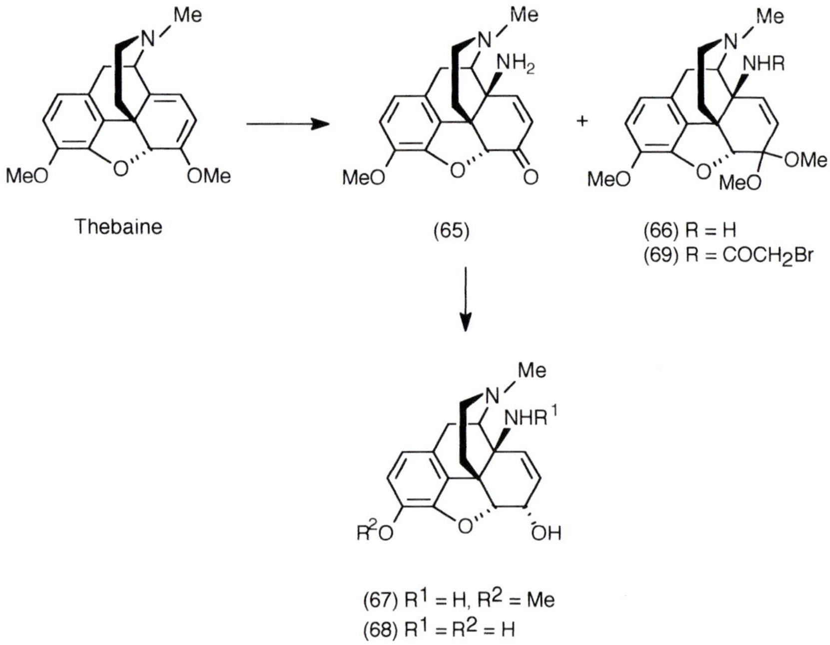

Thebaine

(65)

(66) R = H
(69) R = COCH₂Br

(67) R¹ = H, R² = Me
(68) R¹ = R² = H

Scheme 3.9

mide to afford BAMO (63) [112] (*Scheme 3.9*). Treatment of ketal (66) with boron tribromide yielded 14β-aminomorphinone (70), which after catalytic hydrogenation to (71) and bromoacetylation, yielded H₂BAMO (64) [114] (*Scheme 3.10*).

Archer and co-workers also prepared and evaluated the disulphides (72)-(75). Incubation of bovine striatal membranes with TAMO (72), *N*-CPM-TAMO (73), MET-TAMO (74) and *N*-CPM-MET-TAMO (75) resulted in wash-resistant inhibition of the binding of the μ selective peptide DAMGO. TAMO had no effect on κ or δ binding while *N*-CPM-TAMO moderately inhibited κ binding and weakly inhibited δ binding. MET-TAMO and *N*-CPM-MET-TAMO inhibited μ, κ and δ binding. Thus, TAMO was the most selective of the four ligands and *N*-CPM-TAMO appeared to be the next most selective [118, 119]. TAMO exhibited a short antinociceptive effect in the mouse tail flick test [116].

The synthesis of TAMO is outlined in *Scheme 3.11*. The 3-hydroxy group of 14β-amino-7,8-dihydromorphinone (71) [114] was protected with tert-bu-

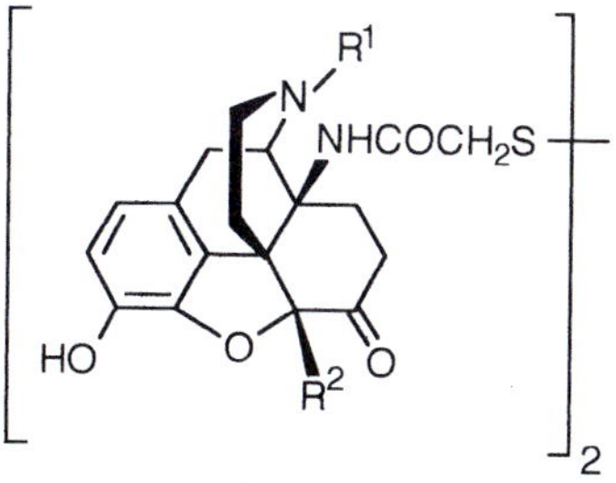

(72) TAMO, R^1 = Me, R^2 = H
(73) N-CPM-TAMO, R^1 = $CH_2CH(CH_2)_2$, R^2 = H
(74) MET-TAMO, R^1 = R^2 = Me
(75) N-CPM-MET-TAMO, R^1 = $CH_2CH(CH_2)_2$, R^2 = Me

tyldimethylsilyl chloride to give the silyl ether (76). Disulphide (77) was
formed by reaction of (76) with dithioglycolyl chloride [120]. Deprotection
was accomplished with tetrabutylammonium fluoride to yield TAMO (72)
[118]. Disulphides (73)-(75) were prepared similarly [118,119].

κ OPIOID RECEPTOR SELECTIVE ANTAGONISTS

PEPTIDES

Several attempts were made to develop κ selective antagonists through
structural modification of dynorphin A. [Ala^2,Trp^4]dynorphin A-(1–13)

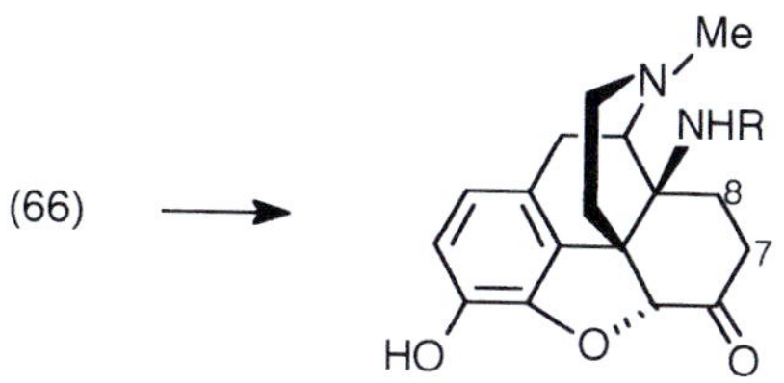

(66) $\longrightarrow$

(70) R = H, $\Delta^{7,8}$
(71) R = H

Scheme 3.10

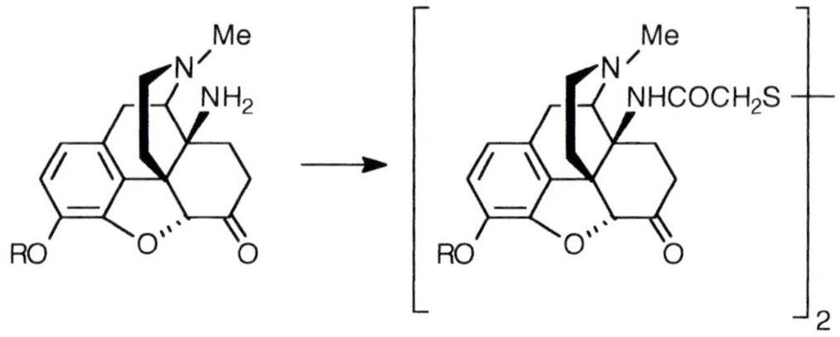

(76) R = t-BuMe$_2$Si

(77) R = t-BuMe$_2$Si
(72) R = H

Scheme 3.11

has been claimed to be a κ selective opioid antagonist [121], but an accurate opioid receptor binding selectivity profile has not been determined. The three 11-peptide analogues [D-Trp[2,8], D-Pro[10]], [D-Trp[5,8], D-Pro[10]]- and [D-Trp[2,4,8], D-Pro[10]]-dynorphin A-(1–11) showed weak antagonism against dynorphin A and low κ versus μ selectivity [122] . [N,N-Diallyl-Tyr[1], D-Pro[10]]dynorphin A-(1–11) and [N,N-diallyl-Tyr[1], Aib[2,3], D-Pro[10]]dynorphin A-(1–11) were reported to act as pure but not very κ selective opioid antagonists *in vitro* [25, 123].

NON-PEPTIDIC LIGANDS

Competitive antagonists

The first selective κ opioid receptor antagonist documented to have a significant degree of selectivity for κ over μ and δ receptors was TENA (78) [124]. The β-naltrexamine derivative (78) was developed out of the double pharmacophore 'bivalent ligand' approach of Portoghese and Takemori [124, 125]. TENA consists of two naltrexone-derived pharmacophores connected to a spacer obtained from triethylene glycol. Extensive studies of the length and conformational flexibility of the spacer [126–128] led to the synthesis of bivalent ligands that have a pyrrole moiety as a very short and rigid spacer [129]. The most potent and selective members of this series, norbinaltorphimine (nor-BNI, 79) and binaltorphimine (BNI, 80) possess very high κ opioid receptor antagonist potency and κ antagonist selectivity *in vitro* and *in vivo* [130, 131]. Its selective antagonist properties have been demonstrated in smooth muscle bioassays (*Table 3.7*), radioligand binding assays (*Table*

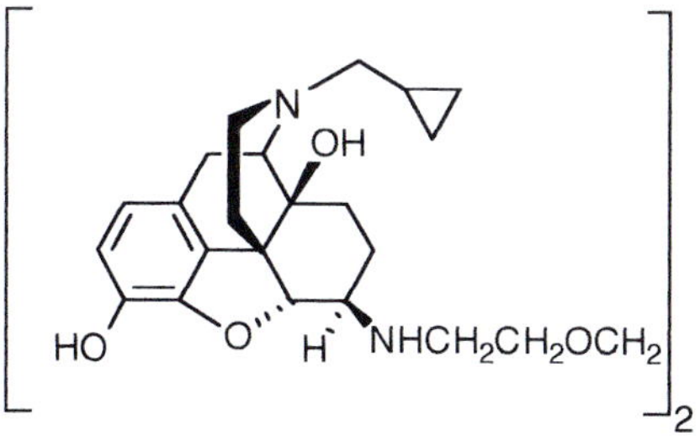

(78) TENA

3.8) [132] and *in vivo*. Nor-BNI has been found to selectively block κ mediated antinociception [133] and diuresis [134], although nor-BNI is weakly potent following systemic administration. Another research group found that nor-BNI was about 400-fold selective for κ versus μ receptors and about 250-fold for κ versus δ receptors. The same group also found that nor-BNI has only limited selectivity for κ versus μ receptors *in vivo* [135]. The reasons for the apparent differences observed *in vivo* appear to relate to the unusually long time following nor-BNI administration to reach the peak κ selective antagonist effect [26].

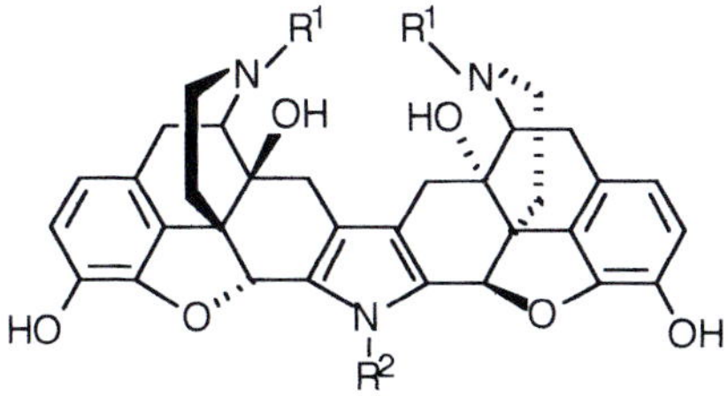

(79) nor-BNI, R^1 = CPM, R^2 = H
(80) BNI, R^1 = CPM, R^2 = Me
CPM = cyclopropylmethyl

It was found that only one pharmacophore is required for the κ opioid antagonist selectivity of nor-BNI [136]. In smooth muscle preparations, the meso isomer (81) (derived from (−)-naltrexone and its inactive (+)-enantiomer [27]) of nor-BNI was more potent than nor-BNI and about half as selective as κ antagonist. Since (81) contains one antagonist pharmacophore but yet retains some κ selectivity, it was concluded that κ selectivity is not de-

Table 3.7.　OPIOID ANTAGONIST ACTIVITY (K_e^a IN NM) OF NOR-BNI (79), BNI (80) AND (81) IN THE GPI AND MVD [129, 136]

Antagonists	$EK^a(\kappa)$ K_e	$IC_{50}\ r^d$	$M^a(\mu)$ K_e	$IC_{50}\ r^d$	$DADLE^b(\delta)$ K_e	$IC_{50}\ r^c$	K_e ratio μ/κ	δ/κ
Nor-BNI(79)	0.41	49.8	12.5	2.6	10	2.0	30	24
BNI (80)	–	128	–	5.4	–	4.5	–	–
(81)	0.08	252	1.1	18.8	1.3	16.8	14	16
Naltrexone	5.5	19	1.0	98	24	5.1	0.2	4.4

[a]Ethylketazocine (EK) and morphine (M) in the GPI. [b][D-Ala^2,D-Leu^5]enkephalin (DADLE) in the MVD. [c]The ID_{50} ratio is the ID_{50} of the agonist in the presence of antagonist divided by the control IC_{50} in the same preparation.

pendent on the presence of two (−)-naltrexone-derived pharmacophores of nor-BNI. It was suggested that the κ selectivity of nor-BNI and (81) is derived from the portions of the second halves of these molecules in that they mimic key 'address' components of dynorphin at κ opioid receptors [136].

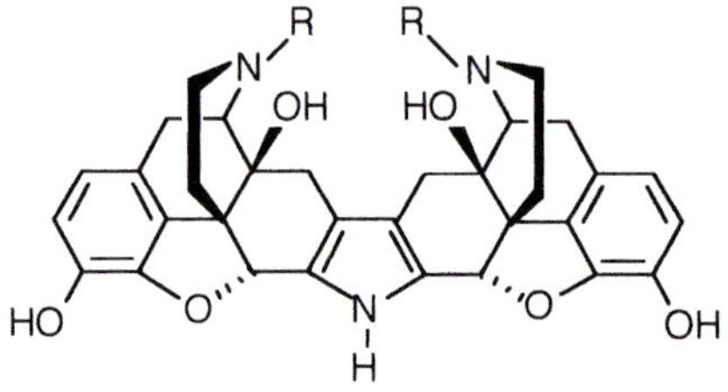

(81) R = cyclopropylmethyl

Bimorphinans (79) and (80) were synthesized [129, 137] by heating the hydrochlorides of naltrexone and the appropriate hydrazine ($RNHNH_2$; R = H, Me) under conditions similar to that reported for the Piloty-Robinson

Table 3.8.　BINDING AFFINITIES (K_i IN nM) AND SELECTIVITIES OF NOR-BNI AND BNI IN COMPETITION WITH [^{3}H]DAMGO (μ), [^{3}H]DADLE (δ) AND [^{3}H]EK (κ) USING GUINEA-PIG BRAIN MEMBRANES [132]

Antagonists	[^{3}H] DAMGO(μ)	[^{3}H]DADLE(δ)	[^{3}H]EK(κ)	K_i ratio μ/κ	ratio μ/κ
Nor-BNI (79)	47	39	0.26	181	150
BNI (80)	18	58	0.41	44	141

synthesis [138, 139]. A procedure was found for the preparation of BNI and analogues having an *N*-methyl group at the pyrrole moiety, which takes place under milder conditions and gives better results in comparison with the procedure reported earlier [137]. Thus, naltrexone and analogues are being treated with *N*-methylhydrazine sulphate in glacial acetic acid at room temperature to give the corresponding bimorphinans in good yields [140].

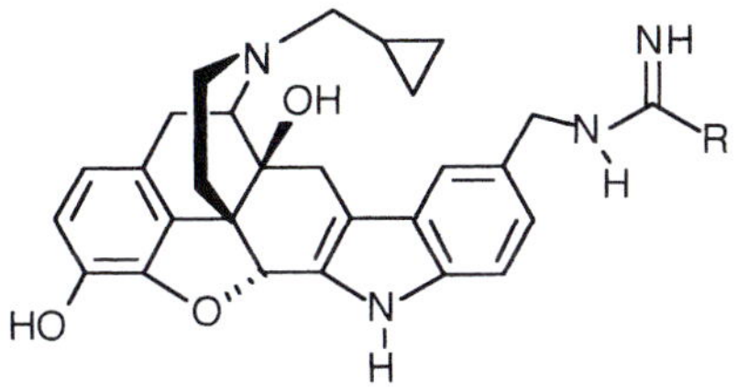

(82) R = (CH$_2$)$_3$Me
(83) R = (CH$_2$)$_2$Me
(84) R = Et

Recently, Portoghese and his collaborators developed a series of 5'-[(*N^2*-alkylamidino)methyl]naltrindole derivatives (82)-(84) as a novel class of κ opioid receptor antagonists [141]. The design rationale for the synthesis of this series involved the attachment of a basic group to the 5'-position of the δ receptor selective antagonist naltrindole (NTI, 85; *Scheme 3.12*) in order to approximate the distance between one of the antagonist pharmacophores of nor-BNI and its second basic group which has been suggested

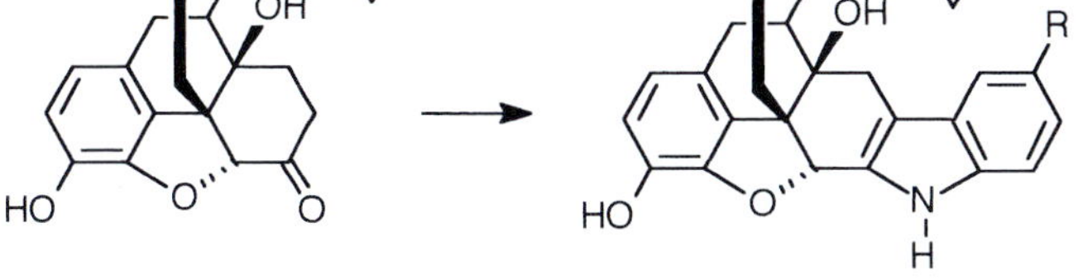

(2) naltrexone

(85) NTI, R = H
(86) R = CN
(87) R = CH$_2$NH$_2$

Scheme 3.12

Table 3.9.   BINDING AFFINITIES AND SELECTIVITIES OF NOR-BNI AND
COMPOUND (83) IN COMPETITION WITH [^{3}H]DAMGO(μ), [^{3}H]U69593(κ) AND
[^{3}H]DADLE(δ) USING GUINEA-PIG BRAIN MEMBRANES [141]

Antagonists	K_i, nM [^{3}H]DAMGO(μ)	[^{3}H]DADLE(δ)	[^{3}H]U69593(κ)	K_i μ/κ	ratio δ/κ
Nor-BNI(79)	47	43	0.28	168	154
(83)	3.5	5.5	0.061	57	90

[136] to function as a κ address mimic. Thus, the indole moiety functions as a
rigid spacer to hold the amidine group in a location similar to that of the
right-hand basic group of nor-BNI. Compound (82), which is the most potent derivative of this series, possesses greater *in vitro* κ antagonist potency
and selectivity in smooth muscle preparations than does nor-BNI. In opioid
binding, compound (82) shows a higher κ affinity but somewhat less selectivity than nor-BNI (*Table 3.9*).

Compounds (82)-(84) were synthesized as outlined in *Scheme 3.12*. Reaction of naltrexone with 4-hydrazinobenzonitrile [143] under Fischer indole
conditions afforded the 5'-nitrile (86) which was catalytically hydrogenated
to the primary amine (87) using Raney nickel. The amidines (82)-(84) were
prepared by reacting amine (87) with the appropriate iminoester [144, 145].

Irreversible ligands

The first site-directed irreversible inhibitors of κ receptors, compounds (88)
and (89) (UPHIT) were described by Rice and his collaborators [146, 147].

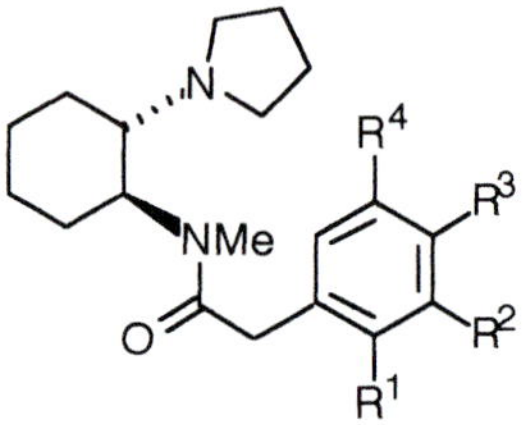

(88) R^1 = NCS, R^2 = R^3 = R^4 = H
(89) R^1 = NCS, R^2 = H, R^3 = R^4 = Cl
(90) U50488, R^1 = R^2 = H, R^3 = R^4 = Cl
(91) R^1 = R^3 = R^4 = H, R^2 = NCS

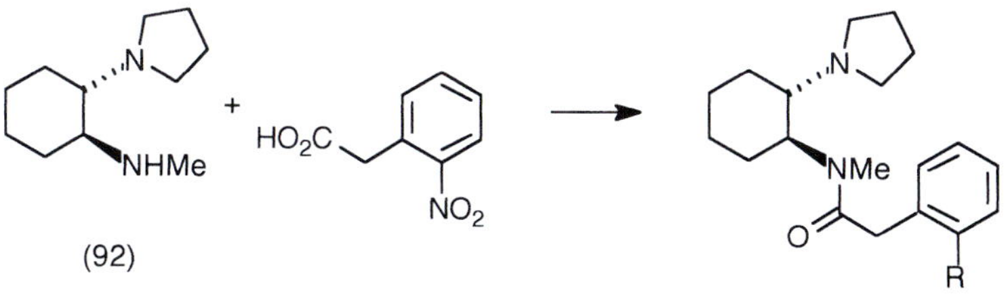

(92)

(93) R = NO$_2$
(94) R = NH$_2$
(88) R = NCS

Scheme 3.13

They designed these acylators of κ receptors through structural modifications of the U50488 (90; κ agonist) molecule. Compound (88) is able to specifically and irreversibly inhibit κ receptors labelled by [^{3}H]U69593 (κ agonist) while i.c.v. administration into guinea-pig brain failed to produce any
irreversible inhibition of κ receptors. In contrast, i.c.v. injection of compounds (89) and (91) resulted in a significant reduction in κ receptors that
bind [^{3}H]U69593 [147, 148].

The synthesis of compound (88) is outlined in *Scheme 3.13*. DCC coupling of enantiomerically pure (1S,2S)-(+)-*trans*-2-pyrrolidinyl-N-methylcy-
clohexylamine (92) with 2-nitrophenylacetic acid gave the nitro derivative
(93). Catalytic hydrogenation over Pd-C catalyst afforded amine (94) which
was treated with thiophosgene to yield (88).

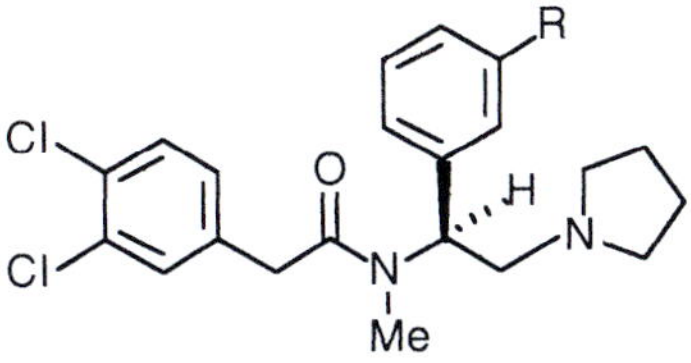

(95) DIPPA, R = NCS
(96) R = H

Another U50488-derived irreversible κ receptor antagonist, DIPPA (95)
was described by Portoghese and his collaborators [149, 150]. The design of
DIPPA (which has also an isothiocyanate acylating group) as an affinity label was based on the report that arylacetamide (96) was a potent, κ selective

agonist [151, 152]. Unlike compounds (88), DIPPA possessed selective κ opioid receptor antagonism *in vivo*. This antagonism was measured in the mouse tail flick assay and lasted 48 h. In the abdominal stretch assay in mice, DIPPA showed a short-term antinociceptive effect and in bioassays (GPI and MVD) it exhibited full agonism. This agonist effect in bioassays was antagonized by nor-BNI indicating interaction with κ receptors. Receptor binding studies showed that DIPPA binds selectively and with high affinity to κ receptors. Wash studies have suggested that this involves covalent binding. Short term agonism, followed by long term antagonism, has also been reported for β-CNA [58].

The synthesis of DIPPA (*Scheme 3.14*) involved the nitration of optically pure (97) [151] to give a regioisomeric mixture consisting of 88% *m*- and

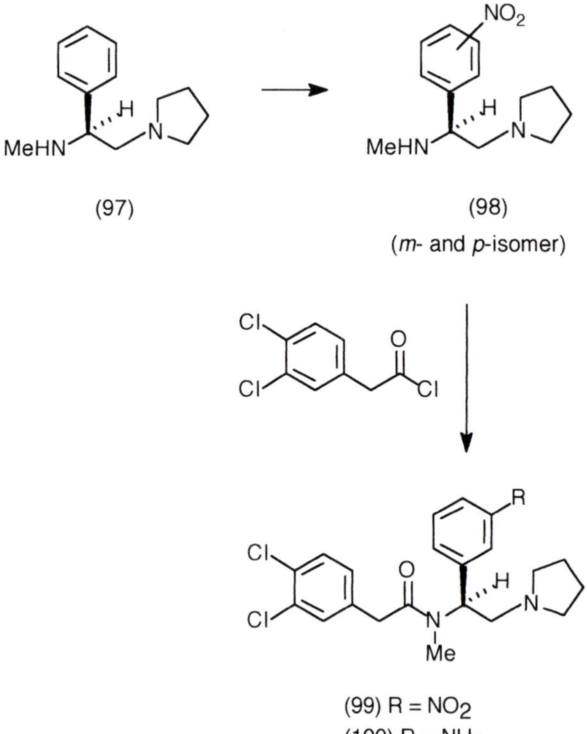

(97)

(98)
(*m*- and *p*-isomer)

(99) R = NO_2
(100) R = NH_2

Scheme 3.14

12% *p*-nitro isomers of (98) which were acylated with 3,4-dichlorophenylace-
tyl chloride to yield enantio- and regio-isomerically pure (99) after chromatography and crystallization. Raney nickel-hydrazine reduction of the nitro
group afforded the amine precursor (100) which was treated with thiophosgene to give DIPPA (95) [149, 150].

δ OPIOID RECEPTOR SELECTIVE ANTAGONISTS

PEPTIDES

Competitive antagonists

Several δ selective opioid antagonists have been obtained through diallyla-
tion of the α-amino group of enkephalin-related peptides [25]. The design of
these analogues was based on analogy with the known *N*-allyl substituted
morphinan antagonists. *N,N*-Diallylated leu-enkephalin (101) was shown
to be a moderately potent δ selective antagonist in the MVD [153]. Replacement of the 3,4 position peptide bond in (101) with a thiomethylene moiety
resulted in a compound, *N,N*-diallyl-Tyr-Gly-Glyψ-[CH$_2$S]Ph-Leu-OH [ICI
154129, (102)] which also was a δ selective antagonist with moderate δ receptor affinity [154, 155]. Subsequently, a conformationally restricted enkephalin analogue, *N,N*-diallyl-Tyr-Aib-Aib-Phe-Leu-OH [ICI 174864, (103)]
(Aib = aminoisobutyric acid), was found to be a more potent and selective
δ antagonist [25, 156, 157].

   In 1992, Schiller and his collaborators reported the discovery of a new
class of opioid peptide-derived δ antagonists which contain a 1,2,3,4,-tetra-
hydroisoquinoline-3-carboxylic acid (Tic) residue (22b) in the 2-position of
the peptide sequence [158]. The two prototype antagonists were the tetrapeptide H-Tyr-Tic-Phe-Phe-OH (TIPP, 104) and the tripeptide H-Tyr-Tic-Phe-
OH (TIP, 105). TIPP showed high antagonist potency against various δ agonists in the MVD (K$_e$ = 3–5 nM), high δ affinity (K$_i^\delta$ = 1.22 nM), and extraordinary μ/δ selectivity ratio (K$_i^\mu$/K$_i^\delta$ = 1410). No agonism was found in
the GPI at concentrations as high as 10 μM. The compound was also prepared in tritiated form and [^{3}H]TIPP was shown to be a good radioligand
for the study of δ opioid receptor interactions [159]. In comparison with
TIPP, TIP was a somewhat less potent and less selective δ receptor antagonist.

   Both TIPP and TIP were stable in the aqueous buffer solution (pH 7.7)
used for biological testing for periods of up to 6 months. However, these
peptides were shown to undergo slow spontaneous Tyr-Tic diketopiperazine

formation with concomitant cleavage of the Tic-Phe peptide bond in dimethyl sulphoxide and methanol [160, 161]. This observation prompted
Schiller and his collaborators to design the corresponding peptides containing a reduced peptide bond between the Tic[2] and Phe[3] residues, since this
structural modification eliminates the possibility of Tyr-Tic diketopiperazine formation. Thus, they prepared H-Tyr-Ticψ-[CH$_2$NH]Phe-Phe-OH
(TIPP[ψ], 106) and H-Tyr-Ticψ-[CH$_2$NH]Phe-OH (TIP[ψ], 107) [161]. In
comparison with their respective parent peptides [compounds (104) and
(105)], both pseudopeptide analogues showed increased δ antagonist potency in the MVD, higher δ receptor affinity and improved δ receptor selectivity. The more potent compound, TIPP[ψ], displayed subnanomolar δ receptor affinity and in direct comparisons with other selective δ ligands was
shown to have an unprecedented μ/δ ratio (K$_i^\mu$/K$_i^\delta$ = 10,500). TIPP[ψ] had
no agonist effect at concentrations up to 10 μM either in the MVD or in the
GPI, and showed no antagonist activity at μ and κ receptors in the GPI at
concentrations as high as 10 μM. Most interestingly, TIPP[ψ] was able to decrease naloxone-precipitated withdrawal symptoms in rats and attenuated
the development of morphine tolerance in the rat tail-flick test. These results
suggest that δ opioid receptors are critically involved in the development of
morphine tolerance and dependence [162]. TIPP[ψ] turned out to be stable
against enzymatic degradation. [^{3}H]TIPP[ψ] has been prepared and was
found to be a valuable tool for *in vitro* and *in vivo* studies [163].

Both peptides (106) and (107) were synthesized by the solid-phase method
using *tert*-butyloxycarbonyl (Boc) protected amino acids and 1,3-di-isopro-
pylcarbodi-imide (DIC)/1-hydroxybenzotriazole (HOBt) as coupling
agents. Introduction of the reduced peptide bond between the Tic[2] and
Phe[3] residues required a reductive alkylation reaction [164] between 2-Boc-
1,2,3,4-tetrahydroisoquinoline-3-aldehyde and the amino group of the re-
sin-bound phenylalanine or H-Phe-Phe dipeptide. 2-Boc-1,2,3,4-tetrahy-
droisoquinoline-3-aldehyde was synthesized *via* preparation of Boc-1,2,3,4-
tetrahydroisoquinoline-3-(*N*-methoxy-*N*-methylamide) by using a published
procedure [165]. Peptides were cleaved from the resin by HF-anisole treatment in the usual manner. Crude products were purified by reversed-phase
chromatography.

Replacement of Phe[3] in TIPP with Leu, Ile or norvaline (Nva) resulted in
compounds that retained potent δ antagonist activity in the MVD and
showed high δ selectivity in opioid receptor binding assays. Obviously, an
aromatic residue in position 3 of the peptide sequence is not absolutely necessary for high δ antagonist potency. Interestingly, saturation of the Phe[3]
aromatic ring in TIPP, achieved through substitution of cyclohexylamine
(Cha), led to a compound (H-Tyr-Tic-Cha-Phe-OH [TICP], 108) with in-

creased δ antagonist potency and higher δ selectivity than the parent peptide. The corresponding pseudopeptide H-Tyr-Ticψ[CH$_2$NH]Cha-Phe-OH (TICP[ψ], 109) showed a further improvement in δ antagonist activity [166].

Rónai and his collaborators demonstrated that the Tyr-Pro-Gly-Phe-Leu-Thr hexapeptide sequence accepts N-terminal substituents such as N-t-Boc (110), N-phenylacetyl (111) and N-diphenylacetyl (112) where the nitrogen cannot become protonated, as well as 'traditional' substituents such as N,N-diallyl (113) where protonation is likely under physiological conditions. Compounds (110)-(113) were pure opioid antagonists of medium δ affinity in the MVD (K$_e$ values against [Met^5]-enkephalin range from 0.6 to 4 μM) and showed high μ/δ selectivity ratios (K$_e^{\mu}$/K$_e^{\delta}$ = 50–350 in MVD) [167].

Irreversible antagonists

The hexapeptide analogue [D-Ala^2,Leu^5,Cys^6]enkephalin (DALCE) contains a single sulphydryl group and has been reported to bind covalently to δ receptors by forming a disulphide bond with a sulphydryl group present at the binding site [168, 169]. DALCE is moderately selective and it appears to possess a pharmacological profile *in vivo* consistent with nonequilibrium δ opioid receptor antagonism.

NON-PEPTIDIC LIGANDS

Competitive antagonists

The rationale for the design of naltrindole (NTI, 82) by Portoghese and his collaborators [170] was based on the 'message-address' concept [171, 172]. This design strategy for nonpeptide δ selective antagonists employed the naltrexone pharmacophore for the message moiety and a key element in the leu-

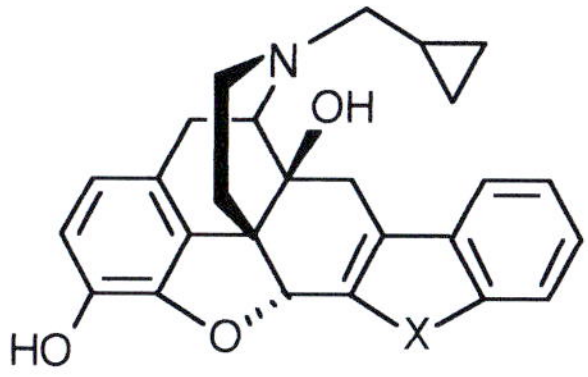

(85) NTI, X = NH
(114) X = NMe
(115) NTB, X = O

cine-enkephalin δ address [142, 173]. The key element, which was hypothesized to be the phenyl group of Phe[4] of leucine-enkephalin, was attached to the morphinan structure of naltrexone through a rigid spacer. The first target compound synthesized, NTI (82), contained a pyrrole spacer because it was easily accessible from naltrexone through a Fisher indole synthesis. NTI is a δ opioid receptor antagonist with high δ affinity and good selectivity as found in bioassays. Its N-methyl (114) and benzofuran (NTB, 115) analogues were also δ receptor opioid antagonists in bioassays (*Table 3.10*). All three compounds displayed relatively moderate antinociceptive potency in the writhing assay [174, 175]. NTB is able to distinguish between δ receptor subtypes and is selective for the δ_2 site [176].

Interestingly, NTI was found to possess immunosuppressant properties while being less toxic (NTI does not show any cytotoxic effect) than cyclosporin [177–179]. Since the immunosuppressive mechanism of NTI is different from that of other immunosuppressants, the use of NTI or other δ opioid receptor antagonists in combination with other immunosuppressive agents could be more beneficial than treatment with a single compound (e.g., cyclosporin) after organ transplantation. There is increasing evidence of improvement after renal transplantation when cyclosporin is combined with other immunosuppressive agents.

Development of morphine tolerance and physical dependence is markedly suppressed by the administration of NTI or its isothiocyanate analogue naltrindole 5'-isothiocyanate (NTII, 116) before and during morphine treatment [180]. These effects are produced by NTI and NTII at doses that do not block the antinociceptive effects due to interaction at μ receptors. These data are of interest from the standpoint of preventing tolerance and physical dependence in patients who receive morphine on a chronic basis [173]. NTI

Table 3.10. ANTAGONIST K_e VALUES OF NTI AND ANALOGUES (114) AND (115) IN THE MVD AND GPI [170]

| Antagonists | $K_e{}^a$ (nM) | | | Selectivity ratio | |
	$M^b(\mu)$	$EK^b(\kappa)$	$DADLE^c(\delta)$	μ/δ	κ/δ
NTI (82)	32	58	0.21	152	276
(114)	11	20	1.0	11	20
NTB (115)	27	48	1.1	25	44
Naloxone	2.2	16	40	0.06	0.4

[a]Derived from the Schild relationship and calculated from an average of at least three IC$_{50}$ ratio determinations by using K_e = [antagonist]/(IC$_{50}$ ratio -1). [b]Morphine (M) or ethylketazocine (EK) in the GPI. [c][D-Ala^2,D-Leu^5]enkephalin (DADLE) in the MVD.

seems also to block the ability of cocaine to produce positive reinforcement in rats [181, 182].

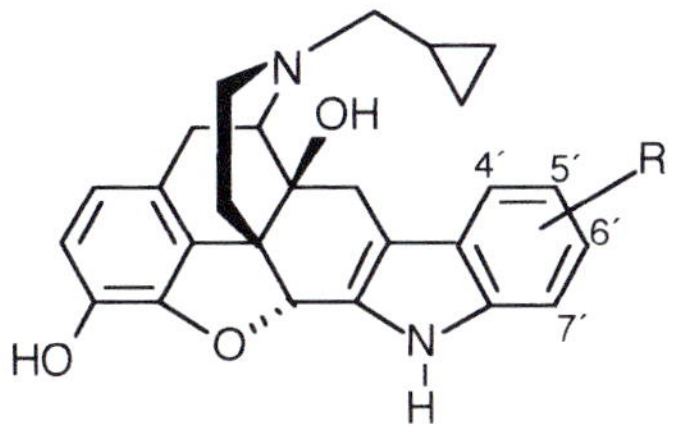

(116) NTII, R = 5'-NCS
(117) R = 6'-CO$_2$Me
(118) R = 7'-CO$_2$Et
(119) R = 7'-SO$_2$Me

Aside from the demonstrated antinociceptive effect produced by agonist interactions at δ receptors [174], NTI has been employed to demonstrate that δ opioid receptors are involved in the antinociceptive effects of cholecystokinin octapeptide in mice [183] and in swim stress-induced antinociception in adult rats [184, 185]. NTI was found to produce a marked and long-lasting antitussive effect in mice and rats which was not antagonized by the irreversible μ antagonist β-FNA [186].

NTI (82) and its N'-methyl analogue (114) were prepared from naltrexone through a Fischer indole synthesis by treatment with phenylhydrazine and methylphenylhydrazine respectively [170, 187]. NTB (115) was synthesized by reaction of naltrexone with O-phenylhydroxylamine [188].

A series of heterocyclic analogues related to NTI and 6-arylnaltrexone derivatives has been synthesized in order to determine the role of the spacer and the address moieties in conferring δ opioid receptor antagonist activity [188]. It was found that heterocycles other than pyrrole could serve as spacers for the address moiety. However, pyrrole seems to be superior to other heterocycles. It was also found that the aromatic address component is important for δ antagonist potency. The role of the indole in enhanced δ recognition is likely to be an interaction with a lipophilic site in the receptor [189]. The indole moiety confers δ receptor selectivity by decreasing μ affinity and enhancing δ affinity. It is likely that the reason for decreased affinity of μ is the loss of the C-6 carbonyl oxygen as the proton-accepting centre which is believed to be important for recognition of the μ receptor [189].

Nagase and his collaborators have synthesized and evaluated a series of NTI analogues substituted in the indole benzene moiety [190]. Three compounds showed higher δ antagonist potency than NTI in the MVD against the δ agonist DPDPE ([D-Pen^2,D-Pen^5]enkephalin). The K_e values (nM) of indoles (117), (118), (119) and NTI were 0.13, 0.16, 0.14 and 0.21 respectively. Selectivities were not determined. Compound (120) exhibited higher immunosuppressive activity than NTI.

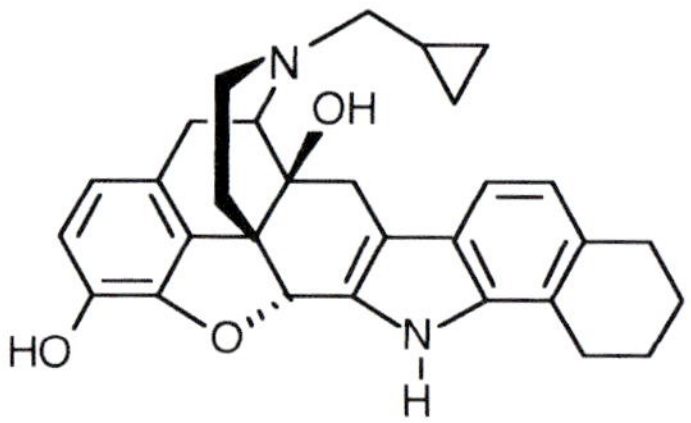

(120)

Introduction of 14β-ethoxy and 5β-methyl groups into the NTI molecule resulted in the pure opioid antagonist (121) with somewhat lower δ potency but much higher δ selectivity in the MVD due to very low μ and κ affinities (*Table 3.11*) [191]. Indole (121) was prepared by reacting the μ receptor-preferring opioid antagonist 14-*O*-ethyl-5-methylnaltrexone [192] with phenylhydrazine under conditions used for the Fisher indole synthesis.

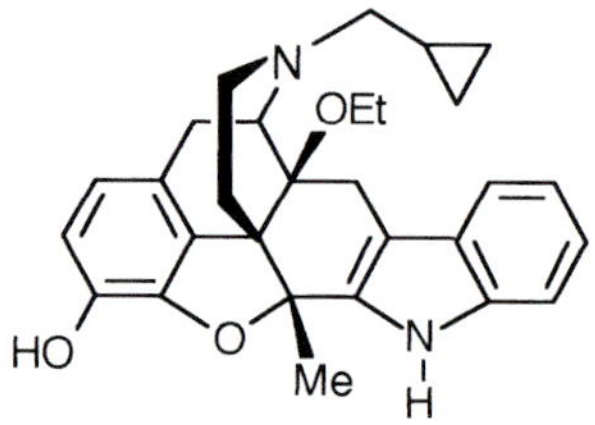

(121)

Table 3.11. ANTAGONIST K_e^a VALUES (NM) OF (121) AND NTI IN THE MVD [191]

Antagonists	DAMGO(μ)	CI977(κ)	DPDPE(δ)	selectivity ratio μ/δ	κ/δ
(121)	133	529	1.3	102	407
NTI	5.25	32.4	0.18	29	180

$^a K_e$ = [antagonist]/DR-1, where DR is the dose ratio (i.e., ratio of equiactive concentrations of the test agonist in the presence and absence of the antagonist).

Portoghese and his coworkers prepared a series of amino acid conjugates of NTI in order to obtain δ antagonists that would have limited access to the central nervous system (CNS) upon peripheral administration [193]. Two of the more δ selective conjugates, the glycinate (122) and the aspartate (123), were evaluated by the i.v. and i.c.v. routes in mice against the δ agonist DPDPE ([D-Pen^2,D-Pen^5]enkephalin). The i.v./i.c.v. dose ratios of (122) and (123) to produce equivalent antagonism of DPDPE-induced antinociception were very high ($> 49,000$) which is consistent with poor CNS penetration.

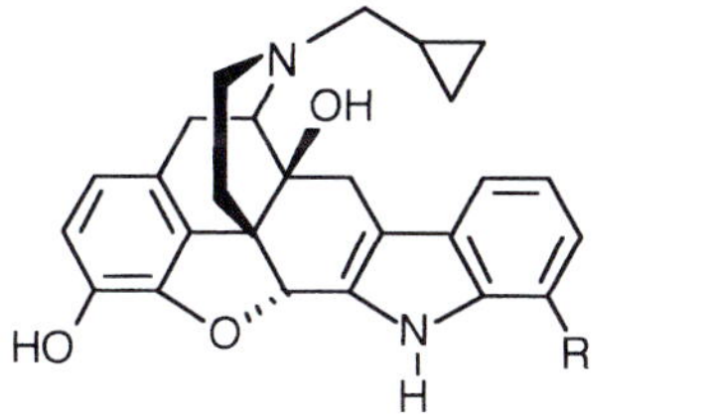

(122) R = CONHCH$_2$CO$_2$H
(123) R = CONHCH(CO$_2$H)CH$_2$CO$_2$H
(124) R = CO$_2$H
(125) R = CONHCH$_2$CO$_2$CH$_2$Ph
(126) R = CONHCH(CO$_2$CH$_2$Ph)CH$_2$CO$_2$CH$_2$Ph

Compounds (122) and (123) have been prepared from 7'-carboxynaltrindole (124) which is available from naltrexone *via* the Fischer indole synthesis using 2-hydrazinobenzoic acid in acetic acid. The coupling of (124) with suitable protected amino acids using the Bop reagent [benzotriazolyloxytris-(dimethylamino)phosphonium hexafluorophosphate] afforded the corresponding intermediates (125) and (126). Catalytic hydrogenation yielded compounds (122) and (123) [193].

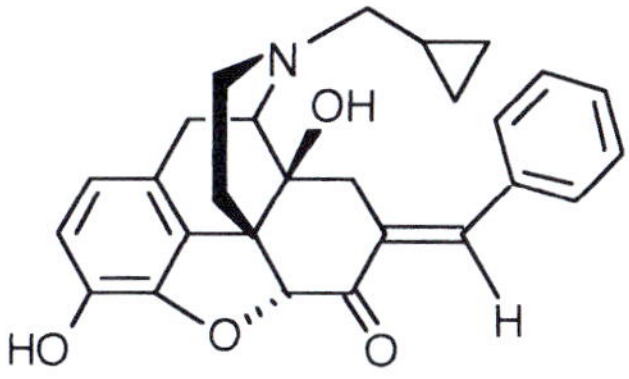

(127) BNTX

7-Benzylidenenaltrexone (BNTX, 127), developed by Portoghese and his coworkers [194], was found to be a δ_1 opioid receptor antagonist. In the mouse tail-flick assay, BNTX effectively antagonized the δ_1 agonist DPDPE but did not significantly change the ED_{50} values of the δ_2 selective agonist DSLET ([D-Ser^2,Leu^5]enkephalin-Thr^6), morphine or U50488. Like NTI, BNTX and NTB produced a marked antitussive effect in mice which was not antagonized by the irreversible μ antagonist β-FNA [195].

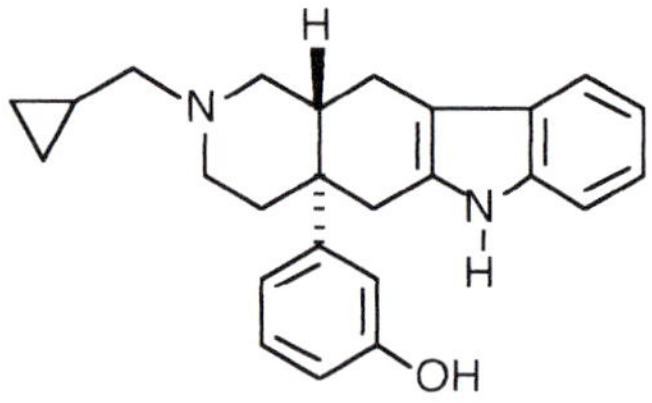

(128)

Nagase and his collaborators have developed a series of indolo[2,3-g]octa-hydroisoquinoline derivatives mimicking the structure of NTI [196]. The compounds show in the MVD and radioligand binding studies antagonism and selectivity at δ receptors. Racemic compound (128), for instance, exhibited in binding a δ K_i value of 3.5 nM and a μ/δ and k/δ selectivity ratio of 205 and 272 respectively. Compound (128) did not affect the inhibition of contraction in the GPI and showed a K_e value in the MVD against DPDPE of 4.8 nM.

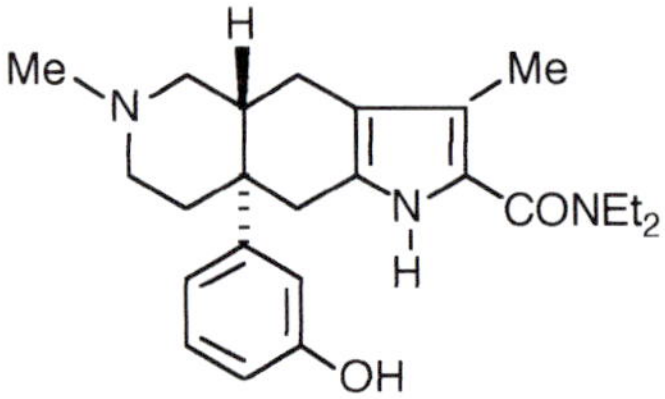

(129)

A similar approach to δ selective antagonists was undertaken by Dondio and Ronzoni [197]. They prepared pyrrolo-octahydroisoquinoline deriva-tives. Studies of the binding of the racemic form of compound (129) showed a δ K_i value of 2.15 nM and μ/δ and k/δ selectivity ratios of 45 and 403 re-spectively. In the MVD, compound (129) exhibited a K_e value of 7 nM against DADLE.

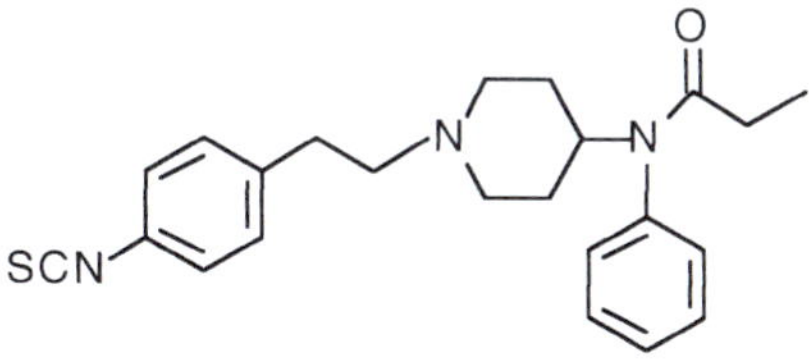

(130) FIT

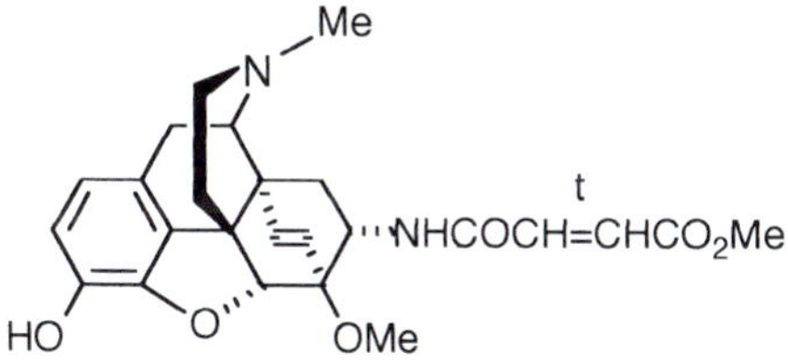

(131) FAO

Long acting and irreversible antagonists

Rice and his coworkers reported on the first δ selective irreversible ligands FIT (fentanyl isothiocyanate, 130) and FAO (7a-methylfumaramido-6,14-*endo*-ethenotetrahydro-oripavine, fumaramido-oripavine,131) [95, 96]. In binding experiments, FIT and FAO were found to bind irreversibly to δ receptors while *in vivo* experiments show that they have morphine-like antinociceptive potencies. The 3-methyl analogue of FIT, (+)-enantiomer of compound (132, SUPERFIT) was shown to be highly potent and specific for acylation of δ opioid receptors in rat brain membranes like its achiral prototype FIT, and was about 10 times as potent as FIT in this assay. SUPERFIT was about 5 times as potent as FIT in acylation of δ receptors in NG108–15 neuroblastoma X glioma hybrid cells and about 50 times as potent as its enantiomer. Both FIT and SUPERFIT behaved as partial agonists in inhibiting δ receptor-coupled adenylate cyclase in NG108–15 membranes.

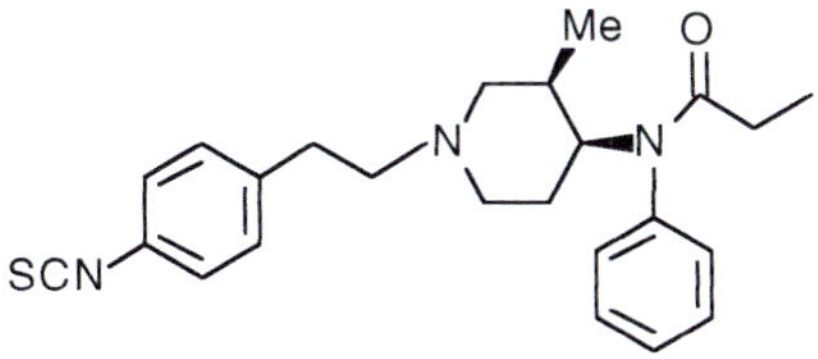

(132) SUPERFIT

SUPERFIT was 5–10 times more potent than FIT and about 100 times more potent than its enantiomer in this assay [198].

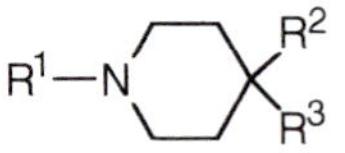

$$R^1-N\left(\underset{R^3}{\overset{R^2}{\diagdown}}\right)$$

(133) R^1 = H, R^2, R^3 = O
(134) R^1 = 4-NO_2-$C_6H_4CH_2CO$, R^2, R^3 = O
(135) R^1 = 4-NO_2-$C_6H_4CH_2CO$, R^2, R^3 = NC_6H_4
(136) R^1 = 4-NO_2-$C_6H_4CH_2CO$, R^2 = H, R^3 = NHC_6H_4
(137) R^1 = 4-NO_2-C_6H_4-CH_2CH_2, R^2 = H, R^3 = NHC_6H_4
(138) R^1 = 4-NO_2-$C_6H_4CH_2CH_2$, R^2 = H, R^3 = NC_6H_4 COEt
(139) R^1 = 4-NH_2-$C_6H_4CH_2CH_2$, R^2 = H, R^3 = NC_6H_4 COEt

The synthesis of FIT (130) was a modification of the previously reported route [199] to fentanyl and congeners (133)-(139). Reaction of 4-piperidone (133) with (4-nitrophenyl)acetyl chloride afforded amide (134) which was treated with aniline to give the Schiff base (135). Reduction with sodium borohydride to (136) followed by borane reduction of the amide function gave diamine (137). Acylation with propionic anhydride gave the known [200] nitrofentanyl (138). Catalytic hydrogenation over Pd-C provided aminofentanyl (139) which was converted into the isothiocyanate (132) by treatment with thiophosgene [96].

FAO (131) was synthesized from the 7α-amino-6,14-*endo*-ethenomorphinan (140) which is available from thebaine in three steps as described by Bentley and his coworkers [201]. Selective O-demethylation of the phenolic ether function of (140) with boron tribromide afforded the phenol (141) which was treated with methylfumaroyl chloride to yield FAO (131) [96].

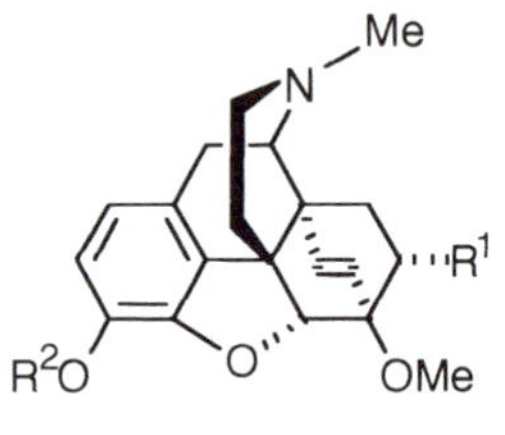

(140) R^1 = NH_2, R^2 = Me
(141) R^1 = NH_2, R^2 = H

Naltrindole 5'-isothiocyanate (NTII, 116) exhibited high δ antagonist effects and δ selectivity in the MVD [202]. The irreversible antagonism was confirmed by the fact that there were no significant differences between pre- and post-wash IC_{50} ratios. NTII was also a very potent δ selective antagonist *in vivo* (mouse abdominal stretch assay, i.c.v. administration), causing a more than 50-fold increase of the ED_{50} for the δ_2 agonist [D-Ser^2,D-Leu^5]enkephalin-Thr^6 (DSLET) and no increase for the δ_1 agonist [D-Pen^2,D-Pen^5]enkephalin (DPDPE) [202, 203]. In binding experiments, it was found that NTII only changed the binding characteristics of [^{3}H]DSLET and not that of [^{3}H]DPDPE which supports the postulated existence of δ opioid receptor subtypes [204].

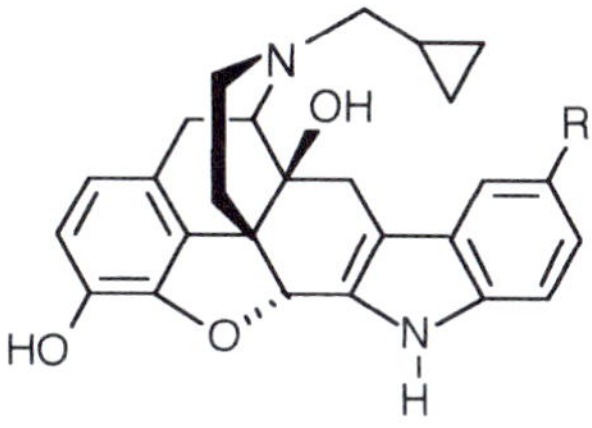

(142) R = NO$_2$
(143) R = NH$_2$

NTII was prepared by reacting naltrexone with (4-nitrophenyl)hydrazine to form 5'-nitroindole (142) which was reduced by catalytic hydrogenation over Raney nickel to the 5'-amino derivative (143). Treatment with thiophosgene yielded NTII [203].

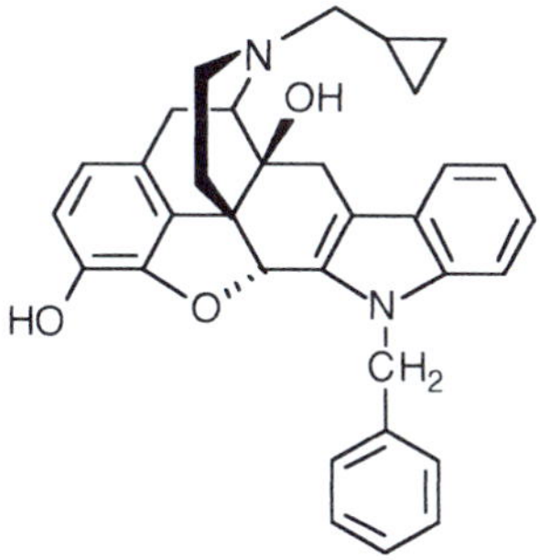

(144) BNTI

Another δ opioid receptor antagonist developed by Portoghese and his

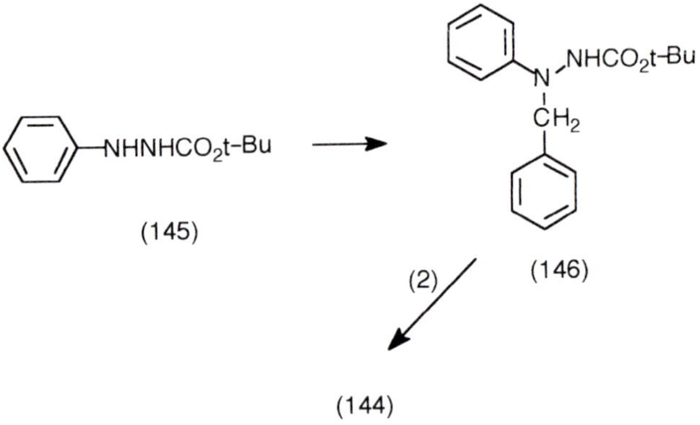

Scheme 3.15

coworkers is N^1-benzylnaltrindole (BNTI, 144) which is very long lasting and selective for δ_2 receptors [205, 206]. BNTI was a potent δ antagonist in bioassays and *in vivo*. In the abdominal stretch assay in mice, BNTI antagonized the antinociceptive effect of the δ_2 agonist DSLET for five days while no significant antagonism of the antinociceptive effect of the δ_1 selective agonist DPDPE was observed.

BNTI was prepared as outlined in *Scheme 3.15*. Condensation of Boc-phenylhydrazine (145) with benzyl bromide gave Boc-N^1-benzyl-N_1-phenyl-hydrazine (146) which was treated with naltrexone to give BNTI [205].

Radiolabelled opioid antagonists (see [207] for further details) are of great value in opioid research.

CONCLUSION

In recent years, substantial progress has been made towards the development of opioid receptor antagonists which exhibit high selectivity for μ, κ and δ receptors or receptor subtypes. These highly selective antagonists have advantages over the universal opioid antagonists (for example, naloxone and naltrexone) because they are of value in probing the interaction of endogenous opioid peptides with opioid receptors. They are also useful in evaluating the selectivity of new opioid agonists. Selective opioid antagonists also have therapeutic potential in the treatment of a variety of disorders where endogenous peptides play a modulatory role. These include, for instance, food intake, immune function, shock, constipation, drug addiction, alcoholism and certain mental disorders.

REFERENCES

1   Fischer, E. (1894) Ber. Dtsch. Chem. Ges. 27, 2985–2993.
2   Langley, J. N. (1909) J. Physiol. 39, 235–295.
3   Ehrlich, P. (1913) Lancet 2, 445–451.
4   Beckett, A.H. and Casy, A.F. (1954) J. Pharm. Pharmacol. 6, 986–1001.
5   Beckett, A.H. and Casy, A.F. (1954) Nature (London) 173, 1231–1232.
6   De Stevens, G. (1965) in Analgetics, p. 475, Academic Press, New York.
7   Jacobson, A.E., May, E.L. and Sargent, L.J. (1970) in Burger's Medicinal Chemistry (Burger, A., ed.) 3rd edn, Part II, John Wiley, New York.
8   Janssen, P.A.J., Hellerbach, J., Schnider, O., Besendorf, L.T. and Pellmont, B. (1960) in Synthetic Analgesics Part 1, Pergamon, New York.
9   Jannsen, P.A.J., Hellerbach, J., Schnider, O., Besendorf, L.T. and Pellmont, B. (1960) in Synthetic Analgesics, Part II, Pergamon, New York.
10 Portoghese, P.S. (1966) J. Pharm. Sci. 55, 865–887.
11 Portoghese, P.S. (1970) Annu. Rev. Pharmacol. 10, 51–76.
12 Pert, C.B. and Snyder, S.H. (1973) Science (Washington, D.C.) 179, 1011–1014.
13 Terenius, L. (1973) Acta Pharmacol. Toxicol. 33, 377–384.
14 Simon, E.J., Hiller, J.M. and Edelmand, I. (1973) Proc. Natl. Acad. Sci. U.S.A. 70, 1947–1949.
15 Martin, W.R. (1967) Pharmacol. Rev. 19, 463–521.
16 Portoghese, P.S. (1965) J. Med. Chem. 8, 609.
17 Hughes, J., Smith, T.H., Kosterlitz, H.W., Fothergill, L.A., Morgan, B.A. and Morris, H.R. (1975) Nature (London) 258, 577–579.
17a Zadina, J.E., Hackler, L., Ge, L.J. and Kastin, A.J. (1997) Nature (London) 386, 499–502.
17b Akil, H., Watson, S.J., Young, E., Lewis, M.E., Khachaturian, H. and Walker, J.M. (1984) Annu. Rev. Neurosci. 7, 223–255.
18 Reisine, T. and Bell, G.I. (1993) Trends Neurosci. 16, 506–510.
19 Herz, A. (1987) in Trends in Medicinal Chemistry (Mutschler, E. and Winterfeldt, E., eds.), pp. 337–350, VCH Verlagsges, Weinheim.
20 Portoghese, P.S. (1991) J. Med. Chem. 34, 1757–1762.
21 Mansour, A., Hoversten, M.T., Raylor, L.P., Watson, S.J. and Akil, H. (1995) Brain Res. 700, 89–98.
22 Rothman, R.B., Xu, H., Wang, J.B., Partilla, J.S., Kayakiri, H., Rice, K.C. and Uhl, G.R. (1995) Synapse 21, 60–64.
23 Abood, M.E., Noel, M.A., Carter, R.C. and Harris, L.S. (1995) Biochem. Pharmacol. 50, 851–859.
24 Emmerson, P.J., Clark, M.J., Mansour, A., Akil, H., Woods, J.H. and Medzihradsky, F. (1996) J. Pharmacol. Exp. Ther. 278, 1121–1127.
24a Giershik, P., Bouillon, T. and Jakobs, K.H. (1994) Methods Enzymol. 237, 13–26.
24b Traynor, J.R. and Nahorski, S.R. (1995) Mol. Pharmacol. 47, 848–854.
24c Befort, K., Tabbara, L. and Kieffer, B. (1996) Neurochem. Res. 21, 1301–1307.
25 Schiller, P.W. (1991) Prog. Med. Chem. 28, 301–340.
26 Zimmerman, D.M. and Leander, J.D. (1990) J. Med. Chem. 33, 895–902.
27 Iijima, I., Minawikawa, J., Jacobson, A.E., Brossi, A., Rice, K. C. and Klee, W.A. (1978) J. Med. Chem. 21, 398–400.
28 Aceto, M.D., Harris, L.S., Dewey, W.L. and May, E.L. (1979) National Institute on Drug Abuse Research Monograph 27, 341

29 Gurll, N.J., Reynolds, D.G., Vargish, T. and Lechner, R. (1982) J. Pharmacol. Exp. Ther. 220, 621–624.
30 Holaday, J.W. and Faden, A.J. (1978) Nature (London) 275, 450–451.
31 Faden, A.I. and Holaday, J.W. (1979) Science (Washington, D.C.) 205, 317–318.
32 Faden, A.I. and Holaday, J.W. (1980) J. Pharmacol. Exp. Ther. 212, 441–447.
33 DeBoecky, C., Van Reemts, P., Rigatto, H. and Chernick, V. (1984) J. Appl. Physiol. 56, 1507–1511
34 Kreek, M.-J., Schaefer, R.A., Hahn, E.F. and Fishman, J. (1983) Lancet i, 261–262.
35 Martin, W.R., Jasinski, D.R. and Mansky, P.A. (1973) Arch. Gen. Psychiat. 28, 784.
36 Gold, M.S., Dackis, C.A. Pottash, A.L.C., Sternbach, H.H., Annitto, W.J., Martin, D. and Dackis, M.P. (1982) Med. Res. Rev. 3, 211–246.
37 McIntosh, T.K. and Faden, A.I. (1988) Pharmacological Approaches to the Treatment of Brain and Spinal Injury (Stein, G. and Sabel, B.A., eds.) pp. 89–102, Plenum Publishing Corporation, New York.
38 Kreek, M.-J. (1996) in Pharmacological Aspects of Drug Dependence: Handbook of Experimental Pharmacology (Schuler, C.R. and Kuhar, M.J., eds.) Volume 118, pp. 563–599, Springer, Berlin.
39 Cone, E.J. (1973) Tetrahedron Lett. 2607–2610.
40 Gonzales, J.P. and Brogden, R.N. (1988) Drugs 35, 192–213.
41 Chatterjie, N., Inturrisi, C.E., Dayton, H.B. and Blumberg, H. (1975) J. Med. Chem. 18, 490–492.
42 Freund, M. and Speyer, E. (1916) J. Prakt. Chem. 94, 135–178.
43 Hauser, F.M., Chen, T.-K. and Carroll, F.I. (1974) J. Med. Chem. 17, 1117.
44 Krassnig, R., Hederer, C. and Schmidhammer, H. (1996) Arch. Pharm. Med. Chem. 329, 325–326.
45 Weiss, U. (1957) J. Org. Chem. 22, 1505–1506.
46 Sankyo (1963) Br. Pat. 939,287; (1964) Chem. Abstr. 60, 12070h.
47 Lewenstein, M.J. and Fishman, J. (1967) U. S. Pat. 3,320,262; (1967) Chem. Abstr. 67, 90989q.
48 Kobylecki, R.J., Carling, R.W., Lord, J.A.H., Smith, C.F.C. and Lane, A.C. (1982) J. Med. Chem. 25, 116–120.
49 Olofson, R.A., Schnur, R.C., Bunes, L. and Pepe, J.P. (1977) Tetrahedron Lett. 1567–1570.
50 Rice, K.C. (1977) J. Med. Chem. 20, 164–165.
51 Michel, M.E., Bolger, G. and Weissman, B.A. (1984) Pharmacologist 26, 201.
52 Rzeszotarski, W.J. and Mavunkel, B.J. (1989) U.S. Pat. 4,889,860; (1990) Chem. Abstr. 112, 235663.
53 Nash, C.B., Caldwell, R.W. and Tuttle, R.R. (1984) Fed. Proc. Abstr. 43, 3987.
54 Key Pharmaceuticals (1984) Drugs Future 9, 518–519.
55 Faden, A.I., Sacksen, I. and Noble, L.J. (1988) J. Pharmacol. Exp. Ther. 245, 742–748.
56 Vink, R., McIntosh, T.K., Rhomhanyi, R. and Faden, A.I. (1990) J. Neurosci. 10, 3524–3530.
57 Hahn, E.F., Fishman, J. and Heilman, R.D. (1975) J. Med. Chem. 18, 259–262.
58 Portoghese, P. S., Larson, D.L., Jiang, J.B., Takemori, A.E. and Caruso, T.P. (1978) J. Med. Chem. 21, 598–599.
59 Portoghese, P.S., Larson, D.L., Jiang, J.B., Takemori, A.E. and Caruso, T.P. (1979) J. Med. Chem. 22, 168–173.
60 Casy, A.F. and Parfitt, R.T. (1986) Opioid Analgesics: Chemistry and Receptors, pp. 61–65 Plenum Press, New York and London.
61 Schoenecker, J.W., Takemori, A.E. and Portoghese, P.S. (1987) J. Med. Chem. 30, 1040–1044.

62 Wiley, G.A. Hershkowitz, R.L., Rein, B.M. and Chung, B.C. (1964) J. Am. Chem. Soc. 86, 964–965.
63 Pelton, J.T., Gulya, K., Hruby, V.J., Duckles, S.P. and Yamamura, H.I. (1985) Proc. Natl. Acad. Sci. U.S.A. 82, 236–239.
64 Pelton, J.T., Kazmierski, W., Gulya, K., Yamamura, H.I. and Hruby, V.J. (1986) J. Med. Chem. 29, 2370–2375.
65 Kazmierski, W., Wire, W.S., Lui, G.K., Knapp, R.J., Shook, J.E., Burks, T.F., Yamamura, H.I. and Hruby, V.J. (1988) J. Med. Chem. 31, 2170–2177.
66 Venn, R.F. and Barnard, E.A. (1981) J. Biol. Chem. 256, 1529–1532.
67 Szücs, M., Belcheva, M., Simon, J., Benyhe, S., Tóth, G., Hepp, J., Wollemann, M. and Medzihradszky, K. (1987) Life Sci. 41, 177–184.
68 Newman, E.L. and Barnard, E.A. (1984) Biochemistry 23, 5385–5389.
69 Benyhe, S., Hepp, J., Simon, J., Borsodi, A., Medzihradszky, K. and Wollemann, M. (1987) Neuropeptides 9, 225–235.
70 Schmidhammer, H. (1989) Trends in Medicinal Chemistry '88 (van der Goot, H., Domany, G., Pallos, L. and Timmerman, H., eds.) pp. 483–490, Elsevier Science, Amsterdam.
71 Schmidhammer, H., Burkard, W.P., Eggstein-Aeppli, L. and Smith, C.F.C (1989) J. Med. Chem. 32, 418–421.
71a Freye, E., Latasch, L. and Schmidhammer, H. (1992) Anaesthesist 41, 527–533.
71b Chen, Y., Mestek, A., Liu, J., Hurley, J.A. and Yu, L. (1993) Mol. Pharmacol. 4, 8–12.
71c Caudle, R.M., Chavkin, C. and Dubner, R. (1994) J. Neurosci.. 14, 5580–5589.
71d Márki, A., Monory, K., Ötvös, F., Tóth, G., Krassnig, R., Schmidhammer, H., Traynor, J.R., Maldonado, R., Roques, B.P. and Borsodi, A. Br. J. Pharmacol., submitted.
71e Hutcheson, D., Garzon, J., Garcia-España, A., Sanchez-Blasquez, P., Schmidhammer, H., Borsodi, A., Roques, B.P. and Maldonado, R. Br. J. Pharmacol., submitted.
71f Ötvös, F., Tóth, G. and Schmidhammer, H. (1992) Helv. Chim. Acta 75, 1718–1720.
72 Schmidhammer, H., Aeppli, L., Atwell, L., Fritsch, F., Jacobson, A.E., Nebuchla, M. and Sperk, G. (1984) J. Med. Chem. 27, 1575–1579.
73 Krassnig, R. and Schmidhammer, H. (1994) Heterocycles 38, 877–881.
74 Musliner, W.J. and Gates, J.W. (1966) J. Am. Chem. Soc. 88, 4271–4273.
75 Schmidhammer, H., Smith, C.F.C., Erlach, D., Koch, M., Krassnig, R., Schwetz, W. and Wechner, C. (1990) J. Med. Chem. 33, 1200–1206.
76 Schmidhammer, H., Jennewein, H. and Smith, C.F.C. (1991) Arch. Pharm. (Weinheim) 324, 209–211.
77 Schmidhammer, H., Jennewein, H.K., Krassnig, R., Traynor, J.R., Patel, D., Bell, K., Froschauer, G., Mattersberger, K., Jachs-Ewinger, C., Jura, P., Fraser, G.L. and Kalinin, V.N. (1995) J. Med. Chem. 38, 3071–3077.
78 Olofson, R.A., Marts, J.T., Senet, J.-P., Piteau, M. and Malfroot, T.A. (1984) J. Org. Chem. 49, 2081–2082.
79 Portoghese, P.S., Larson, D.L., Sayre, L.M., Fries, D.S. and Takemori, A.E. (1980) J. Med. Chem. 23, 233–234.
80 Ward, S.J., Portoghese, P.S. and Takemori, A.E. (1982) J. Pharmacol. Exp. Ther. 220, 494–498.
81 Ward, S.J., Portoghese, P.S. and Takemori, A.E. (1982) Eur. J. Pharmacol. 80, 377–384.
82 Takemori, A.E., Larson, D.L. and Portoghese, P.S. (1981) Eur. J. Pharmacol. 70, 445–451.
83 Jiang, J.B., Hanson, R.N., Portoghese, P.S. and Takemori, A.E. (1977) J. Med. Chem. 20, 1100–1102.
84 Sayre, L.M. and Portoghese, P.S. (1980) J. Org. Chem. 45, 3366–3368.
85 Pasternak, G.W. and Hahn, E.F. (1980) J. Med. Chem. 23, 674–676.
86 Hahn, E.F., Carroll-Buatti, M. and Pasternak, G.W. (1982) J. Neurosci. 2, 572–576.

87 Itzhak, Y. (1988) in The Opiate Receptors (Pasternak, G.W., ed.) pp. 95–142, Humana Press, Clifton, N.J.

88 Ling, G.S.F. and Pasternak, G.W. (1983) Brain Res. 271, 152–156.

89 Ling, G.S.F., Spiegel, K., Nishimura, S. and Pasternak, G.W. (1983) Eur. J. Pharmacol. 86, 487–488.

90 Ling, G.S.F., Spiegel, K., Lockhart, S.H. and Pasternak, G.W. (1985) J. Pharmacol. Exp. Ther. 232, 149–155.

91 Ling, G.S.F., MacLeod, J.M., Lee, S., Lockhart, S.H. and Pasternak, G.W. (1984) Science (Washington, D.C.) 226, 462–464.

92 Johnson, N. and Pasternak, G.W. (1984) Mol. Pharmacol. 26, 477–483.

93 Cruciani, R.A., Lutz, R.A., Munson, P.J. and Rodbard, D. (1987) J. Pharmacol. Exp. Ther. 242, 15–20.

94 Garzon-Aburbeh, A., Lipkowski, A.W., Larson, D.L. and Portoghese, P.S. (1989) Neurochem. Int. 15, 207.

95 Rice, K.C., Jacobson, A.E., Burke Jr., T.R., Bajwa, B.S., Streaty, R.A. and Klee, W.A. (1983) Science (Washington, D.C.) 220, 314–316.

96 Burke Jr., T.R., Bajwa, B.S., Jacobson, A.E., Rice, K.C., Streaty, R.A. and Klee, W.A. (1984) J. Med. Chem. 27, 1570–1574.

97 Rothman, R.B., Bowen, W.D., Herkenham, M., Jacobson, A.E., Rice, K.C. and Pert, C.B. (1985) Mol. Pharmacol. 27, 399–409.

98 McLean, S., Rothman, R.B., Jacobson, A.E., Rice, K.C. and Herkenham, M. (1987) J. Comp. Neurol. 255, 497–510.

99 Tocque, B., Jacobson, A.E., Rice, K.C. and Frey, E.A. (1987) Eur. J. Pharmacol. 143, 127–130.

100 Danks, J.A., Tortella, F.C., Long, J.B., Bykov, V., Jacobson, A.E., Rice, K.C., Holaday, J.W. and Rothman, R.B. (1988) Neuropharmacology 27, 965.

101 Hunger, A., Kebrle, J., Rossi, A. and Hoffmann, K. (1960) Helv. Chim. Acta 43, 1032–1046.

102 Lewis, J.W., Smith, C.F.C., McCarthy, P.S., Walter, D.S., Kobylecki, R.J., Myers, M., Haynes, A.S., Lewis, C.J. and Waltham, K. (1989) in Problems of Drug Dependence (Harris, L.S., ed.) pp. 136–143, National Institute on Drug Abuse Research Monograph No. 90, U.S Government Printing Office, Washington, D.C.

103 Aceto, M.D., Bowman, E.R., May, E.L., Harris, L.S., Woods, J.H., Smith, C.B., Medzih-radsky, F. and Jacobson, A.E. (1989) Arzneim.-Forsch. Drug Res. 39, 570–575.

104 Comer, S.D., Burke, T.F., Lewis, J.W. and Woods, J.H. (1992) J. Pharmacol. Exp. Ther. 262, 1051–1056.

105 Burke, T.F., Woods, J.H., Lewis, J.W. and Medzihradsky, F. (1994) J Pharmacol. Exp. Ther. 271, 715–721.

106 Sebastian, A., Bidlack, J.M., Jiang, Q., Deecher, D., Teitler, M., Glick, S.D. and Archer, S. (1993) J. Med. Chem. 36, 3154–3160.

107 Jiang, Q., Sebastian, A., Archer, S. and Bidlack, J.M. (1993) Eur. J. Pharmacol. 230, 129–130.

108 Jiang, Q., Sebastian, A., Archer, S. and Bidlack, J.M. (1994) J. Pharmacol. Exp. Ther. 268, 1107–1113.

109 Boden, R.M., Gates, M., Ho, S.P. and Sundararaman, P. (1982) J. Org. Chem. 47, 1347–1349.

110 Schmidhammer, H., Fritsch, F., Burkard, W.P., Eggstein-Aeppli, L., Hefti, F. and Holck, M.I. (1988) Helv. Chim. Acta 71, 642–647.

111 Kirby, G.W. and McLean, D. (1985) J. Chem. Soc., Perkin Trans. 1, 1443–1445.

112 Archer, S., Seyed-Mozaffari, A., Osei-Gyimah, P., Bidlack, J.M. and Abood, L.G. (1983) J. Med. Chem. 26, 1775–1777.
113 Bidlack, J.M., Frey, R.A., Seyed-Mozaffari, A. and Archer, S. (1989) Biochemistry 28, 4333–4339.
114 Bidlack, J.M., Frey, D.K., Kaplan, R.A., Seyed-Mozaffari, A. and Archer, S. (1990) Mol. Pharmacol. 37, 50–59.
115 Bidlack, J.M., Kaplan, R.A. Subbramanian, R.A., Seyed-Mozaffari, A. and Archer, S. (1993) Biochemistry 32, 6703–6711.
116 Jiang, Q., Seyed-Mozaffari, A., Archer, S. and Bidlack, J.M. (1992) J. Pharmacol. Exp. Ther. 262, 526–531.
117 Allen, R.M., Kirby, G.W. and McDougall, D.J. (1981) J. Chem. Soc., Perkin Trans.1 1143–1147.
118 Archer, S., Seyed-Mozaffari, A. Jiang, Q. and Bidlack, J.M. (1994) J. Med. Chem. 37, 1578–1585.
119 Bidlack, J.M., Kaplan, R.A., Sebastian, A., Seyed-Mozaffari, A., Hutchinson, I. and Archer, S. (1995) Bioorg. Med. Chem. Lett. 5, 1695–1700.
120 Benary, E. (1913) Ber. Dtsch. Chem. Ges. 46, 1021–1027.
121 Lemaire, S. and Turcotte, A. (1986) Can. J. Physiol. Pharmacol. 64, 673–678.
122 Gairin, J.E., Mazarguil, H., Alvinerie, P., Saint-Pierre, S., Meunier, J.-C. and Cros, J. (1986) J. Med. Chem. 29, 1913–1917.
123 Gairin, J.E., Mazarguil, H., Alvinerie, P., Botanch, C., Cros, C. and Meunier, J.-C. (1988) Br. J. Pharmacol. 95, 1023–1030.
124 Portoghese, P.S. and Takemori, A.E. (1985) Life Sci. 36, 801–805.
125 Erez, M., Takemori, A.E. and Portoghese, P.S. (1982) J. Med. Chem. 25, 847–849.
126 Botros, S., Lipkowski, A.W., Takemori, A.E. and Portoghese, (1986) J. Med. Chem. 29, 874–876.
127 Portoghese, P.S., Ronsisvalle, G., Larson, D.L. and Takemori, A.E. (1986) J. Med. Chem. 29, 1650–1653.
128 Portoghese, P.S., Larson, D.L., Yim, C.B., Sayre, L.M., Ronisvalle, G., Tam, S.W. and Takemori, A.E. (1986) J. Med. Chem. 29, 1855–1861.
129 Portoghese, P. S., Lipkowski, A.W. and Takemori, A.E. (1987) J. Med. Chem. 30, 238–239.
130 Portoghese, P.S., Lipkowski, A.W. and Takemori, A.E. (1987) Life Sci. 40, 1287–1292.
131 Takemori, A.E., Ho, B.Y., Naeseth, J.S. and Portoghese, P.S. (1988) J. Pharmacol. Exp. Ther. 246, 255–258.
132 Portoghese, P.S., Garzon-Aburbeh, A., Nagase, H., Lin, C.-E. and Takemori, A.E., (1991) J. Med. Chem. 34, 1292–1296.
133 Czlonkowski, A., Millan, M. and Herz, A. (1987) Eur. J. Pharmacol. 142, 183–184.
134 Takemori, A.E., Schwartz, M.M. and Portoghese, P.S. (1988) J. Pharmacol. Exp. Ther. 247, 971–974.
135 Birch, P.J., Hayes, A.G., Sheehan, M.J. and Tyers, M.B. (1987) Eur. J. Pharmacol. 144, 405–408.
136 Portoghese, P.S., Nagase, H. and Takemori, A.E. (1988) J. Med. Chem. 31, 1344–1347.
137 Portoghese, P.S., Nagase, H., Lipkowski, A.W., Larson, D.L. and Takemori, A.E. (1988) J. Med. Chem. 31, 836–841.
138 Piloty, O. (1910) Ber. Dtsch. Chem. Ges. 43, 489–498.
139 Robinson, G.M. and Robinson, R. (1918) J. Chem. Soc. 639–645.
140 Schmidhammer, H. and Smith, C.F.C. (1989) Helv. Chim. Acta 72, 675–677.
141 Olmsted, S.L., Takemori, A.E. and Portoghese, P.S. (1993) J. Med. Chem. 36, 179–180.
142 Portoghese, P.S. (1989) Trends Pharmacol. Sci. 10, 230–235.
143 Rivett, D.E., Rosevear, J. and Wilshire, J.F.K. (1979) Aust. J. Chem. 32, 1601–1612.

144   Sandler, S.R. and Karo, W. (1972) Organic Chemistry, pp. 268–299, Academic Press, New York.
145   Cereda, E., Ezhaya, A., Quintero, M.G., Bellora, E., Dubini, E., Micheletti, R., Schiavone, A., Brambille, A., Schiavi, G.B. and Donnetti, A.J. (1990) J. Med. Chem. 33, 2108–2118.
146   de Costa, B.R., Rothman, R.B., Bykov, V., Jacobson, A.E. and Rice, K.C. (1989) J. Med. Chem. 32, 281–283.
147   de Costa, B.R., Band, L., Rothman, R.B., Jacobson, A.E., Bykov, V., Pert, A. and Rice, K.C. (1989) FEBS Lett. 249, 178–182.
148   de Costa, B.R., Rothman, R.B., Bykov, V., Band, L., Pert, A., Jacobson, A.E. and Rice, K.C. (1990) J. Med. Chem. 33, 1171–1176.
149   Chang, A.-C., Takemori, A.E. and Portoghese, P.S. (1994) J. Med. Chem. 37, 1547–1549.
150   Chang, A.-C., Takemori, A.E., Ojala, W.H., Gleason, W.B. and Portoghese, P.S. (1994) J. Med. Chem. 37, 4490–4498.
151   Costello, G.F., James, R., Shaw, J.S., Slater, A.M. and Stutchbury, N.C. (1991) J. Med. Chem. 34, 181–189.
152   Barlow, J.J., Blackburn, R.P., Costello, G.F., James, R., Le Count, D.J., Main, B.G., Pearce, R.J., Russell, K. and Shaw, J.S. (1991) J. Med. Chem. 34, 3149–3158.
153   Belton, P., Cotton, R., Giles, M.B., Gormley, J.J., Miller, L., Shaw, J.S., Timms, D. and Wilkinson, A. (1983) Life Sci. 33 (Supp.1), 443–446.
154   Shaw, J.S., Miller, L., Turnbull, M.J., Gormley, J.J. and Morley, J.S. (1982) Life Sci. 31, 1259–1262.
155   Corbett, A.D., Gillan, M.G.C., Kosterlitz, H.W., McKnight, A.T., Paterson, S.J. and Robson, L.E. (1984) Br. J. Pharmacol. 83, 271–279.
156   Cotton, R., Giles, M.G., Miller, L., Shaw, J.S. and Timms, D. (1984) Eur. J. Pharmacol. 97, 331–332.
157   Hirning, L.D., Mosberg, H.I., Hurst, R., Hruby, V.J., Burks, T.F. and Porreca, F. (1985) Neuropeptides 5, 383.
158   Schiller, P.W., Nguyen, T.M.-D., Weltrowska, G., Wilkes, B.C., Marsden, B.J., Lemieux, C. and Chung, N.N. (1992) Proc. Natl. Acad. Sci. U.S.A. 89, 11871–11875.
159   Nevin, S.T., Tóth, G., Nguyen, T.M.-D., Schiller, P.W. and Borsodi, A. (1993) Life Sci. 53, PL 57–62.
160   Marsden, B.J., Nguyen, T.M.-D. and Schiller, P.W. (1993) Int. J. Pept. Protein Res. 41, 313–316.
161   Schiller, P.W., Weltrowska, G., Nguyen, T.M.-D., Wilkes, B.C., Chung, N. N. and Lemieux, C. (1993) J. Med. Chem. 36, 3182–3187.
162   Fundytus, M.E., Schiller, P.W., Shapiro, M., Weltrowska, G. and Coderre, T.J. (1995) Eur. J. Pharmacol. 286, 105–108.
163   Nevin, S.T., Tóth, G., Weltrowska, G., Schiller, P.W. and Borsodi, A. (1995) Life Sci., 56, PL 225–230.
164   Sasaki, Y, and Coy, D.H. (1987) Peptides 8, 119–121.
165   Fehrentz, J.-A. and Castro, B. (1983) Synthesis 676–678.
166   Schiller, P.W., Weltrowska, G., Nguyen, T.M.-D., Lemieux, C., Chung, N.N., Zelent, B., Wilkes, B.C. and Carpenter, K.A. (1996) in Peptides: Chemistry, Structure and Biology (Kaumaya, P.T.P. and Hodges, R.S., eds.) pp. 609–611, Mayflower Scientific Ltd.
167   Rónai, A.Z., Botyanszki, J., Hepp, J. and Medzihradszky, K. (1992) Life Sci. 50, 1371–1378.
168   Bowen, W.D., Hellewell, S.B., Keleman, M., Huey, R. and Stewart, D. (1987) J. Biol. Chem. 262, 13434–13439.
169   Calcagnetti, D.J., Fenselow, M.S., Helmstetter, F.J. and Bowen, W.D. (1989) Peptides 10, 319–326.

170 Portoghese, P.S., Sultana, M., Nagase, H., and Takemori, A.E. (1988) J. Med. Chem. 31, 281–282.

171 Schwyzer, R. (1977) Ann. N.Y. Acad. Sci. 297, 3–26.

172 Chavkin, C. and Goldstein, A. (1981) Proc. Natl. Acad. Sci. U.S.A. 78, 6543–6547.

173 Portoghese, P.S. (1992) J. Med. Chem. 35, 1927–1937.

174 Takemori, A.E., Sofuoglu, M., Sultana, M., Nagase, H. and Portoghese, P.S. (1990) in New Leads in Opioid Research (van Ree, J.M., Mulder, A.H., Wiegant, V.M. and van Wimersma Greidanus, T.B., eds.) pp. 277–278, Excerpta Medica Elsevier Science, Amsterdam.

175 Takemori, A.E. and Portoghese, P.S. (1992) Annu. Rev. Pharmacol. Toxicol. 32, 239–269.

176 Sofuoglu, M., Portoghese, P.S. and Takemori, A.E. (1991) J. Pharmacol. Exp. Ther. 267, 676–680.

177 Arakawa, K., Akami, T., Okamoto, M., Oka, T., Nagase, H. and Matsumoto, S. (1992) Transplantation 53, 951–953.

178 Arakawa, K., Akami, T., Okamoto, M., Nakajima, H., Mitsuo, M., Naka, I., Oka, T. and Nagase, H. (1992) Transplant Proc. 24, 696–697.

179 Arakawa, K., Akami, T., Okamoto, M., Akioka, K., Akai, I., Oka, T and Nagase, H. (1993) Transplant. Proc. 25, 738–740.

180 Abdelhamid, E.E., Sultana, M., Portoghese, P.S. and Takemori, A.E. (1991) J. Pharmacol. Exp. Ther. 258, 299–303.

181 Menkens, K., Bilsky, E.J., Wild, K.D., Portoghese, P.S., Reid, L.D. and Porreca, F. (1992) Eur. J. Pharmacol. 219, 345–346.

182 Reid, L.D., Hubbell, C.L., Glaccum, M.B., Bilsky, E.J., Portoghese, P.S. and Porreca, F. (1993) Life Sci. 52, PL 67–71.

183 Hong, E.K. and Takemori, A.E. (1989) J. Pharmacol. Exp. Ther. 251, 594–598.

184 Jackson, H.C., Ripley, T.L. and Nutt, D.J. (1989) Neuropharmacology 28, 1427–1430.

185 Kitchen, I. and Pinker, S.R. (1990) Br. J. Pharmacol. 100, 685–688.

186 Kamei, J., Iwamoto, Y., Suzuki, T., Misawa, M., Nagase, H. and Kasuya, Y. (1993) Eur. J. Pharmacol. 249, 161–165.

187 Portoghese, P.S., Sultana, M. and Takemori, A.E. (1990) J. Med. Chem. 33, 1714–1720.

188 Portoghese, P.S., Nagase, H., MaloneyHuss, K.E., Lin, C.-E. and Takemori, A.E. (1991) J. Med. Chem. 34, 1715–1720.

189 Maguire, P.A., Perez, J.J., Tsai, N.F., Rodriguez, L., Beatty, M.F., Villar, H.O., Kamal, J.J., Upton, C., Casy, A.F. and Loew, G.H. (1993) Mol. Pharmacol. 44, 1246–1251.

190 Nagase, H., Mizusuna, A., Kawai, K. and Nakatani, I. (1994) PCT Int. Appl. WO 94 07,896; (1994) Chem. Abstr. 121, P134540b.

191 Schmidhammer, H., Daurer, D., Wieser, M., Monory, K., Borsodi, A., Elliott, J. and Traynor, J.R. (1997) Bioorg. Med. Chem. Lett. 7, 151–156.

192 Schmidhammer, H., Schratz, A., Schmidt, C., Patel, D. and Traynor, J.R. (1993) Helv. Chim. Acta 76, 476–480.

193 Portoghese, P.S., Farouz-Grant, F., Sultana, M and Takemori, A.E. (1995) J. Med. Chem. 38, 402–407.

194 Portoghese, P.S., Sultana, M., Nagase, H. and Takemori, A.E. (1992) Eur. J. Pharmacol. 218, 195–196.

195 Kamei, J., Iwamoto, Y., Suzuki, T., Misawa, M., Nagase, H. and Kasuya, Y. (1994) Eur. J. Pharmacol. 251, 291–294.

196 Nagase, H., Onoda, Y., Kawai, K., Matsumoto, S. and Endo, T., (1991) PCT Int.Appl. WO 91 18,901; (1992) Chem. Abstr. 116, P255598g.

197 Dondio, G. and Ronzoni, S. (1995) PCT Int, Appl. WO 95 04734; (1995) Chem. Abstr. 122, P314536j.

198 Burke, Jr., T.R., Jacobson, A.E., Rice, K.C., Silverton, J.V., Simmonds, W.F., Streaty, R.A. and Klee, W.A. (1986) J. Med. Chem. 29, 1087–1093.
199 Lobbezoo, M.W., Soudjin, W. and van Wijngaarden, I. (1980) Eur. J. Med. Chem. 15, 357–361.
200 Janssen, C. (1964) French Pat. M2430; (1965) Chem. Abstr. 62, 1463e.
201 Bentley, K.W., Hardy, D.C. and Smith, A.C.B. (1969) J. Chem. Soc. 2235–2236.
202 Portoghese, P.S., Sultana, M. and Takemori, A.E. (1990) J. Med. Chem. 33, 1547–1548.
203 Portoghese, P.S., Sultana, M., Nelson, W.L., Klein, P. and Takemori, A.E. (1992) J. Med. Chem. 35, 4086–4091.
204 Chakrabati, S., Sultana, M., Portoghese, P.S. and Takemori, A.E. (1993) Life Sci. 53, 1761–1765.
205 Korlipara, V.L., Takemori, A.E. and Portoghese, P.S. (1994) J. Med. Chem. 37, 1882–1885.
206 Korlipara, V.L., Takemori, A.E. and Portoghese, P.S. (1995) J. Med. Chem. 38, 1337–1343.
207 Borsodi, A. and Tóth, G. (1995) Ann. N.Y. Acad. Sci. 757, 339–352.

Progress in Medicinal Chemistry – Vol. 35
Edited by G.P. Ellis, D.K. Luscombe and A.W. Oxford

4 Mechanisms of Bacterial Resistance to Antibiotics and Biocides

A.D. RUSSELL, D.Sc.

Welsh School of Pharmacy, University of Wales, Cardiff, CF1 3XF, UK

INTRODUCTION

DEFINITIONS

Although originally defined as 'a substance produced by one micro-organism that inhibits the growth of another micro-organism', an antibiotic is now considered in a broader context. The term refers to 'a substance produced by one micro-organism or a similar substance that, in low concentrations, inhibits the growth of another micro-organism' [1]. As such, the term encompasses chemotherapeutic drugs such as the older penicillins, streptomycin and some tetracyclines as well as chloramphenicol (originally obtained from *Streptomyces venezuelae* but now produced synthetically) and the semi-synthetic penicillins and cephalosporins, together with those drugs, such as 4-quinolones, sulphonamides and trimethoprim that are made by synthetic means. The term 'antibiotic' as used in this review refers to the broader definition.

Other antimicrobial compounds consist of antiseptics, disinfectants and preservatives [2]. *Antisepsis* is defined as the destruction or inhibition of micro-organisms on living tissues, thereby limiting or preventing the harmful results of infection. *Disinfection* is the process of removing micro-organisms, including potentially pathogenic ones, from the surfaces of inanimate objects. Bacterial spores are not necessarily inactivated. Chemical disinfectants are capable of different levels (high, intermediate or low) of activity [3]. *Preservation* is a term used to denote prevention of spoilage of a pharmaceutical, food, cosmetic or other product [4, 5].

Some compounds are employed in only one of these areas, for example, the parabens (esters of 4-hydroxybenzoic acid) are used as preservatives in some types of cosmetic and pharmaceutical products, whereas chlorhexidine salts are utilized in all three areas. The collective term 'biocide' (*bios*, life; *cido*, to kill) is used widely nowadays to denote a compound used in one or more of these fields [2].

Three words (resistance, insusceptibility, tolerance) and their corresponding adjectives find widespread use, often interchangeably, to signify the lack of effect of an antimicrobial agent on a microbial cell. 'Tolerance' is usually used to denote lack of lysis in some penicillin-treated streptococci. 'Resistance' in terms of insusceptibility to biocides is often used to denote an increase of approximately 4–8 fold in the inhibitory concentration which, as will be discussed later, is not without difficulty because lethal concentrations are much more relevant.

SIGNIFICANCE OF BACTERIAL RESISTANCE

The introduction of antibiotics into clinical medicine has brought about a whole range of unforeseen problems. Apart from potential toxicity and, with some antibiotics, such as penicillins, of hypersensitivity reactions, several micro-organisms (especially many types of bacteria) show high levels of resistance. Furthermore, transferability of resistance from resistant to sensitive cells of the same or different species or genus is a well-known phenomenon with potentially serious clinical consequences [6–8].

β-Lactamase-producing staphylococci resistant to benzylpenicillin emerged shortly after the introduction into clinical practice of this antibiotic and similar events have occurred when other antibiotics have been newly used [9], thereby complicating therapy. Currently, there is considerable concern about methicillin-resistant *Staphylococcus aureus* (MRSA) [10, 11], vancomycin-resistant enterococci [12, 13], multi-drug resistant *Mycobacterium tuberculosis* (MDRTB) [14, 15], antibiotic-resistant Gram-negative bacteria such as *Serratia marcescens, Acinetobacter* spp. and *Klebsiella* spp [16–18], antibiotic-resistant pneumococci [19, 20] and 'emerging' pathogens such as *Stenotrophomonas (Xanthomonas) maltophilia, Burholderia (Pseudomonas) cepacia* and *Enterobacter* spp. [21, 22].

Bacteria may also be resistant to some types of biocides. Gram-negative bacteria, such as *Pseudomonas aeruginosa, Proteus* spp. and *Providencia stuartii*, are much less susceptible to many biocides than are Gram-positive cocci [23–25]. Some MRSA strains demonstrate plasmid-mediated resistance to cationic biocides [26–28] and mycobacteria possess above-average resistance to biocides [3, 29]. The least sensitive of all types of bacteria to biocides are bacterial spores [8, 29–31] and germinating and outgrowing spores, for example, of some *Bacillus* and *Clostridium* spp. are sometimes implicated as the aetiological agents of food poisoning. Clearly, it is essential to understand how bacterial resistance arises and what measures can be taken to overcome it.

GENERAL MECHANISMS OF RESISTANCE

Bacterial resistance to both antibiotics and biocides is essentially of two types [6–8, 23–25]:

(i) intrinsic resistance (intrinsic insusceptibility [32]), a natural property of an organism, and

(ii) acquired resistance, which results from mutation or via the acquisition of generic material (such as a plasmid or transposon (Tn)). Some bacteria are naturally insusceptible to some antibacterial compounds, but sensitive to others, and some organisms have acquired the ability to circumvent the action of these agents (Table 4.1).

Much is known about the two general mechanisms, especially with regard to antibiotics. In this review, current aspects of the mechanisms of resistance of several different types of bacteria to a broad range of antibiotics and biocides will be discussed. Where relevant, the clinical significance will be addressed, and a final section will consider possible ways of overcoming bacterial resistance.

It must be emphasized that the subject of bacterial resistance to antibacterial agents is a huge one. Consequently, as far as possible, only recent theories and practicalities will be described here, and reviews will be cited frequently, although occasionally some older work will perforce be referred to when necessary.

Table 4.1.   GENERALIZED RESISTANCE MECHANISMS [#] TOWARDS ANTIBIOTICS AND/OR BIOCIDES

Mechanism	Comment and examples
Alteration of antibacterial agent (inactivation/modification)	Occurs with several types of antibiotics, rarely with biocides (except for mercurials)
Impaired uptake	Several antibiotics and biocides*
Modified target site	β-Lactams, 4-quinolones, macrolides, etc
Efflux of antibacterial agent	Tetracyclines, QACs, others (see Tables 4.11 & 4.14)
By-pass of sensitive step	Duplication of target enzymes, second version being insusceptible to action of a drug, e.g. trimethoprim
Overproduction of target	Higher drug concentrations needed to saturate target enzymes, e.g. trimethoprim and DHFR
Absence of an enzyme or metabolic pathway	Applies to some antibiotics and certain bacterial species, e.g. folate auxotrophs, trimethoprim and sulphonamides

[#]   Some are intrinsic in nature, others acquired: see also Tables 4.5, 4.6, 4.7, 4.10, 4.11, 4.18
*   Biocides have been considered in relation to Gram-negative bacteria, mycobacteria and bacterial spores
QACs, quaternary ammonium compounds; DHFR, dihydrofolate reductase

MECHANISMS OF INTRINSIC RESISTANCE (INTRINSIC INSUSCEPTIBILITY)

The first stage in the action of an antibiotic or biocide on a bacterial cell involves interaction between the chemical and the biological entity. Adsorption of a variety of biocides into bacterial cells has been described [33] but this, *per se*, does not necessarily provide information about the mechanism or site of action of the antibacterial compound. However, resistant cells would usually (but not necessarily) be expected to adsorb less of a chemical than sensitive cells. In non-sporulating bacteria, changes to the outer layers of cells may follow the initial binding to the cell surface or there may be diffusion across the cell envelope; in either case, an antibiotic or biocide will penetrate the cell to reach the primary site of action at the cytoplasmic membrane or within the cytoplasm. Little is known about the penetration of antibacterial agents into bacterial spores.

Antibiotics are selectively toxic drugs and their action on non-sporulating bacteria is usually highly specific (Table 4.2) [34, 35]; exceptions to this general statement may be found. For example, the quinolones and fluoroquinolones act on DNA gyrase (topoisomerase II), yet their lethal effect is claimed to result from a 'cascade' of events [36, 37]. In contrast, with biocides, effects on the primary target site may lead to additional, secondary, changes elsewhere in the organism. These secondary events may also contribute to the bacteriostatic or bactericidal activity of biocides. Further, some biocides are highly reactive chemical molecules that will combine with specific receptors at different sites within the cell. For example, glutaraldehyde interacts with amino groups in the cell envelope, cytoplasmic membrane and cytosol [38]. For these reasons, it is not always an easy matter to define the precise mode of action of a biocidal agent [39]. A summary of biocidal mechanisms is presented in Table 4.3.

A knowledge of the site and action of antibacterial agents is important in understanding at least some of the resistance mechanisms presented by bacteria.

INTRINSIC RESISTANCE IN GRAM-NEGATIVE BACTERIA

Any impaired uptake of an antibiotic or biocide means that fewer molecules of the antibacterial agent are able to reach their target site(s) thereby reducing their effect on the cell [23–25, 40–42]. Decreased uptake in Gram-negative bacteria is an important reason for the fact that these organisms are often less sensitive than Gram-positive bacteria, such as staphylococci, to a variety of chemically unrelated molecules [23–25]. This intrinsic insuscep-

Table 4.2.   SUMMARY OF MECHANISMS OF ACTION OF ANTIBIOTICS AGAINST
NON-SPORULATING BACTERIA

Effect	Antibiotic(s)	Action
Inhibition of cell wall peptidoglycan synthesis	D-cycloserine	Early biosynthetic stage: competitive inhibition of (i) alanine racemase and (ii) D-alanine synthetase (ligase)
	Glycopeptides (vancomycin, teicoplanin)	Intermediate biosynthetic stage: binding to peptidoglycan precursor (D-alanyl-D-alanine)
	β-lactams	Late biosynthetic stage: binding to specific PBPs*, inhibition of transpeptidases
	Isoniazid	Mycolic acid biosynthesis (?) in *M. tuberculosis*
Cytoplasmic membrane disruption	Polymyxins	Also damage outer membrane of Gram-negative bacteria
Inhibition of protein synthesis	Streptomycin	Inhibits initiation stage
	Tetracyclines	Inhibits binding of aminoacyl-tRNA to 30S ribosomal subunit
	Chloramphenicol	Inhibits peptidyl transferase
	Erythromycin	Inhibits translocation
	Mupirocin	Inhibits isoleucyl-tRNA synthetase
Inhibition of RNA synthesis	Actinomycin D	Binds to double stranded DNA
	Rifampicin	Inhibits DNA-dependent RNA polymerase
Inhibition of DNA synthesis	Mitomycin C	Covalent linking to DNA
	Quinolones	Effect on DNA gyrase
Inhibition of tetrahydrofolate synthesis	Sulphonamides	Competitive inhibitors of dihydropteroate synthetase
	Trimethoprim	Inhibits dihydrofolate reductase

*PBPs, penicillin-binding proteins

Table 4.3.   SUMMARY OF MECHANISMS OF ACTION OF BIOCIDES AGAINST
NON-SPORULATING BACTERIA

Effect	Biocide(s)*	Action*
Cell envelope (cell wall, outer membrane)	Glutaraldehyde	Cross-links proteins
	EDTA, other permeabilizers	Gram-negative bacteria: removal of Mg^{2+}, release of some LPS
	Lysozyme	Breaks down peptidoglycan (β1-4 sites)
Cytoplasmic (inner) membrane	QACs Chlorhexidine	Generalized membrane damage Low concentrations affect membrane integrity, high concentrations cause congealing of cytoplasm
	PHMB, alexidine	Phase separation and domain formation of membrane lipids
	Hexachlorophane	Inhibition of membrane-bound electron transport chain
	Sorbic acid	Transport inhibitor, effect on PMF
	Parabens	Low concentrations inhibit transport, high concentrations affect membrane integrity
	Mercurials, silver compounds	Interaction with thiol groups
Alkylation	Ethylene oxide	Alkylation of amino, carboxyl, hydroxyl and mercapto groups in proteins
Cross-linking	Formaldehyde	Cross-linking of proteins, RNA and DNA
	Glutaraldehyde	Cross-linking of proteins in cell envelope (q.v.) and elsewhere in cell
Intercalation	Acridines	Intercalation of an acridine molecule between two layers of base pairs in DNA
Interaction with thiol groups	Mercurials (q.v.)	Membrane-bound enzymes
	Isothiazolones	Membrane-bound enzymes
Oxidizing agents	Halogens	Oxidize thiol groups to disulphides, sulphoxides or disulphoxides
	Peroxygens	Hydrogen peroxide: activity results from formation of free hydroxyl radicals ($^{\bullet}OH$) which oxidize thiol groups in enzymes and proteins Peracetic acid: disrupts thiol groups in proteins and enzymes

* LPS, lipopolysaccharide; QACs, quaternary ammonium compounds

Table 4.4. TRANSPORT OF ANTIBIOTICS AND BIOCIDES INTO GRAM-NEGATIVE BACTERIA

Antibiotic(s) or biocide(s)	Passage across OM	Passage across IM
β-Lactams*	Passive diffusion, OmpF and OmpC porins	Not applicable (interact with penicillin-sensitive enzymes on outer face of IM)
D-cycloserine	Probably via porins	D-alanine transport system
Polymyxins	Self-promoted uptake, damage to OM	Not applicable: IM is a major target site
AGACs	Self-promoted uptake (in *Ps.aeruginosa*); porins (other organisms?)	Energy-dependent phase I (EDP-1) for streptomycin
Vancomycin	Hydrophilic, but too large to traverse OM	Not applicable
Chloramphenicol	Porins	Active transport
Tetracyclines	Porins (especially OmpF)	Energy-independent (passive) and energy-dependent (active) transport systems
4-Quinolones	Porins (OmpC, OmpF), also self-promoted uptake	Probably active transport
Chlorhexidine	Self-promoted uptake?	IM is a major target site; damage to IM enables biocide to enter cytosol where further interaction occurs
QACs	Self-promoted uptake? Also, OM might present barrier	IM is a major target site; damage to IM enables biocide to enter cytosol where further interaction occurs
Parabens	OM presents barrier	IM is probably a major target; also inhibit intracellular biosynthetic processes

* Depends on a number of factors; generally, rate of penetration through porins decreases with increase in hydrophobicity, size and molecular weight and net negative charge, but overall rate is influenced by gross physicochemical properties.

OM, outer membrane; IM, inner membrane

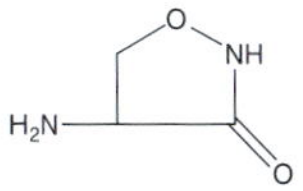

(1) 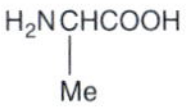   (2)

tibility is associated with the nature and composition of the cell envelope of these organisms [43, 44] and is most marked in *Pseudomonas aeruginosa, Proteus* spp., *Providencia stuartii, Burkholderia* (formerly *Pseudomonas*) *cepacia* and *Serratia marcescens* [7, 25].

In brief, both Gram-positive and Gram-negative bacteria contain a basic cross-linked peptidoglycan structure. The former bacteria, such as staphylococci, also contain glycerol and ribitol teichoic acids. In contrast, in addition to peptidoglycan, the Gram-negative cell envelope contains lipoprotein molecules attached covalently to this backbone, together with lipopolysaccharide and protein attached by hydrophobic interactions and divalent cations (Ca^{2+}, Mg^{2+}). On the inner side is a layer of phospholipid.

The Gram-negative cell envelope is considerably more complex than the cell wall of non-sporulating Gram-positive bacteria, mycobacteria excepted [7, 8, 45, 46]. The outer membrane of Gram-negative bacteria plays an important, though passive, role in acting as a permeability barrier. Antibiotics

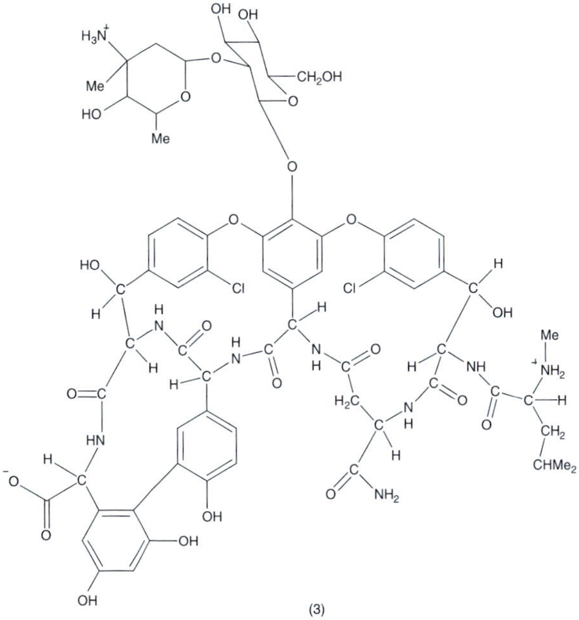

(3)

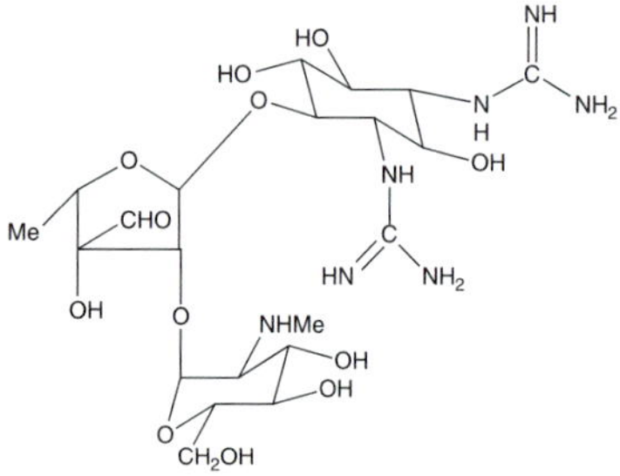

(4)

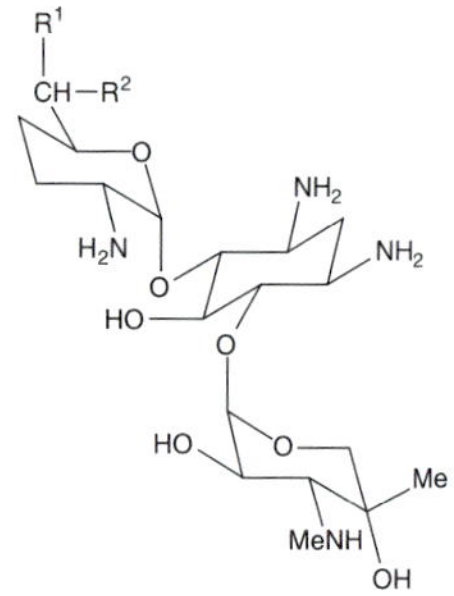

Compound	R^1	R^2
Gentamycin C_{12}	H	NH_2
Gentamycin C_1	Me	NHMe
Gentamycin C_2	Me	NH_2

(5)

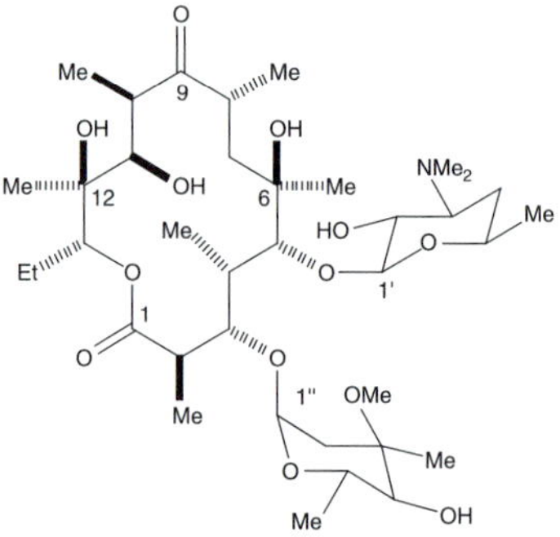

(6)

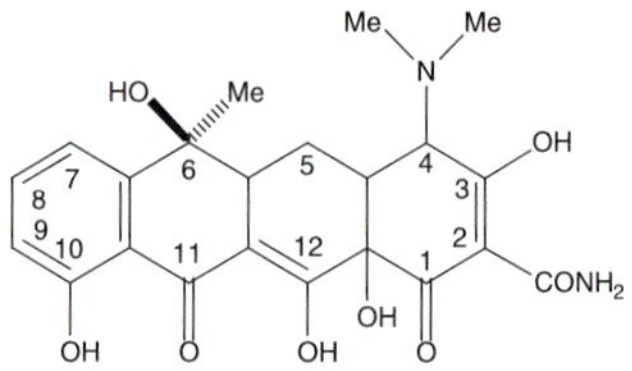

(7)

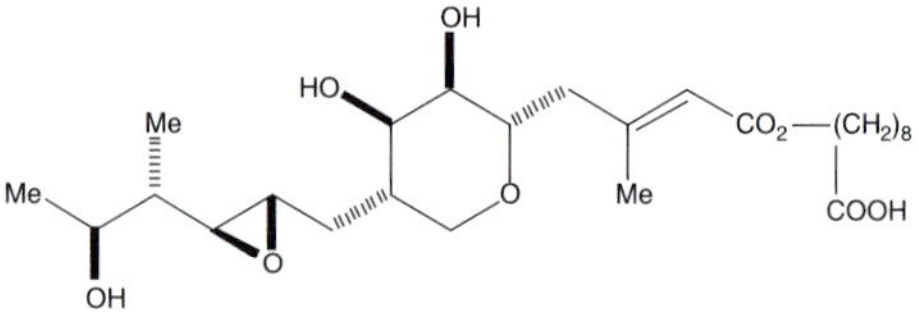

(8)

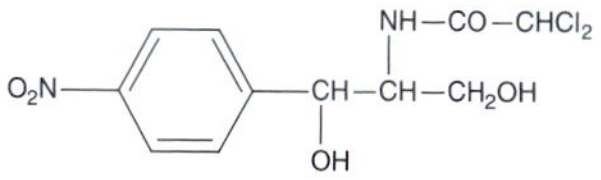

(9)

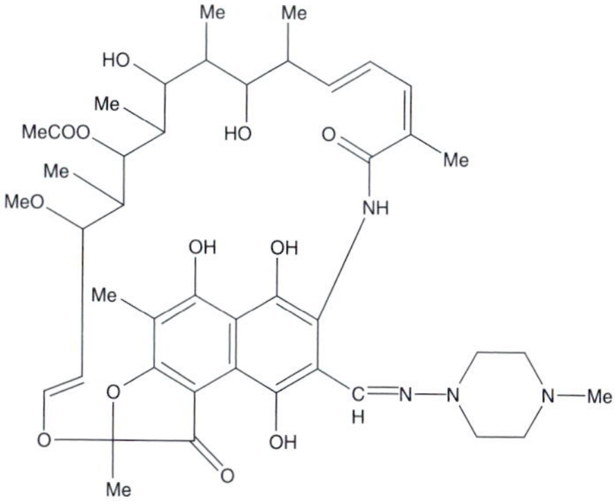

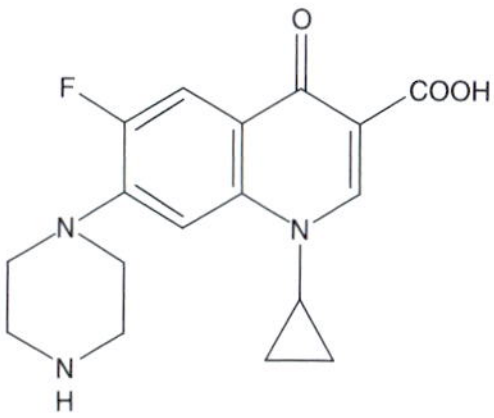

(11)

(10)

leu——dab

D–phe   dab

dab   thr

dab

dab

thr

dab—FA

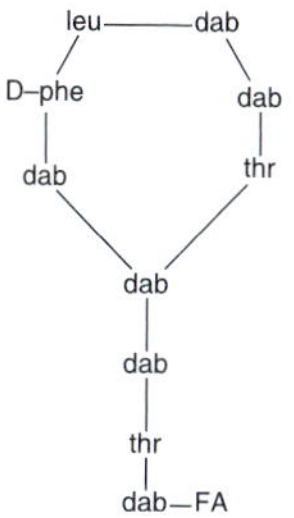

Polymyxin B$_2$: fatty acyl (FA) residue is (+)-6-methyloctanoic acid; dab, L–diaminobutyric acid

(12)

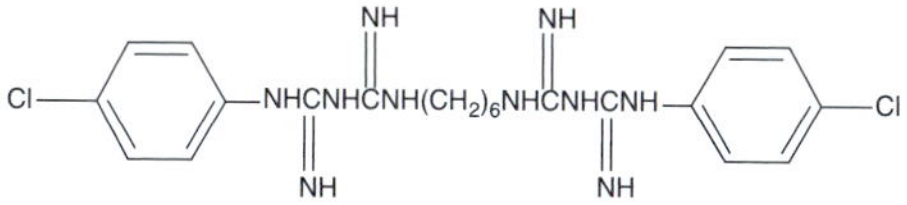

(13)

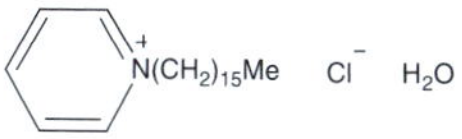

(14)

and biocides (Table 4.4) must traverse the outer membrane, and in some cases the inner (cytoplasmic) membrane also, to reach their target sites. β-Lactam antibiotics interact with penicillin-binding proteins (PBPs) on the outer face of the inner membrane [47, 48] and thus do not need to cross the latter membrane. In contrast, another (but early) inhibitor of peptidoglycan biosynthesis, D-cycloserine (1) has as its site of action the interconversion of L-alanine to D-alanine (2) and the formation of the D-alanyl-D-alanine dipeptide in the cytosol [49, 50] and thus must cross both the outer and inner membranes. The glycopeptide antibiotic vancomycin (3) acts at an intermediate stage in peptidoglycan biosynthesis by interfering with glycan unit insertion [49] and consequently must pass through the outer membrane. Inhibitors of protein biosynthesis, e.g. the aminoglycoside-aminocyclitol antibiotics (AGACs) such as streptomycin (4) and gentamicin (5) [51, 52], the macrolides such as erythromycin (6) [53, 54], the tetracyclines exemplified by tetracycline itself (7) [55, 56], mupirocin (8) [57] and chloramphenicol (9) [58], and inhibitors of RNA synthesis (rifampicin (10) [59]) and DNA synthesis (4-quinolones such as ciprofloxacin (11) [36, 37]) must all traverse both the outer and inner membranes in order to reach their target within the cell. The so-called 'membrane-active' agents such as the polymyxin antibiotics (12) and cationic biocides like chlorhexidine (13) and quaternary ammonium compounds (QACs, e.g. cetylpyridinium chloride (14)) cause severe damage to the inner membrane, although changes are also induced in the outer membrane [60–63]. The nature of the action of all these agents is summarized in Tables 4.2 and 4.3.

The membrane-active agents are believed to enter Gram-negative cells (Table 4.4) by a self-promoted entry mechanism [7, 60–62] in which damage to the outer membrane first occurs, and as a consequence, these chemicals are able to reach the inner membrane. Deleterious effects on this membrane [60–62] (Table 4.3) presumably enables the compounds to enter the cytoplasm where interaction with proteins and nucleic acids can occur [33, 60–62]. Studies with outer membrane mutants suggest that chlorhexidine enters readily some types of Gram-negative bacteria such as *Escherichia coli* [64–66] but to a lesser extent with other Gram-negative organisms, such as *Providencia stuartii* [67, 68]. The outer membrane is likely to act as a significant barrier to QACs in Gram-negative cells [65, 69]. Investigations have also been conducted with a homologous series of the methyl, ethyl, propyl and butyl esters (15; the parabens) of 4-hydroxybenzoic acid [70] and with the phenolics, phenol (16), cresol (17) and chlorocresol (18) [69] on the basis of which the following conclusions have been reached:

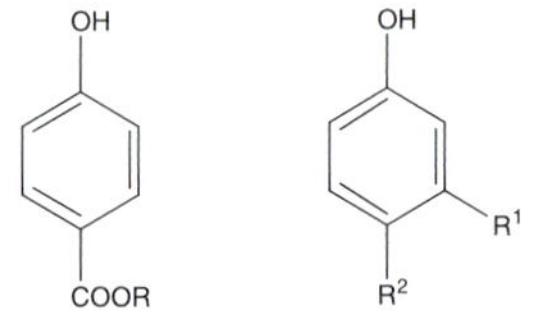

OH   OH

COOR   R²

R = Me; Et; Pr; Bu etc.   (16) R¹ = R² = H
                                (17) R¹ = Me, R² = H
(15)   (18) R¹ = Me, R² = Cl

(i) activity increases in the paraben series in the order Me, Et, Pr and Bu ester, but this is accompanied by a corresponding decrease in aqueous solubility;

(ii) wild-type strains of *E.coli* and *Salmonella typhimurium* are intrinsically more susceptible to parabens than rough and especially deep rough mutants;

(iii) the greatest effect in the paraben series is shown by the Bu ester against the deep rough strains;

(iv) in the phenolic series, phenol is the least active, chlorocresol (4-chloro-3-methyl phenol) the most active, and the greatest effect is shown by chlorocresol against the deep rough strains.

Uptake of the parabens probably proceeds by a general dissolution of an ester into the cell with no specific binding sites at the cell surface. In the deep rough mutants, the increased sensitivity to the parabens probably results from the appearance of phospholipid patches at the cell surface which aid in penetration of the parabens and especially of the most hydrophobic one (the Bu ester). A similar mechanism is likely to account for the uptake of phenolics across the outer membrane into the cells.

At the time when these experiments were undertaken in the mid-1980s, the possibility of biocide efflux had not been described. Since then, evidence that efflux of cationic biocides from Gram-negative bacteria has been presented [41, 42, 71] and it would be instructive to re-examine some of these mutant strains to determine whether an efflux mechanism is operating here, and to what extent it is responsible for the differing susceptibilities noted. Some aspects such as multidrug resistance in Gram-negative bacteria are considered later in this review.

Microbial degradation of biocides has been described by Hugo [72] who points out that soil organisms are able to break down substances such as phenols added as fumigants. He also reviewed the utilization by bacteria of aromatic compounds (including the preservatives cresol, phenol, benzoic acid and esters of 4-hydroxybenzoic acid). Several types of preservatives and disinfectants, such as the QACs (e.g. cetrimide, cetylpyridinium chloride, benzalkonium chloride), chlorhexidine and phenylethanol can also be inactivated. Significantly, this only occurs at concentrations well below inhibitory or 'in-use' concentrations [33] and thus cannot be responsible for insusceptibility. A further comment about chlorhexidine is given below.

Even in bacteria such as *Providencia stuartii, Ps. aeruginosa* and MRSA, there is no evidence to suggest that chlorhexidine and other biocides are broken down [25, 73, 74]. However, an aldehyde hydrogenase expressed constitutively and at high levels in some *Pseudomonas* spp. is capable of reducing formaldehyde and other aldehydes, but not glutaraldehyde, to the corresponding alcohol [75].

In general, it appears that biocide molecules are more recalcitrant than chemotherapeutic agents to inactivation by bacterial enzymes. It has been suggested [76] that chlorhexidine might be inactivated; if this is confirmed, then it adds another dimension to mechanisms of bacterial insensitivity to biocides [76].

Intrinsic resistance to biocides as a consequence of bacterial degradative activities is thus not a major mechanism of insusceptibility. There are, of course, examples of plasmid-mediated enzymes that confer resistance to inorganic (and sometimes organic) mercurials and these will be discussed later (see 'Plasmid-mediated resistance to biocides'). The mechanisms of intrinsic resistance to antibotics and biocides are summarized in Table 4.5.

INTRINSIC RESISTANCE IN MYCOBACTERIA

Mycobacteria comprise a fairly diverse group of acid-fast, Gram-positive bacteria. The best-known members are *Mycobacterium tuberculosis*, *M. bovis, M. leprae* and *M. avium intracellulare* (MAI). Others include *M. terrae,M. fortuitum* and *M. smegmatis,* whilst of increasing medical concern is *M. chelonae (M. chelonei),* some strains of which show high resistance to glutaraldehyde [77–79]. Tuberculosis is on the increase and multi-

Table 4.5.   INTRINSIC RESISTANCE MECHANISMS

Type of resistance	Examples	Mechanisms of resistance
Impermeability Gram-negative bacteria	QACs Triclosan Diamidines Rifampicin Some β-lactams (e.g. methicillin cloxacillin) Fucidin Vancomycin	Barrier presented by OM preventing biocide or antibiotic uptake; glycocalyx may also be involved
Impermeability (Mycobacteria)	Chlorhexidine QACs Organomercurials Glutaraldehyde	Waxy cell wall prevents adequate biocide entry Reason for high resistance of some strains of *M.chelonae*?
Impermeability (Bacterial spores)	Chlorhexidine QACs Organomercurials Phenolics	Spore coat(s) and cortex present a barrier to biocide entry
Impermeability (Gram-positive bacteria)	β-lactams Chlorhexidine	Glycocalyx/ mucoexopolysaccharide may be associated with reduced biocide or antibiotic diffusion
Inactivation (chromosomally mediated)	β-Lactamase-susceptible β-Lactams	Antibiotic inactivation

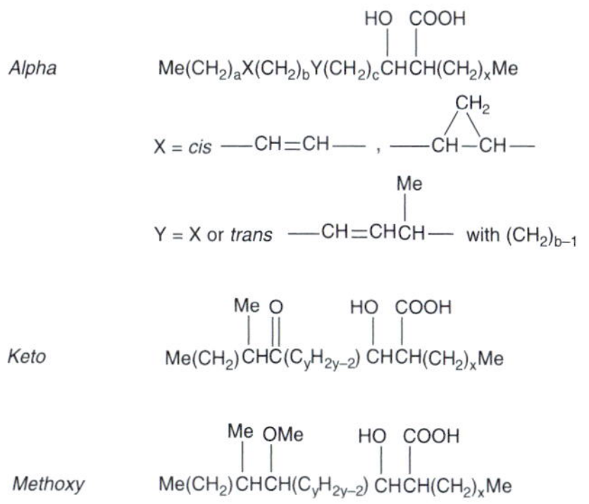

Alpha $Me(CH_2)_aX(CH_2)_bY(CH_2)_c\overset{\underset{\displaystyle |}{HO}}{C}\overset{\underset{\displaystyle |}{COOH}}{C}H(CH_2)_xMe$

$X = cis$ ——CH=CH—— , $\overset{\overset{\displaystyle CH_2}{\diagup\diagdown}}{—CH—CH—}$

$Y = X \text{ or } trans$ ——CH=CH$\overset{\underset{\displaystyle |}{Me}}{C}$H—— with $(CH_2)_{b-1}$

Keto $Me(CH_2)\overset{\underset{\displaystyle |}{Me}}{C}H\overset{\underset{\displaystyle \|}{O}}{C}(C_yH_{2y-2})\,\overset{\underset{\displaystyle |}{HO}}{C}H\overset{\underset{\displaystyle |}{COOH}}{C}H(CH_2)_xMe$

Methoxy $Me(CH_2)\overset{\underset{\displaystyle |}{Me}}{C}H\overset{\underset{\displaystyle |}{OMe}}{C}H(C_yH_{2y-2})\,\overset{\underset{\displaystyle |}{HO}}{C}H\overset{\underset{\displaystyle |}{COOH}}{C}H(CH_2)_xMe$

a = 15, 17, 19; b = 14, 16; c = 11, 13, 15, 17; x = 19, 21, 23; y = 31 – 39

(19)

drug resistant *M. tuberculosis* (MDRTB) strains have appeared [80–82]. It is thus important to consider the mechanisms responsible for resistance. Mycobacterial resistance to antibiotics may be intrinsic (considered here) or acquired (considered later), whereas resistance to biocides is mainly intrinsic in nature.

The mycobacterial cell wall

The mycobacterial cell wall is complex and differs considerably from the cell wall of other (non-mycobacterial) Gram-positive organisms. It consists of peptidoglycan covalently linked to a polysaccharide co-polymer (arabinogalactan), made up of arabinose and galactose esterified to the fatty acids, mycolic acids (19); also present are complex lipids, lipopolysaccharides and proteins [83–86]. Porin channels are also present through which hydrophilic nutrients can diffuse into the cell [86, 87]. Similar cell wall structures exist in all the mycobacterial species examined to date [85, 86]. The cell wall composition of a particular species may be influenced by its environmental niche; pathogenic bacteria such as *M. tuberculosis* exist in a relatively nutrient-rich environment whereas saprophytic mycobacteria living in soil or water are exposed to natural antibiotics and tend to be more intrinsically resistant to antibiotics [88]. The mycobacterial cell wall may act as a barrier (Table 4.6) to the entry of antibiotics or biocides.

Intrinsic resistance of mycobacteria to antibiotics

The porins present in mycobacterial cell walls probably permit the entry of

Table 4.6.   RESISTANCE MECHANISMS IN MYCOBACTERIA

Type of resistance	Postulated mechanism
Intrinsic	Cell wall barrier
Several antibiotics	Arabinogalactan
	Mycolic acids
Acquired	
Isoniazid	Deletion of katG in *M. tuberculosis*
Ethambutol	Mutation of target enzyme involved in arabinogalactan synthesis
Ethionamide	Single amino acid substitution in protein InhA
Rifampicin	Alterations in β-subunit of RNA polymerase
Quinolones	Point mutations in *gyr*A

only very small hydrophilic antibiotics and this is undoubtedly an important factor in the intrinsic resistance of mycobacteria to such drugs [86–89]. Many of the antibiotics effective against mycobacteria are hydrophobic in nature and the probable uptake pathway involves passive diffusion following dissolution into lipid-containing regions of the cell wall. For example, derivatives of isoniazid (20) that are more hydrophobic than isoniazid itself have a greater antimycobacterial action [89, 90].

The component(s) of the mycobacterial cell wall responsible for conferring intrinsic resistance to antibiotics are not yet known with any certainty. However, it has been shown [91–93] that inhibitors of arabinogalactan synthesis increase mycobacterial sensitivity to antibiotics. When *M. avium* is treated with inhibitors of mycolic acid biosynthesis there are significant alterations in outer cell wall layers and the cells show an increased antibiotic susceptibility [93]. Thus, arabinogalactan and mycolic acids are components of the wall associated with intrinsic resistance of mycobacteria to chemotherapeutic drugs and alteration of these structural building blocks leads to increased intracellular penetration of antibiotics [88, 94, 95].

Little is known about the uptake of specific antibiotics into mycobacteria, although the low permeability of mycobacteria to relatively hydrophilic molecules such as β-lactam antibiotics is a major factor in the resistance of organisms to these drugs. Clearly, further work is needed to obtain a better understanding of antibiotic uptake in general with the overall intention of increasing uptake into the mycobacterial cell and thereby improving the efficacy of antibiotic therapy against mycobacteria.

Intrinsic resistance of mycobacteria to biocides

The responses of mycobacteria to biocides have been comprehensively re-viewed [44, 95–97]. Three levels of germicidal activity have been described [3, 98, 99]: (i) high-level activity, inactivating bacterial spores, vegetative bacteria (including mycobacteria), fungi and viruses; (ii) intermediate activity, lethal to all those in (i) except spores; (iii) low-level activity, lethal only to vegetative bacteria (excluding mycobacteria), lipid-enveloped viruses and some fungi. Mycobacteria are thus more resistant to biocides than other non-sporulating bacteria but less resistant than bacterial spores.

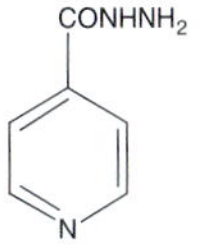

Glutaraldehyde (21), peracetic acid , phenol (16) and ethanol are amongst those agents with mycobactericidal activity [95,100–103]. Other well-known biocidal-type agents such as chlorhexidine (13), QACs such as CPC (14), or-ganomercurials such as phenylmercuric nitrate (22) and parabens (15) are mycobacteriostatic rather than mycobactericidal [95, 104]. *M. chelonae* sub-species *abscessus*, isolated from endoscope washers, was not killed after a 60-minute exposure to alkaline glutaraldehyde whereas a reference strain showed a five log reduction after a contact period of 10 minutes [77]; this glu-taraldehyde-resistant strain was also resistant to peracetic acid but not to sodium dichloroisocyanurate (NaDCC) (23) or to a phenolic [77]. Other workers have also noted an above-average resistance of *M. chelonae* to glu-taraldehyde [78, 79] although not to peracetic acid [78]. This organism has a particular propensity for adhering to smooth surfaces [95] which might contribute to its resistance.

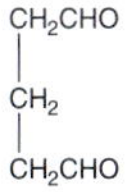

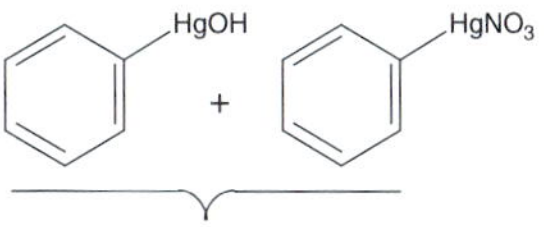

Many years ago, it was proposed [105] that resistance of mycobacteria to QACs was related to the lipid content of the cell wall; thus, *M. phlei* (with low total lipid) was more sensitive than *M. tuberculosis*, which possessed a higher total cell lipid content. It was also pointed out [105] that the resist-ance of various species of mycobacteria was related to the content of waxy material in the wall. As noted above, the mycobacterial cell wall is highly hy-drophobic with a mycoylarabinogalactanpeptidoglycan skeleton. Hydrophi-lic type biocides are generally unable to penetrate the cell wall in sufficiently high concentration to produce a lethal effect. However, low concentrations of biocides such as chlorhexidine must traverse this permeability barrier because minimal concentrations inhibiting growth are generally of the same order as those found with other non-mycobacterial strains [104, 106], although *M. avium* may be particularly resistant [104]. The activity of chlorhexidine can be potentiated in the presence of ethambutol (24), an inhibitor of arabinogalactan biosynthesis [106]. Thus, arabinogalactan is one component of the cell wall that acts as a permeability barrier. Other components have yet to be investigated. More information and studies on newer inhibitors of mycobacterial wall synthesis are needed [107] and further data on wall structural damage will be useful [108].

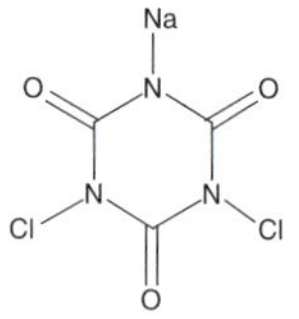

$$MeCH_2CHNH(CH_2)_2NHCHCH_2Me$$
$$CH_2OH \qquad CH_2OH$$

. 2HCl

(23) 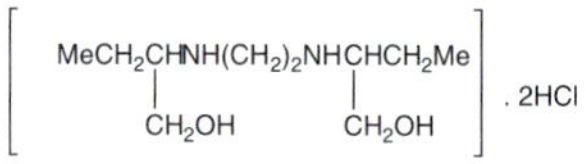   (24)

INTRINSIC RESISTANCE IN OTHER NON-SPORULATING GRAM-POSITIVE BACTERIA

The cell wall of Gram-positive bacteria such as staphylococci consists of a highly cross-linked peptidoglycan network which incorporates teichoic and teichuronic acids [45, 46]. The peptidoglycan moiety acts as a molecular sieve, a property that depends upon its thickness, and the degree of cross-linking. These can vary depending on the physiological status of the cells being affected by growth rate and any growth-limiting nutrient [109]. The susceptibility of *B. megaterium* vegetative cells to chlorhexidine and phenoxyethanol varies when growth rate is altered and nutrient limitation introduced [109]. However, lysozyme-induced protoplasts remain sensitive to these cytoplasmic membrane-active agents, from which it may be inferred that changes in the cell wall must be responsible for the modified response in whole cells [109].

    Other studies [110] have demonstrated that when staphylococci are repeatedly subcultured in media containing glycerol they become 'fattened' with thicker cell walls. Such fattened cells have enhanced levels of cellular lipid that protects them from higher phenols but not from lower phenols, tetrachlorosalicylanilide (25) or hexachlorophene (26) [33, 110].

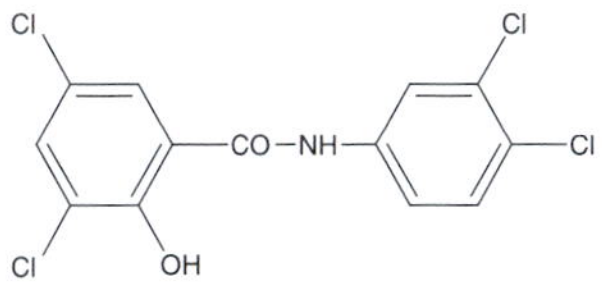

(25)

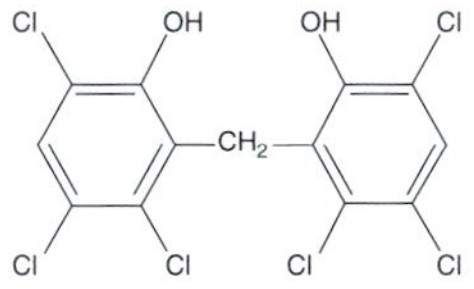

(26)

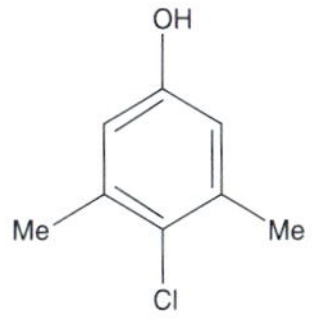

(27)

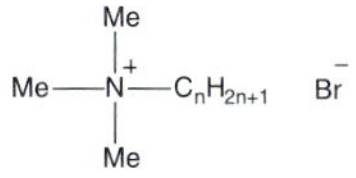

(28)

(n= 12, 14 or 16)

In nature, *S. aureus* may occur as mucoid strains, in which a slime layer surrounds the cells. Such mucoid strains are less susceptible than non-mucoid strains to chloroxylenol (27), cetrimide (28) or chlorhexidine (13) although there is little difference in response with phenols or chlorinated phenols [111]. Removal of slime by washing renders cells as phenotypically sensitive to biocides as non-mucoid cells [111]. The protective nature of slime could be achieved by its acting as either (i) a physical barrier to biocide penetration, or (ii) a loose layer that interacts with, or absorbs, the biocide.

INTRINSIC RESISTANCE IN BACTERIAL SPORES

A complex sequence of events takes place during the life-cycle of cells of the genera *Bacillus* and *Clostridium* [30, 112, 113]. The sporulation process comprises some seven stages and several stages also occur from the activation of a mature spore via germination and outgrowth to the production of a vegetative cell [114, 115]. It is not surprising, therefore, that changes in sensitivity to antibacterial agents occur during the overall life-cycle.

Chemotherapeutic antibiotics vs. spores

The effects of antibiotics on sporulation, germination and outgrowth have been reviewed [116]. The inhibitory action depends upon the stage of development of the sporulating cell when the antibiotic is added. The two peaks of penicillin binding to sporulating cultures are associated with the appearance of specific penicillin-binding proteins (PBPs). The inner primordial layer of the spore cell wall develops, after germination, into the cell wall of the outgrowing vegetative cell [117], and retains the chemical structure of the vegetative wall peptidoglycan. In contrast, the outer (cortex) layer of peptidoglycan has a different, unique muramic lactam structure. The primordial cell wall is synthesized at the inner forespore membrane and the cortex at the outer forespore membrane. Penicillin inhibits spore cortex synthesis, thus transpeptidase and carboxypeptidase activities are essential [113].

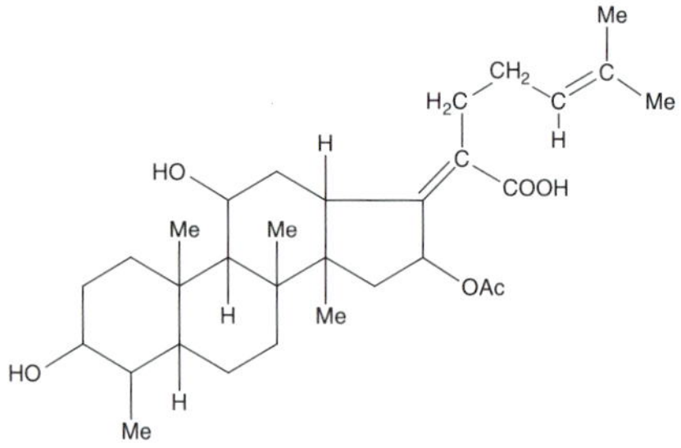

(29)

Other antibiotics such as chloramphenicol (9), fucidin (29), tetracycline (7) and rifampicin (10) inhibit sporulation; the underlying mechanisms involved have been discussed [116]. These chemotherapeutic drugs have no sporicidal activity. This is hardly surprising since spores are considered to be metabolically inert [118]. Polymyxin affects the germination of spores [116] but most antibiotics do not inhibit this process [116]. For example, even at high concentrations, penicillin has no effect on the germination of *Bacillus* or *Clostridium* spores [116] whereas *Cl. terani* cells are injured considerably during the early steps of outgrowth [116]. *B. cereus* spores synthesize cell wall material immediately after germination; in this organism, the cells are relatively stable to penicillin in the early stages of post-germinative development but become highly sensitive to the penicillin-induced blockage of peptidoglycan synthesis during the later phase of elongation and first division [116].

Peptidoglycan, protein and nucleic acid syntheses are not an essential feature of germination but are shown during outgrowth. The germination process is thus insusceptible to inhibitors of these biosyntheses which occur during outgrowth; for example, actinomycin D (30) inhibits mRNA synthesis, protein synthesis inhibitors such as chloramphenicol (9) and tetracycline (7) prevent the incorporation of amino acids into protein and mitomycin C (31) inhibits DNA synthesis [116].

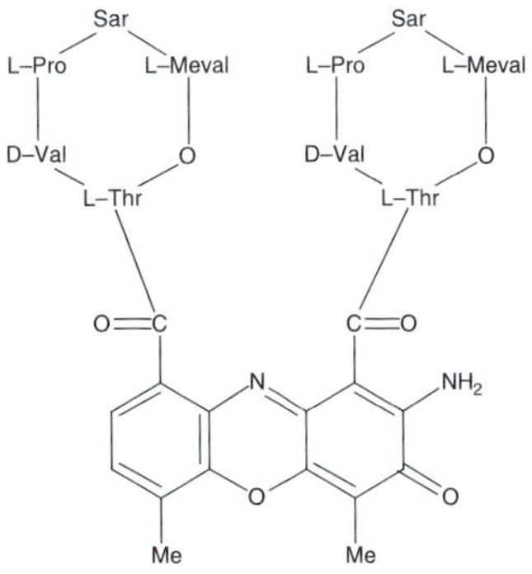

Sar = sarcosine; Meval = N–methylvaline

(30)   (31)

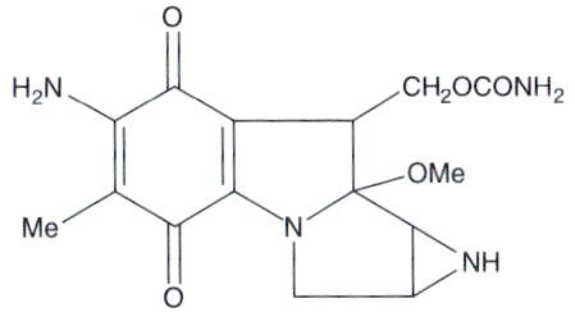

Antibiotics used as food preservatives vs. spores

Nisin is a bacteriocin produced by Group N streptococci such as *Lactobacillus lactis* [119, 120]; it is present naturally in milk and in fermented dairy products and its use as a food preservative (in dairy products and especially processed cheeses) is permitted in several European countries and in the USA. Nisin is active only against Gram-positive bacteria and is considered as a bacteriocin because it is a polypeptide with inhibitory activity towards closely related species. Its mechanism of action has not been fully elucidated. It appears that cytoplasmic membrane disruption is the primary target in non-sporulating Gram-positive bacteria [119] although peptidoglycan synthesis is inhibited [121] and it is likely that Gram-negative organisms are insusceptible by virtue of the outer membrane acting as a permeability barrier. Nisin does not activate bacterial spores *per se*; used in combination with heat, however, it enhances the thermal sensitivity of spores and inhibits the outgrowth of surviving spores [116]. It does not prevent germination but germinated spores are sensitive to the bacteriocin [122]. The reasons for these differing susceptibilities to nisin are unknown; it is possible that the intact spore is impermeable to nisin thereby exhibiting a form of intrinsic resistance.

Other bacteriocins are known, including subtilin [123]. This also inhibits outgrowth and enhances heat processes against some types of spores [124]. Its mechanism of action and the basis for differing responses of spores, and of germinating and outgrowing spores, remain speculative.

Tylosin is a macrolide produced by *Streptomyces fradiae* [125]. It is not sporicidal; it increases the sensitivity of spores to heat and especially to ionizing radiation and inhibits outgrowth although at a later stage than nisin or subtilin [126].

Biocidal agents vs. spores

Many commonly used biocides such as QACs, biguanides, phenolics, parabens and organomercurials are bactericidal but not sporicidal. Others, including glutaraldehyde, peracetic acid, hydrogen peroxide and ethylene oxide are bactericidal and sporicidal, although higher concentrations for longer periods may be necessary to produce the latter effect [30, 31, 127]. Members of both categories may inhibit germination and/or outgrowth usually at concentrations equivalent to those that affect non-sporulating bacteria [128–133]. Clearly, changes must occur during sporulation, germination and outgrowth that are responsible for differences in susceptibility to various compounds. Data are now available that enable us to begin to pin-point these changes [30, 128].

The response of a developing spore to a biocide depends upon its stage of development [30, 128–132]. Resistance increases during sporulation and may be an early, intermediate or late event [128–132]. Resistance to chlorhexidine takes place at an intermediate stage and appears to be associated with the development of the cortex [133]. In contrast, decreasing susceptibility to glutaraldehyde is a very late event [131, 132] and is linked to the biosynthesis of the spore coats.

Coatless forms of spores produced by treatment of mature spores under alkaline conditions with urea plus dithiothreitol plus sodium lauryl sulphate (UDS), have been of considerable use in estimating the role of the spore coats in limiting access of biocides to their target sites within cells [8, 30]. This treatment does, however, remove a certain amount of spore cortex also [31, 127, 134]; the amount of cortex remaining can be further reduced by the subsequent use of lysozyme [134, 135]. The spore coats play an important part in conferring intrinsic resistance of spores to many biocides (Table 4.7) but the cortex also is an important barrier especially since UDSL-treated spores are much more sensitive to chlorine- and iodine-releasing agents than are UDS-exposed spores (Table 4.8) [136]. There is little information about the uptake of biocides by mature and modified spores, although such experiments could provide additional evidence to support the current concept of the barrier roles for the coats and cortex. Consideration should also be given to the substantial amounts of low molecular weight basic proteins, the small acid-soluble spore proteins (SASPs), which are present in the spore core [137] and which are rapidly degraded during germination [137, 138]. SASPs are essential for expression of spore resistance to ultraviolet radiation [139] and also appear to be involved in resistance to hydrogen peroxide [139]. Spores ($\alpha^- \beta^-$) which are deficient in α/β-type, but not γ-type, SASPs are much more sensitive to peroxide than are wild-type (normal) spores [139]. In wild-type spores, it is likely that DNA is saturated with α/β-type

Table 4.7.   MECHANISMS OF SPORE RESISTANCE TO BIOCIDAL AGENTS

Antibacterial agent	Spore component (s)	Comment
Alkali	Cortex	Alkali-resistant
Lysozyme	Coat(s)	Cortex is site of action
Hypochlorites, chlorine dioxide	Coat(s), cortex	UDS and UDSL spores highly sensitive
Glutaraldehyde	Coat(s), cortex (?)	UDS spores highly sensitive
Iodine	Coat(s), cortex	UDS spores highly sensitive
Hydrogen peroxide	Coat(s)	Varies with strain
Chlorhexidine	Coat(s), cortex (?)	
Ethylene oxide	Coat(s)?	Exact relationship unclear
Octanol	Cortex	Findings based on Dap⁻ mutants of
Xylene	Cortex	B.sphaericus
Ozone	Coat(s), cortex (?)	UDS spores highly sensitive

UDS,   urea + dithiothreitol + sodium lauryl sulphate (SLS)
UDSL,   urea + dithiothreitol + SLS + lysozyme

SASPs and is thus protected from free radical damage. The $\alpha^-\beta^-$ mutants do not appear to have been tested against other biocides. In view of their postulated role in protecting DNA, it is interesting to speculate that α/β-type SASPs would not be associated with spore resistance to those biocides that have little or no effect on DNA.

During germination and/or outgrowth, spores become sensitive to biocidal agents [140–143]. Some inhibitors (e.g. phenolics and parabens) prevent

Table 4.8.   PRETREATMENT OF BACTERIAL SPORES AND SUBSEQUENT RESPONSE TO HALOGEN-RELEASING AGENTS (HRAS)

| Pretreatment | HRA | |
	Chlorine	Iodine
Sodium hydroxide	NaOCl and NaDCC: increased sporidical activity Chloramine: no effect	Slight increase
UDS	Increased activity to all	Increased activity
UDS then lysozyme (UDSL)	Increased activity to all	Marked increased activity

germination [29, 30] whereas others, such as chlorhexidine and QACs [29, 30], do not affect germination but inhibit outgrowth. Several degradative changes occur during germination: these include loss of dry weight, decrease in optical density of suspensions, loss of dipicolinic acid (32) together with an increase in stainability and an increase in oxygen consumption [114, 115]. There is generally little information about the relative amounts of biocides taken up by mature, germinating and outgrowing spores, although it has been shown [130] that the uptake can vary depending upon the stage of the cycle in which cells are treated. At present, it is not known why some biocides act at the germination stage and others on outgrowth. It is possible to revive small numbers of spores treated with high concentrations of sporicidal agents [140–143] and this aspect should certainly be considered when bacteria are being evaluated for susceptibility or resistance.

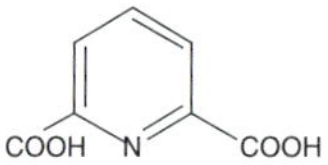

COOH   N   COOH

(32)

PHYSIOLOGICAL (PHENOTYPIC) ADAPTATION AND INTRINSIC RESISTANCE

The association of bacteria or other micro-organisms with solid surfaces leads to the generation of a biofilm, a consortium of organisms organised within an extensive exopolysaccharide exopolymer (glycocalyx) [144, 145]. Biofilms can consist of monocultures, several diverse species or mixed phenotypes of a given species [146–148]. Bacteria in different parts of a biofilm experience different nutrient environments such that their physiological properties are affected [147]. Within the depths of a biofilm, for example, nutrient limitation is likely to reduce growth rates [147]. Consequently, the phenotypes of sessile organisms within biofilms differ considerably from the planktonic cells found in laboratory cultures [148]. This is apposite when considering sensitivity of biofilm cells to antibiotics and other antibacterial agents because nutrient limitation may alter the bacterial cell surface with a resultant change in sensitivity [149]. Slow-growing bacteria are particularly insusceptible [147].

Bacteria within a biofilm may be less sensitive to antibacterial compounds than planktonic cells (reviewed in [150]). Several possible reasons (Table 4.9) can be put forward to account for this, notably reduced access of drug or biocide to cells within a biofilm, chemical reaction with the glycocalyx and modulation of the micro-environment [147]. In addition, the attached cells may produce degradative enzymes although the significance of this

Table 4.9.   BIOFILMS AND BACTERIAL RESPONSE TO BIOCIDES

Mechanism of resistance associated with biofilms	Comment
Exclusion or reduced access of biocide to under-lying bacterial cell	Depends on (i) nature of biocide (ii) binding capacity of glycocalyx towards biocide (iii) rate of growth of microcolony relative to biocide diffusion rate
Modulation of micro-environment	Associated with (i) nutrient limitation (ii) growth rate
Increased production of degradative enzymes by attached cells	Mechanism unclear at present

has yet to be fully assessed [151]. Genetic exchange is likely to occur between cells contained in a biofilm and this could complicate the issue [147]. Extra-cellular β-lactamases within the glycocalyx reinforce its action as a diffusion barrier [150]. Bacteria removed from a biofilm, however, and recultured in ordinary culture media are no more resistant than 'ordinary' planktonic cells of that species [147].

Recently [152], a novel strategy has been described for controlling biofilms through generation of a biocide (hydrogen peroxide) at the biofilm-surface interface rather than simply applied extrinsically. In this procedure, the colo-nized surface incorporated a catalyst that generated active biocide from a treatment agent.

Biofilms provide the most important example of how physiological (phe-notypic) adaptation can play a role in conferring intrinsic resistance [147]. There are, however, other examples which can be cited. Fattened cells of *S. aureus* produced by repeated subculturing in glycerol-containing media were found to be more resistant to benzylpenicillin and higher phenols [110]. Planktonic cultures grown under conditions of nutrient limitation or reduced growth rates have cells with altered sensitivity to biocides [43], prob-ably as a consequence of modifications in the outer membrane of cells [43]. These changes in sensitivity can be considered as the expression of intrinsic resistance brought about by exposure to environmental conditions.

The cell wall of staphylococci is composed essentially of peptidoglycan and teichoic acids. As pointed out earlier (p. 150) substances of higher molec-ular weight can readily cross the wall of these organisms, a property that probably accounts for their greater sensitivity than Gram-negative bacteria

Table 4.10.   ANTIBIOTIC RESISTANCE WHERE THE NORMAL TARGET SITE IS
INSUSCEPTIBLE OR ABSENT

Type of alteration	Examples of drug	Mechanisms of resistance
Target site modification	β-Lactams	Additional PBP(s), e.g. MRSA strains
	β-Lactams	Modified PBP(s) pattern, e.g. mecillinam-resistant *E. coli*
	Chloramphenicol	5OS subunit modification
	Erythromycin	Mono- or di-methylation of adenine residue in 23S rRNA
	Fusidic acid	Modified factor G involved in translocation
	4-Quinolones	Modified DNA gyrase
	Rifampicin	Modified β-subunit of RNA polymerase
	Streptomycin	Modification of S12 in S30 ribosomal subunit
	Sulphonamides	Modified target enzyme (dihydroperoate synthetase)*
	Trimethoprim	Modified target enzyme (dihydrofolate reductase)*
Absence of sensitive target enzyme	Penicillin	Penicillin-tolerant streptococci: peptidoglycan hydrolase absent. Cells thus inhibited but not lysed
Production of second, resistant enzyme**	Mupirocin	Plasmid-mediated mupirocin-resistant isoleucyl tRNA synthetase

*   Chromosomally or plasmid-mediated
** Normal resident mupirocin-sensitive chromosomal enzyme still present.
    PBP, penicillin-binding protein

to a variety of chemical agents [23, 24]. The plasticity of the cell envelope is a well-known phenomenon [146] and growth rate and nutritional limitation affect the physiological state of the cells [109] with alterations to the thickness and degree of cross-linking of peptidoglycan and hence modified sensitivity to antibacterial agents.

Vancomycin-resistant enterococci and staphylococci (see later) are important pathogens. The latter contain rather thicker cell walls than vancomycin-sensitive *S. aureus* [153] and in the absence of a fuller explanation at present, it is tempting to speculate that the cell wall in the resistant organisms could reduce uptake of this glycopeptide antibiotic.

Biofilms are important for several reasons not the least of which are biocorrosion, reduced water quality and foci for the contamination of hygienic products [154–156]. Microbial colonization is also favoured on implanted biomaterials and medical devices resulting in increased infection rates and possible recurrence of infection [150].

Table 4.11. ACQUIRED RESISTANCE MECHANISMS*

Mechanism	Examples	Mechanisms of resistance
Enzymatic detoxification or modification	AGAC antibiotics	Modification by acetyltransferases, adenylylases or phosphotransferases
	β-Lactams	Inactivation (β-lactamases)
	Chloramphenicol	Inactivation (acetyltransferases)
	Erythromycin	Esterases produce anhydroerythromycin
	Tetracyclines	Enzymatic inactivation
	Mercury compounds	Inactivation (hydrolases, lyases)
	Formaldehyde	Dehydrogenase
Enhanced efflux	Arsenate	Efflux by ATPase pump
	Cadmium	Efflux by ATPase pump
	Tetracyclines	Energy-dependent, driven by PMF
	Diamidines	Efflux in MRSA strains
	QACs	Efflux in MRSA strains
Enzymatic trapping	β-Lactams	Trapping via β-lactamase in periplasm
Impaired uptake	Fusidic acid	Plasmid-borne mechanism for decreased uptake
	Penicillins	Mutational alterations to porins
	QACs	Plasmid-encoded porin modifications
	4-Quinolones	Mutational alterations to porins
	Tetracyclines	Failure to accumulate (efflux more important)
Modification of regulation	D-Cycloserine	Mutation in chromosomal gene (cycA) involved in uptake
Pathway compensation	Trimethoprim	Overproduction of DHFR (dihydrofolate reductase)
Ribosomal protection	Tetracyclines	Transposon-specified tetracycline-resistant determinants (tetM, tetO), cytoplasmic proteins acting in conjunction with ribosomes

* See also Table 4.10 (Target site modification)

MECHANISMS OF ACQUIRED RESISTANCE

Unlike intrinsic resistance, which is usually expressed by chromosomal genes, acquired resistance arises as a consequence of mutations in chromosomal genes or by the acquisition of plasmids or transposons [6, 7, 157, 158]. Chromosomal mutations are associated with changes in the base sequence of DNA and the principles involved in conferring resistance to several antibiotics are well understood [6, 7]. Resistance determined by plasmids and transposons are well documented and the underlying basis often elucidated.

CHROMOSOMAL GENE MUTATION

Some bacteria are able to circumvent the action of an inhibitor by virtue of changes in the target site [6, 7, 159] (Tables 4.10 and 4.11). Acquired insusceptibility to 4-quinolones is associated with mutations in chromosomal genes [160–163]. In *E. coli, gyr*A or *nor*A mutations are responsible for changes in the A subunit of DNA gyrase, such that cells become insusceptible to nalidixic acid but not markedly so to the fluoroquinolones. Mutations in *gyr*B producing changes of the B subunit of gyrase account for a reduced susceptibility to nalidixic acid and fluoroquinolones [37, 164]. Mutations in the OM proteins or lipopolysaccharide (LPS) provide an adequate reason for an impermeability mechanism [37]. *nor*B mutants of *E. coli* show low level resistance to some quinolones because of a decreased OmpF porin content [6].

Changes in the nature or expression of penicillin-binding proteins (PBPs) render them insensitive to β-lactam antibiotics although they can still be involved in peptidoglycan synthesis. Laboratory mutants of *E. coli* which are not killed by the mecillinam ester (33) have been described; in these there is a reduced affinity of PBP2 for the antibiotic [165]. More serious is the occurrence of this mechanism, involving PBP2b, in clinical isolates of *Streptococcus pneumoniae* [6].

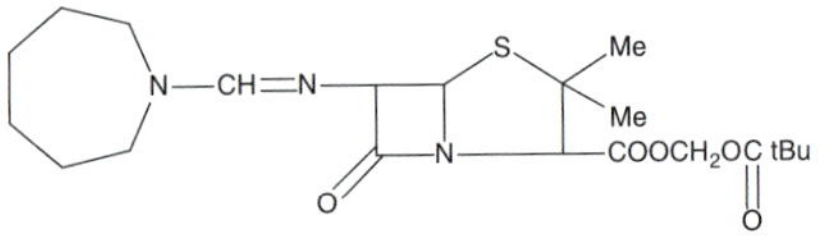

(33) 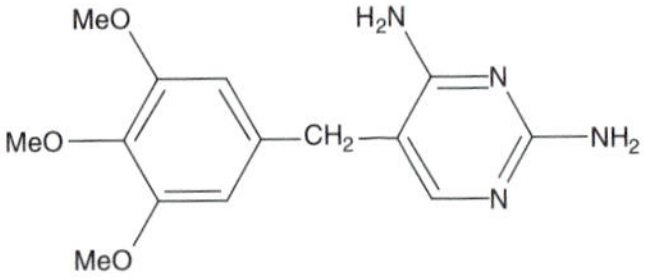   (34)

Alterations in the binding site of ribosomal protein S12, accounts for high-level insusceptibility to streptomycin in *E. coli* and *S. aureus* [166] whereas mutations in RNA polymerase are associated with decreased sensitivity to rifampicin [167]. Other chromosomal mutational examples involving insensitive target enzymes occur with modified dihydrofolate reductases and dihydroperoate synthetases [168]. Chromosomal mutations in *E. coli* cause overproduction of the target enzyme for trimethoprim (34) activity [169]. Mutation of target enzyme in *Strep. pneumoniae* provides lower sulphonamide binding [170], whereas in *S. aureus,* overproduction of 4-amino-benzoic acid overcomes the metabolic block caused by enzyme inhibition [171]. Several examples of target site changes as a consequence of mutation and antibiotic resistance are provided in Table 4.10.

Few studies have determined whether or not chromosomal gene mutation is responsible for conferring resistance to biocides. However, many years ago it was demonstrated that *Ser. marcescens,* normally inhibited by a QAC in broth at a concentration of less than 100 μg/ml , could adapt to grow in this culture medium in the presence of 100 mg/ml of the antibacterial agent [172]. It is unclear how growth was actually monitored because QACs are well known to react with the constituents of culture media to produce a dense opalescence or turbidity. Be that as it may, it was also found that re-sistant and sensitive cells exhibited different surface properties, with the for-mer possessing an increased lipid content. As found by other workers using chlorhexidine, chloroxylenol and other QACs [173, 174], resistance was un-stable being lost when the resistant cells were grown in biocide-free media.

Chromosomal mutants of *Ps. aeruginosa* produced by stepwise exposure to the antibiotic polymyxin are resistant to this drug and to EDTA [175]. The mechanism of this resistance is related to a defective self-promoted up-take pathway in which the cells contain increased amounts of a major outer membrane protein (H1) with a corresponding decrease in envelope Mg^{2+} [175, 176]. Protein H1 may replace Mg^{2+} at cross-bridging sites with LPS; these sites are normally those at which interaction occurs with Mg^{2+}, such as EDTA, or at which displacement of Mg^{2+} takes place, for example, poly-myxin [177, 178].

TRANSFERABLE RESISTANCE TO ANTIBIOTICS: BIOCHEMICAL MECHANISMS

Plasmid- or transposon-encoded resistance to antibiotics in bacteria in-volves drug inactivation, target site changes, decreased antibiotic accumula-tion as a result of impaired uptake or enhanced efflux and duplication of the target site [6] (Table 4.11). Table 4.12 provides examples of the occur-rence of transposons in bacteria and of the types of resistance encoded [6].

Antibiotic inactivation

β-Lactamases hydrolyse the cyclic amide bond in susceptible β-lactam molecules so that antibiotics are unable to bind to PBPs [179]. In Gram-negative bacteria, β-lactamases are intracellular with a periplasmic location, whereas in Gram-positive bacteria they are mainly excreted from the cell and thus are extracellular [180]. Extended-spectrum β-lactamases are plas-mid-enclosed enzymes that confer resistance on those β-lactams (e.g. cefo-

Table 4.12.   EXAMPLES OF THE OCCURRENCE OF TRANSPOSONS IN BACTERIA*

Transposons found in	Types of resistance encoded
Gram-negative bacteria	Ampicillin, kanamycin, streptomycin, trimethoprim, chloramphenicol, tetracycline, sulphonamides
Staphylococci	MLS group*, penicillin, spectinomycin, gentamicin, trimethoprim, kanamycin
Streptococci	Tetracycline
Enterococci	Glycopeptides

* Based on Russell and Chopra [6]
# MLS: macrolides, lincosamides and streptogramins

taxime, cefrazidine and aztreonam) that were designed to resist such enzyme attack [180].

β-Lactamases are heterologous in nature both in terms of their chemical structure and their substrate profile [181] (Table 4.13). The original Richmond-Sykes [182] classification of β-lactamases produced by Gram-negative bacteria envisaged five classes of enzymes. Of these, class I had a cephalo-

Table 4.13.   TYPES AND PROPERTIES OF β-LACTAMASES

Group	Preferred substrate*	Type of enzyme/source
1	Cephalosporins	Chromosomal, Gram-negative bacteria
2a	Penicillins	Gram-positive bacteria
2b	Cephalosporins, penicillins	TEM-1, TEM-2
2b'	Cephalosporins, penicillins	TEM-3, TEM-5
2c	Penicillins (including carbenicillin)	PSE-1, PSE-3, PSE-4
2d	Penicillins (including cloxacillin)	PSE-2, OXA-1
2e	Cephalosporins	*Proteus vulgaris*
3	Variable	*Bacillus cereus* II, *Ps. maltophilia* L1
4	Penicillins	*Burkholderia cepacia*

* For further details, see Bush [181]; see also [180]

sporinase-type activity and was chromosomally located; class II penicillinase activity (chromosomal); class III penicillinase and cephalosporinase
(plasmid-encoded); class IV as for class III but chromosomally-mediated;
and class V as for class III but not inhibited by cloxacillin (35). The enzymes
of classes II to V are constitutive, and class I was the most inducible.

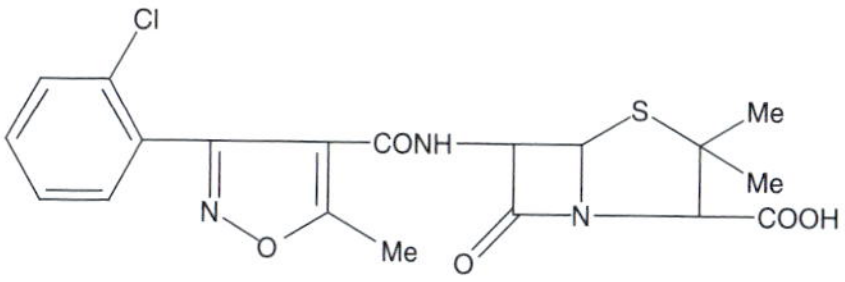

(35)

A more modern general classification scheme for bacterial β-lactamases
[181] places the enzymes into four groups: 1, active against cephalosporins
and consisting of chromosomally encoded enzymes from Gram-negative
bacteria; 2a, Gram-positive penicillinases, 2b, TEM-1 and TEM-2 enzymes,
2b' TEM-3 and TEM-5 enzymes, 2c PSE-1, PSE-3 and PSE-4 β-lactamases,
2d OXA-1 and PSE-2 enzymes, 2e cephalosporinases produced by *Proteus
vulgaris*; 3, enzymes with variable substrate activity; and 4, penicillinases as
exemplified by the enzymes produced by *Burkholderia* (formerly *Pseudomonas*) *cepacia*. The β-lactamases may be constitutive or inducible, extracellular or intracellular (q.v.) and plasmid- or chromosomally-encoded. Many,
but not all, types of β-lactamases are inhibited by clavulanic acid (36), usually at a concentration of 10 μM, but few types are inhibited by EDTA. Class
3 β-lactamases exhibit zinc-dependent activity and are sometimes referred
to as 'carbapenemases' because they hydrolyse carbapenems (37) as well as
penams (38), on which the penicillin structure is based, and cephems (39),
on which the cephalosporins are based.

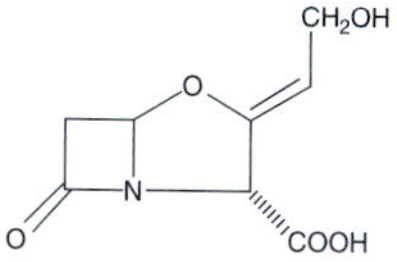

(36) 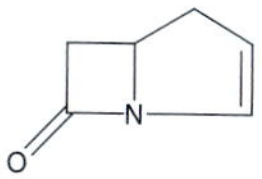   (37) 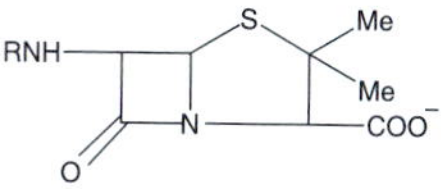   (38)

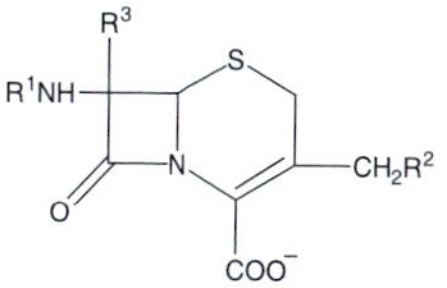

(39) 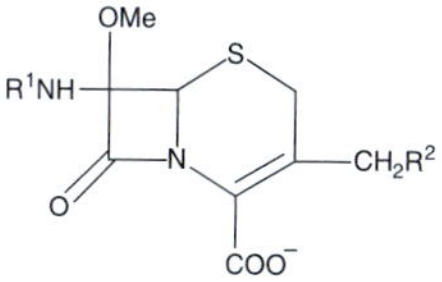   (40)

In *E.coli*, resistance to all β-lactams except cephamycins (40) and carbapenems (37) may be caused by extended-spectrum β-lactamases [183]. These enzymes are, however, sensitive to clavulanic acid and strains producing these β-lactamases are often, but not always, sensitive to β-lactam-β-lactamase inhibitor combinations [183]. In contrast, inhibitor-resistant TEM β-lactamases that are resistant to amoxycillin (41)-clavulanic acid combinations have also been described in *E. coli* [184].

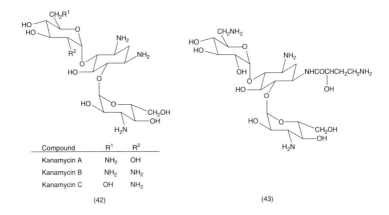

(41)

The chloramphenicol-inactivating enzymes have long been known [185]. These chloramphenicol acetyltransferases (CATs) are usually plasmid-mediated, and are inducible in Gram-positive bacteria and constitutive in Gram-negative organisms. CATs are the most frequent mechanism of chloramphenicol insusceptibility in the Enterobacteriaceae and *Haemophilus influenzae* and are also associated with insensitivity to this antibiotic in Gram-positive genera [186]. Chloramphenicol is degraded to inactive acetoxy-chloramphenicol products. There are several clinically important aminoglycoside-aminocyclitol (AGAC) antibiotics, notably gentamicin, amikacin, netilmicin, sissomicin, kanamycin, tobramycin and the older, now less widely employed, streptomycin and neomycin [6]. One mechanism of bacterial insusceptibility is the production of aminoglycoside-modifying enzymes (AMEs) which can attack different sites on AGACs [6, 187, 188]. The genes which encode AMEs are usually found on plasmids and transposons and are constitutively expressed. The enzymes are placed into three classes, *viz.*

Compound	R¹	R²
Kanamycin A	NH₂	OH
Kanamycin B	NH₂	NH₂
Kanamycin C	OH	NH₂

(42)   (43)

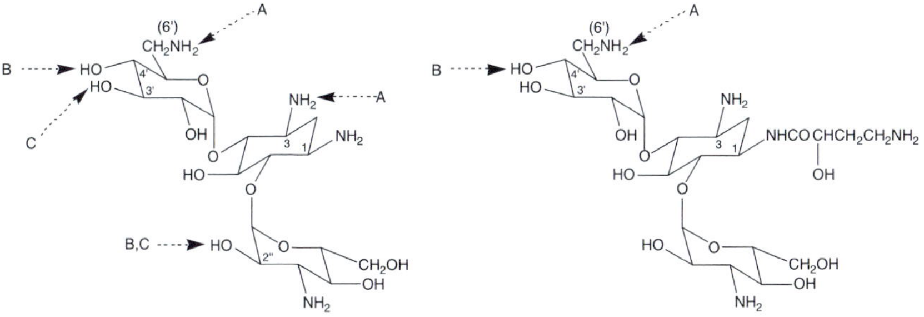

Figure 4.1. Enzymatic sites of modification of (a) kanamycin (42) and (b) amikacin (43). A, acetylation; B, adenylylation; C, phosphorylation

acetyltransferases (AACs) involving *N*-acetylation of susceptible amino groups; phosphotransferases (APHs), the phosphorylation of certain hydroxyl groups; and adenylylases (or nucleotidyltransferases, AAD) which bring about adenylylation of some hydroxyl groups. Examples of susceptible groups on kanamycin (42) and amikacin (43) are provided in Figure 4.1 [187, 188]. The emergence of high-level gentamicin insusceptibility in *Enterococcus faecalis* [189–191] as a consequence of an AME with both 6'-acetyltransferase (AAC-6') and 2"-phosphotransferase (APH-2") activities is clinically significant.

Alteration of the tetracycline molecule (7) is not normally considered as being a mechanism of bacterial insusceptibility to the tetracycline group of antibiotics [192, 193]. Nevertheless, there have been recent examples of tetracycline modification, ostensibly associated with the *tetB* determinant [194, 195].

The macrolides form an important group of clinically useful antibiotics. Older members include erythromycin (6), oleandomycin, triacetyloleandomycin and spiramycin, now joined by clarithromycin (44), roxithromycin (45) and azithromycin (46) [196]. Enterobacteria which show a high level of insusceptibility to erythromycin can inactivate the antibiotic by plasmid-encoded esterases that hydrolyse the lactone ring [197].

The lincosamides, lincomycin and clindamycin are active against Gram-positive bacteria. Plasmid-mediated inactivation from enzymatic nucleotidylation occurs in some staphylococci. Plasmid-encoded enzymes can modify streptogramin A (*O*-acetyltransferase enzyme) and streptogramin B (hydrolase enzyme involved) in *S. aureus* [198, 199]. There is no evidence that bacteria can circumvent the action of other antibiotics; for example, mupirocin is not degraded [200].

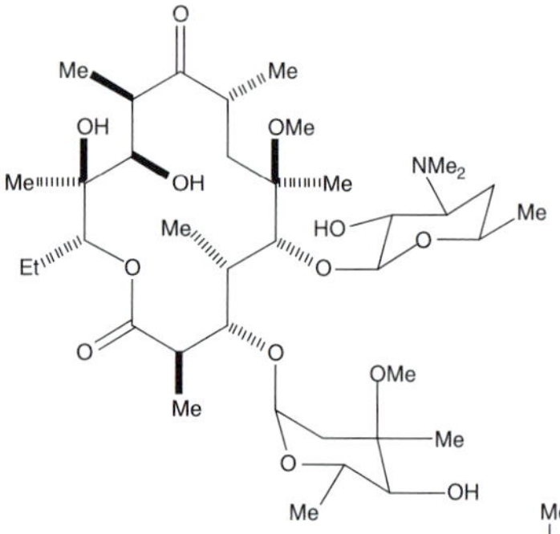

(44)

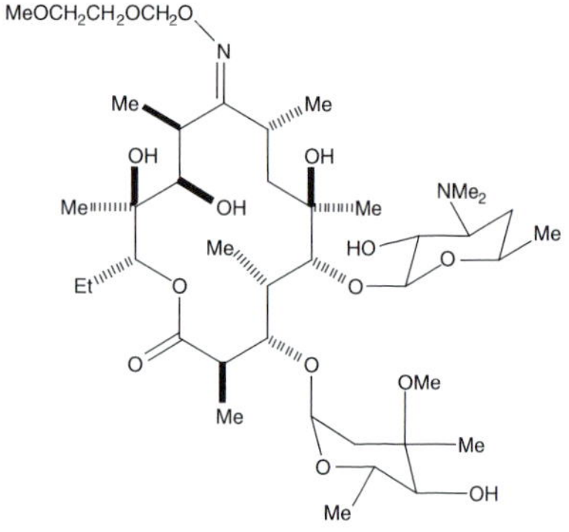

(45)

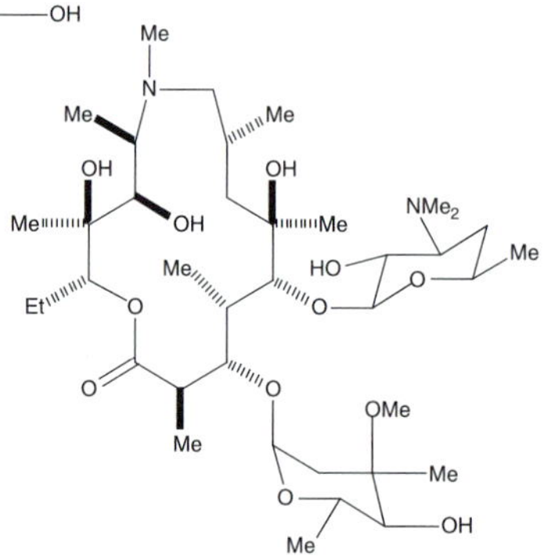

(46)

Plasmid (or transposon)-encoded enzymes are thus responsible for the degradation of several different types of antibiotics. The inactivation of several β-lactams, AGACs, 14-membered macrolides, other macrolides, lincosamides and streptogramis (MLS) and chloramphenicol is a major resistance mechanism; it has yet to be shown that inactivation of other antibiotics falls into this category.

Plasmid-mediated changes in target sites and site duplication

Some bacteria can overcome the action of an antibiotic as a result of changes in the target site [159]; this can be either by chromosomal gene mutation (q.v.) or be plasmid-encoded. High-level mupirocin (8) resistance in staphylococci is associated with the production of a plasmid-encoded, mupirocin-resistant enzyme, isoleucyl tRNA synthetase [201]. The normal, resident chromosomal enzyme is mupirocin-sensitive. Staphylococcal strains that show moderate level resistance to mupirocin do not contain the gene that encodes high-level resistance. They contain a single resistant isoleucyl tRNA synthetase which is believed to have arisen by chromosomal mutation [202].

Plasmid-mediated resistance to sulphonamides results from the duplication of dihydropteroate synthetase (DHPS). The normal DHPS remains sensitive but the plasmid-encoded DHPS, two types (I and II) of which have been found in Gram-negative bacteria, binds considerably less of these drugs [203].

Plasmid- and transposon-mediated resistance to trimethoprim involves a by-pass of the sensitive step by duplication of the chromosomally-encoded dihydrofolate reductase (DHFR) target enzyme [203]. Several trimethoprim-resistant bacterial DHFRs have been identified, resistance ensuing because of altered enzyme target sites [204]. Low-level resistance to tetracyclines arises in *E. coli* as a result of chromosomal mutations leading to loss of the outer membrane porin OmpF through which these drugs normally pass [6, 193].

Plasmid-encoded efflux mechanisms

Decreased antibiotic accumulation has often been attributed to impermeability of a bacterial cell. This is undoubtedly true with many Gram-negative bacteria, notably *Ps. aeruginosa*, and with mycobacteria [75]. It is now realized, however, that an increasing number of energy-dependent efflux systems are responsible for conferring bacterial resistance to many drugs [41, 42, 205]. Such systems are often plasmid-mediated, but chromosomally-controlled effluxing is also known, for example, in Gram-negative bacteria showing multiple antibiotic resistance (Mar: see above and later also) (Table 4.14).

Intracellular concentrations of an antibiotic or other antibacterial agent can be reduced when an organism is capable of pumping out the agent at a rate which is equal to or greater than its uptake [41, 42, 193, 194]. Efflux of

Table 4.14.   EFFLUX AS A RESISTANCE MECHANISM

Efflux system	Resistance(s) conferred
NorA	4-quinolones, chloramphenicol, cationic dyes
TetK	Tetracyclines
MsrA	Erythromycin
Mar*	Tetracycline, chloramphenicol, 4-quinolones
Mdr#	Range of compounds (depends on actual system)

* see also Table 4.18
# see also Table 4.17

tetracyclines is the predominant mechanism of insusceptibility in the Enterobacteriaceae [193, 195]. The *tet* determinants remove tetracyclines from the bacterial cell by encoding membrane-located proteins which mediate energy-dependent efflux of the antibiotics. This efflux is driven by the proton-motive force (PMF) and the end result is a reduced cellular accumulation from that seen in sensitive cells. The membrane-located 'resistance proteins' are plasmid- or transposon-encoded. Eight genetic determinants (*tet*A-*tet*F, *tet*K and *tet*L) have been described in bacteria that encode tetracycline uptake and efflux across the bacterial cytoplasmic membrane is presented in Figure 4.2. In essence, this involves an electrically-neutral proton-tetracycline antiport system with exchange of a monocationic magnesium-tetracycline chelate complex (THMg$^+$) for a proton (H$^+$) [6]. A plasmid-encoded, non-enzymatic resistance involving the gene *cml*A has been described for conferring resistance to chloramphenicol; this involves active efflux of the antibiotic mediated by the membrane-located CmlA protein [6]. Efflux of macrolides has been described in clinical isolates of *S. aureus* [206] in which the gene, *msr*A, is responsible.

Plasmid/transposon-encoded ribosomal protection factors

Ribosomal protection is a mechanism in which a cytoplasmic protein interacts with the ribosome so that the latter becomes insensitive to inhibition by tetracyclines. Several classes of ribosomal protein resistant genes have been characterized, namely *tet*M, *tet*O and *tet*Q [193, 194]. The *tet*M deter-

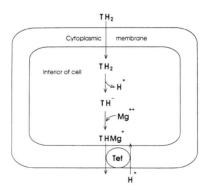

Figure 4.2. Proposed mechanism for tetracycline (TET) uptake and efflux across the bacterial cytoplasmic membrane. Mg^{++}, divalent magnesium cation; TH$_2$ and TH$^-$, protonated and deprotonated tetracycline, respectively; TH Mg$^+$, magnesium-tetracycline chelate complex. Tet proteins confer resistance to tetracycline by mediating expulsion of TH Mg$^+$ from the cell in exchange for a proton (H$^+$).
Reprinted with permission from [6].

minant, for example, is found in various bacterial genera, including *Neisseria, Enterococcus* and *Campylobacter* [193, 194]. The TetM and TetO proteins probably act in a catalytic manner so as to modify the tetracycline binding site on the 30S ribosomal subunit [193].

PLASMID-MEDIATED RESISTANCE TO BIOCIDES

Plasmid- and transposon-encoded bacterial resistance to antibiotics and to anions and cations (including mercury and silver) has been widely studied. Acquired bacterial resistance to chemotherapeutic drugs has been described above. Resistance to non-antibiotic (biocidal) agents is being increasingly investigated [23–25]. Earlier reports [207] concluded that apart from the specific examples of mercury and silver, plasmids were not responsible for the high levels of biocide resistance associated with certain species or strains. However, in recent years there have been numerous reports linking the presence of plasmids in bacteria with resistance to chlorhexidine and QACs and the matter has thus been re-examined [208].

At one time, plasmid-encoded resistance to non-chemotherapeutic antibacterials had been almost exclusively studied with inorganic and organic mercury compounds, silver compounds and other cations and some anions [209–212]. In brief, it has been shown that:

(i) 'narrow-spectrum' plasmids are responsible for resistance to Hg^{2+} (Figure 4.3a) and to the organomercurials (Figure 4.3b) merbromin and fluorescein mercuric acetate, but not to methylmercury, ethylmercury, phenylmercuric nitrate (PMN, 22) or acetate (PMA), *p*-hydroxymercuribenzoate and thiomersal;

(a)

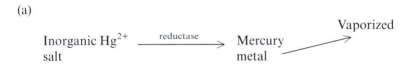

(b)

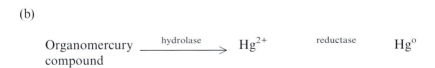

Figure 4.3. Biochemical mechanisms of resistance to (a) inorganic mercury compounds, (b) some organomercurials

(ii) 'broad-spectrum' plasmids encode resistance to all of the compounds listed in (i) by the mechanism described in Figure 4.3a,b;

(iii) plasmid-encoded resistance to Ag^+ occurs in clinical isolates of *Enterobacter* spp., *Pseudomonas* spp. and *Citrobacter* spp. by an unknown mechanism not involving efflux or silver reduction;

(iv) resistance to arsenate (AsO_4^{3-}) by plasmid-located *ars* genes involves an energy-dependent efflux of the inhibitor mediated by an ATPase transport system such that arsenate accumulated in *E. coli* by two constitutive systems, Pi transport (Pit) and phosphate specific transport (Pst), is rapidly pumped out of the cell;

(v) plasmid-mediated chromate (CrO_4^-) resistance in several *Pseudomonas* strains is expressed at the uptake rather than the efflux level, because there is no difference in efflux between sensitive and resistant cells;

(vi) cadmium (Cd^{2+}) resistance is conferred by at least four plasmid-determined systems, of which the *cad*A and *cad*B systems unique to staphylococcal plasmids have been most widely studied. Cd^{2+} is transported into cells by an energy-dependent, chromosomally-determined manganese (Mn) transport system; this is highly specific for Cd^{2+} and Mn^{2+} ions. The *cad*A gene specifies a 100-fold or so increase in Cd^{2+} resistance, the *cad*B gene a 10-fold increase detectable in *cad*A$^-$ mutants. An energy-dependent efflux of Cd^{2+} via a specific efflux ATPase that involves a Cd^{2+}-proton exchange is the *cad*A mechanism of resistance. The *cad*A and *cad*B systems can also confer resistance to other heavy metal ions, whereas the third system that confers Cd^{2+} resistance has a *cad*A-type resistance. In the fourth system, by an unknown mechanism, there is simultaneous plasmid-mediated resistance to Cd^{2+}, Zn^{2+} and Co^{2+};

(vii) in some Gram-negative bacteria such as *E. coli*, copper (Cu^{2+}) resistance may be manifested. The exact mechanism is unknown, but plasmid Rts1 encodes resistance by a decreased accumulation process.

These mechanisms are of considerable microbiological and biochemical interest, although not all of the above agents find current use as biocides. The plasmid-mediated efflux pumps are particularly important, since efflux is one means whereby acquired resistance to antibiotics occurs (see earlier) and can be a mechanism of resistance to some clinically useful biocides (see later). No efflux pump comparable to those described for arsenate and cadmium [212] has yet been detected in silver-resistant bacteria [213]; however, an up-to-date assessment of this subject is available [212].

Increased MICs of some biocides have been observed in *S. aureus* strains possessing a plasmid carrying genes encoding resistance to gentamicin (5) [214–220]. Such biocides are chlorhexidine, QACs, acridines and diamidines together with ethidium bromide which is often studied in a similar manner.

In *S. aureus*, resistance to acridines (AC^R), QACs (QA^R), propamidine isethionate (PI^R) and ethidium bromide (EB^R) is mediated by a common determinant on a group of structurally related plasmids [221–228]. Staphylococci remain the only bacteria in which the genetic aspects of plasmid-mediated biocide resistance have been studied in detail [208]. Resistance to cationic biocides is encoded by at least three separate multidrug resistance determinants; these, their gene location and the resistance phenotypes are summarized in Table 4.15. Resistance to these biocides is widespread among multiresistant strains of *S. aureus* and is specified by two families of determinants, the *qac* A/B and *qac* C/D gene families [222]. The *qac*A gene is present predominantly on the pSK1 family of multiresistance plasmids, but is likely to be present also on the chromosome of clinical *S. aureus* isolates as an integrated pSK1 family plasmid or part thereof [229]. The *qac* A/B gene family encodes proton-dependent export proteins. The *qac*A gene is detected in a large number of clinical *S. aureus* isolates but not in antiseptic-susceptible strains [223, 224]. Genes encoding QA^R may be widespread in food-associated staphylococci species [226].

Coagulase-negative staphylococci (CNS, *S. epidermidis*) contain either or both *qac*A and *qac*C genes, with some 40% of clinical isolates containing both [225]. A selective advantage may be conferred on strains possessing both genes compared with those containing only *qac*A. There is increasing evidence that *S. aureus* and CNS share a common pool of resistance determinants [225].

TABLE 4.15. *qac* GENES AND RESISTANCE TO CATIONIC BIOCIDES IN
STAPHYLOCOCCI

Multidrug resistance determinant	Gene location	Biocide resistance encoded
*qac*A gene	Predominant: pSK1 family of multiresistance plasmids Other: β-lactamase and heavy metal resistance plasmids	Intercalating agents, QACs, diamidines, biguanides
*qac*B gene	β-Lactamase and heavy metal resistance plasmids	Intercalating agents, QACs
*qac*C gene*	Small plasmids (<3 kb) or large conjugative plasmids	Ethidium bromide, some QACs
*qac*D gene*	Large (50 kb), conjugative, multiresistance plasmids	Ethidium bromide, some QACs

* Have identical phenotypes and show restriction site homology

The *ebr* gene is identical to the *qac*C/*qac*D gene family [221]. Sasatsu and his colleagues [230, 231] demonstrated that the nucleotide sequence of an amplified DNA fragment of the *ebr* gene from sensitive and resistant cells of *S. aureus* was identical and thus concluded that antiseptic-resistant cells result from an increase in copy number of a gene whose usual function is to remove toxic substances from normal sensitive cells of staphylococci (and also of enterococci). It is not always possible to demonstrate low-level resistance to cationic biocides in *S. aureus* [232–236]; this could be due to the instability of plasmids in these clinical isolates.

PATHOGENIC RESISTANT GRAM-POSITIVE BACTERIA

MYCOBACTERIA

The possible mechanisms of intrinsic resistance of mycobacteria to antibiotics and biocides were considered earlier and were shown to be associated with the mycobacterial cell wall (Table 4.6). Acquired resistance to antimycobacterial drugs is also well-known [89, 90, 237, 238] (Table 4.6). One such example is the acquired resistance of *M. tuberculosis* to isoniazid (INH,20), one of the most important antitubercular drugs [239, 240]. Despite the fact that INH has been used for many years in the treatment of tuberculosis, its precise mechanism of action remains unknown. However, it is likely that expression of the gene *kat*G is important [89, 90, 241]. This gene encodes an enzyme with catalase and peroxidase activities in *M. tuberculosis* and this enzyme converts INH into an active form. This activated INH interferes with mycolic acid biosynthesis and the INH polypeptide involved is believed to be the molecular target site [238, 241]. Deletion of *kat*G from *M. tuberculosis* results in loss of INH susceptibility, whereas transformation of the gene into INH-resistant mutants restores susceptibility [89, 90, 238].

Other important antitubercular drugs include ethambutol (24), ethionamide (47), rifampicin (10) and the 4-quinolones (e.g. 11) and acquired mycobacterial resistance to each has been described. Acquired resistance to ethambutol might result from mutation of a target enzyme involved in arabinogalactan synthesis [242] (a) to ethionamide from a single amino acid substitution in protein INHA [243], (b) to rifampicin in *M. tuberculosis* and *M. leprae* from alterations in the β-subunit of DNA-dependent RNA polymerase [244], and (c) to quinolones with point mutations in *gyr*A which encodes the A subunit of DNA gyrase [245]. All of these acquired mechanisms are of clinical significance.

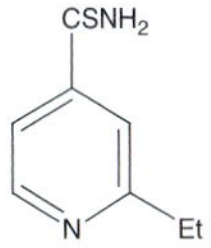

(47) 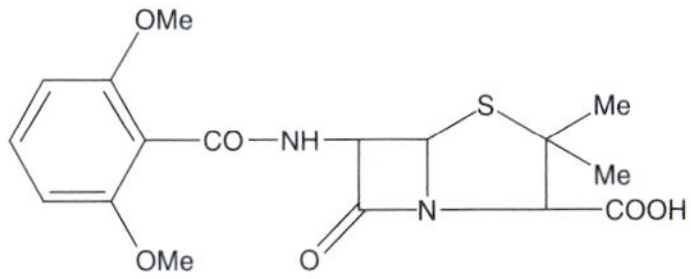   (48)

METHICILLIN-RESISTANT STAPHYLOCOCCI

Methicillin(48)-resistant *S. aureus* (MRSA) strains have been recognized as being a major cause of sepsis in hospitals in the UK and in other countries [246–248], although not all MRSA strains, are of increased virulence. 'Epidemic' MRSA (EMRSA) is a term used to denote the ease with which these strains can spread [249] and those particularly at risk are immunocompromised patients and patients with open sores.

The evolution of MRSA strains is not fully understood, but the same mechanisms of mutation and gene transfer that exist in other species provide a likely reason. The emergence of gentamicin resistance plasmids illustrates the evolutionary potential of translocatable elements [186]. MRSA strains which are also resistant to this aminoglycoside antibiotic are referred to as MGRSA. This evolutionary progression is also responsible for the formation of the β-lactamase-heavy metal resistant plasmids [250]. Some MRSA isolates are penicillin-resistant by virtue of the enzyme β-lactamase, which pre-dates the use of β-lactams [251]. However, the spread of the phenotype has probably arisen as a result of selection caused by the widespread usage of methicillin in hospitals.

Unlike *S. aureus*, *S. epidermidis* is coagulase-negative, and is found as a universal skin commensal, normally present in the resident skin flora; it is also found in the gut and in the upper respiratory tract. *S. epidermidis* is much less pathogenic than *S. aureus* and rarely causes infection in healthy people but it may be responsible for infective endocarditis in people with prosthetic heart valves and for wound infection following hip replacement surgery [252]. Methicillin-resistant *S. epidermidis* (MRSE) strains are known [253], and genetic exchange may occur with *S. aureus* [254].

Increasing numbers, world-wide, of nosocomical (hospital-acquired) infections are caused by MRSA strains [255]. Such strains particularly affect patients in intensive-care units [246]. In the UK, gentamicin resistance suddenly appeared in 1976 and MGRSA strains caused severe major hospital outbreaks. It has been proposed [186, 227, 229] that, since resistance to 'nucleic acid-binding (NAB)' compounds, such as chlorhexidine, amidines,

QACs, ethidium bromide and acridines was prevalent in the staphylococcal population long before gentamicin resistance emerged, then antiseptic resistance could be of particular significance in terms of the potential for the survival of these strains in the hospital environment. This aspect is considered later (see p. 181).

Casewell [256] discussed the evolution of MRSA and in particular of the 'Kettering strain', EMRSA-16, which has an extraordinary ability for hospital-wide spread and which is impossible to eliminate from some hospitals. Another strain, EMRSA-15, has also spread rapidly. β-Lactam resistance in *S. aureus* can be mediated by the *mec* determinant, in which resistance arises from the synthesis of a unique PBP (PBP2′) which has a low affinity for β-lactams [257].

VANCOMYCIN-RESISTANT ENTEROCOCCI

Enterococci, also known as Lancefield Group D streptococci, do not produce lysis on blood agar. *Enterococcus faecium, E. faecalis* and *E. durans*, the three most important species, are found in the human and animal gut [258, 259]. Enterococci may cause urinary and abdominal wound infection and glycopeptide-resistant enterococci are becoming a matter of increasing clinical concern [260–262]. These enterococci show resistance to the glycopeptide antibiotics vancomycin (vancomycin-resistant enterococci, VRE) and, possibly, teicoplanin. In intrinsically resistant genera and in *S. aureus*, glycopeptide resistance is expressed constitutively, as distinct from the inducible high- and low-level resistance in enterococci [262–264].

Inducible resistance to high levels of vancomycin in enterococci is mediated by transposon Tn 1546 or related transposons [262, 263]. The transposition of Tn 1546 into plasmids with a broad host range or into conjugative transposons would enable resistance to spread to *S. aureus* which can exchange genetic information with enterococci [260]. Plasmid-mediated resistance to vancomycin has been successfully conjugated, in laboratory experiments, from enterococci to *S. aureus*, provided that erythromycin (6) and not vancomycin was used as the selective agent [265]. It was only a question of time before such an event was shown to occur clinically [266, 267] and vancomycin resistance has now been found in an MRSA strain [268]. The seriousness of this cannot be over-emphasized.

The mechanisms of vancomycin resistance in enterococci are being unravelled and the following is a summary of what is currently known. Vancomycin inhibits peptidoglycan synthesis in staphylococci by binding to the terminal D-Ala-D-Ala group of the pentapeptide side-chain of peptidoglycan precursors (Table 4.16), thus preventing the transglycosylation and trans-

Table 4.16.   GLYCOPEPTIDE RESISTANCE MECHANISMS IN ENTEROCOCCI

Resistance phenotype	Features	Resistance encoded to	Mechanism [#]
VanA (acquired)*	vanH, vanA and vanX genes; inducible by vancomycin or teicoplanin, plasmid-encoded (Tn 1546 responsible)	Vancomycin (high-level), teicoplanin (high-level)	vanA dehydrogenase vanA ligase vanX dipeptidase vanY carboxypeptidase
VanB (acquired)	vanB gene; inducible by vancomycin, chromosomally-encoded	Vancomycin (moderate/high-level)	Ligase product: D-Ala-D-Lac
VanC (intrinsic)	vanC gene; constitutive? chromosomal?	Vancomycin (moderate level)	vanC ligase (D-Ala-D-Ser produced)

* See Figure 4.4 for further details
[#] D-Ala, D-Alanine; D-Ser, D-serine; D-Lac, D-lactase

peptidation reactions required for polymerization [49]. Resistance to vancomycin in enterococci results from the synthesis of a new ligase that promotes D-Ala-D-Lac (instead of D-Ala-D-Ala) with loss of vancomycin binding [269].

The sequence of events leading to the biosynthesis of D-Ala-D-Lac in highly resistant enterococci (vanA phenotype) has been described by various workers [262, 269] and summarized in Table 4.16 and Figure 4.4. Several genes are involved, of which three (vanH, vanA and vanX) are important in resistance; these are responsible, respectively, for converting pyruvate to lactate, for synthesizing the depsipeptide D-Ala-D-Lac and for hydrolyzing D-Ala-D-Ala, thereby eliminating normal peptidoglycan precursors. High-level resistance is induced by both vancomycin and teicoplanin and is plasmid-mediated. Moderate- to high-level vancomycin resistance (vanB phenotype) also involves the production of D-Ala-D-Lac but is inducible by vancomycin but not teicoplanin and is chromosomally-mediated. Low level resistance to vancomycin (vanC phenotype) involves the vanC gene, which produces D-Ala-D-Ser, and is believed to be constitutive and chromosomally controlled. There does not appear to be any correlation between glycopeptide resistance in enterococci and biocide resistance [270–272]. This aspect will also be considered later in greater detail (p. 180).

(1)   VANCOMYCIN-SENSITIVE ENTEROCOCCI
      Peptidoglycan linear strand based on
      GlcNAc - MurNAc - AGLAA
      Binding of Vancomycin to terminal D-alanyl-D-alanine (AA)

(2)   HIGH LEVEL INDUCIBLE RESISTANCE
            Tn 1546   (vanA type)

      ENCODES
                  ● VanH Dehydrogenase
                    PYRUVATE → LACTATE
Required
for glyco-   ● VanA Ligase (Synthetase)
peptide   SYNTHESIS of *DEPSIPEPTIDE*
resistance   D-ALA-D-LAC
                    REPLACES D-ALA-D-ALA IN PEPTIDOGLYCAN

                  ● VanX Dipeptidase
                    HYDROLYSES D-ALA-D-ALA   ⎤ Eliminate D-ALA-D-ALA
                  ● VanY Carboxypeptidase   ⎦ Precursors

      i.e.   UDP-MurNAc-AGLAA
                              ↗   ↖
                  Dipeptidase   Carboxypeptidase

      Binding of Vancomycin to peptidoglycan no longer possible

Figure 4.4. Glycopeptide resistance in enterococci

MECHANISMS OF MULTIDRUG RESISTANCE IN GRAM-NEGATIVE BACTERIA

Multidrug resistance (Mdr) is a serious problem in enteric and other Gram-negative bacteria [41, 71, 273–283] (Table 4.17). As distinct from plasmid-mediated resistance, described above, Mdr is a term employed to describe a resistance mechanism by genes that comprise part of the normal cell genome. These genes are activated by induction or mutation caused by some types of stress, considered below. Because the genes are ubiquitously distributed, there is no need for genetic transfer.

Exposure to a single drug leads to resistance not only to that drug but also

Table 4.17.   MULTIDRUG RESISTANCE (MDR) SYSTEMS IN GRAM-NEGATIVE
BACTERIA*

Mdr system	Mechanism
Mar (multiple antibiotic resistance)	See Table 4.18
Acr	Mutations at *acr* locus confer hypersensitivity
robA	Overexpression of RobA protein
soxRS	Induced in response to oxidative stress
emr	Transporter system, in *E. coli*, induced by several drugs

* see text for additional comments

to chemically unrelated drugs [273, 274]. For example, Levy and his colleagues [273] demonstrated in 1989 that chromosomal multiple antibiotic resistant (Mar) mutants of *E. coli* selected on agar containing low concentrations of tetracycline or chloramphenicol were much less sensitive to fluoroquinolones than wild-type *E. coli* and that the frequency of emergence was at least 1000-fold higher than with norfloxacin selection. Selection by tetracycline or chloramphenicol produced a relatively high frequency of about 10^{-7} and by norfloxacin of 10^{-9} to 10^{-11} [273]. Resistance induced by the fluoroquinolone was usually attributed to mutations in the *gyrA* gene or to alterations in outer membrane porin OmpF, whereas *marA*-dependent fluoroquinolone resistance was at that time associated with decreased cell permeability. The Mar system is summarized in Table 4.18.

This Mdr regulatory chromosomal locus is widespread among enteric bacteria [71]. Several bacterial isolates show multidrug resistance by chromosomal mechanisms, e.g. mycobacteria (q.v.), *Neisseria gonorrhoeae* and pneumococci. *E. coli* possesses a unique locus (*mar*) that regulates susceptibility to multiple antibiotics [71]; conversely, deletion of the *mar* locus leads to increased susceptibility to these drugs [274]. Mechanisms of resistance do not involve drug modification or inactivation but are demonstrable as decreased uptake, increased efflux or both [41]. Mar-related chloramphenicol resistance, for example, does not inolve inactivation by chloramphenicol acetyltransferase but instead an efflux of the antibiotic in Mar mutants [281].

The *mar* region has been cloned and sequenced; this reveals an operon composed of three putative structural genes, *marR*, *marA* and *marB* [277]. The mar regulon of *E.coli* is a unique antibiotic resistance mechanism in

Table 4.18:   MULTIPLE ANTIBIOTIC RESISTANCE (MAR) IN GRAM-NEGATIVE
BACTERIA

Property	Mar in Gram-negative bacteria
Plasmid-containing genes	Not applicable
Location of resistance	Chromosomal genes comprising part of normal genomes of bacterial cells
Activation of Mar genes	Induction or mutation caused by stress (exposure to xenobiotics*) in natural or clinical environments
Transfer of genes	Unnecessary
Mechanism or resistance (1) Antibiotic inactivation (2) Efflux	Does not occur Occurs by membrane transporters
Types of Mar efflux systems	(a) Single gene-encoded multi-drug exporters (b) Operons and regulons encoding repressors and transcriptional activators
Consequences of Mar	Possible failure of chemotherapy

* Xenobiotic ... foreign inhibitor

that it is inducible not only by the antibiotics (tetracycline, chloramphenicol) listed above but also by weak aromatic acids such as salicylate and acetylsalicylate [273, 76]. The *marRAB* operon, one of two operons (the other is *marC*) in the *mar* locus that are transcribed from a central regulatory region, *marO*, controls intrinsic resistance or susceptibility to multiple antibiotics [276, 277] such as tetracyclines, chloramphenicol and fluoroquinolones. Both operons contribute to the Mar phenotype [277].

Resistance of *E. coli* and other Enterobacteriaceae to a number of chemically unrelated antibiotics can be induced by derepression of the *marRAB* operon. This operon is negatively autoregulated (controlled) by the MarR protein, which binds to the *marRAB* promoter region [282]. The MarR repression is wild-type *E. coli* can be abolished (derepressed) by tetracycline, chloramphenicol and salicylate as noted above. The MarA protein controls a set of genes (*mar* and *soxRS* regulons) thereby providing the bacteria with resistance not only to several antibiotics but also to superoxide-generating reagents [282]. Miller *et al.* [283] had earlier described a new mutant, *marC*, of *E. coli* isolated on the basis of a Mar phenotype which mapped to the soxRS locus that encoded the regulators of the superoxide stress response.

This mutant, given the designation *soxR201*, was used to study the relationship between *mar* and *sox* loci in conferring antibiotic resistance. It was found that full antibiotic resistance resulting from soxR201 mutation was partially dependent on an intact *mar* locus. The *soxRS* regulon in *E. coli* is induced in response to oxidative stress; it consists of two divergently arranged genes encoding two proteins, SoxR and SoxS [282, 283].

The inducible *mar* locus specifies a novel antibacterial mechanism that could well play an important role in the clinical emergence of fluoroquinolone-resistant *E. coli* isolates [274]. The *mar* regulon is one of several Mdr systems present in members of the Enterobacteriaceae [71]. Other Mdr systems include:

(i) the Acr system, wherein mutations at an *acr* locus render *E. coli* more sensitive to hydrophobic antibiotics, dyes and detergents;

(ii) the *robA* gene in which overexpression of the RobA protein confers multiple antibiotic and heavy metal resistance (interestingly, Ag^+ may be effluxed [284, 289];

(iii) the *soxRS* regulon described above;

(iv) the *E. coli* multidrug resistance (emr) locus in *E. coli* that specifies a transporter system which is induced by several drugs.

Mdr systems are also known in other Gram-negative bacteria. For example, *Ps. aeruginosa* is known to possess more than one efflux mechanism [41, 42]. These efflux systems complement the role of the outer membrane in conferring intrinsic resistance on this organism [42, 285]. Overall, Mdr is widespread among Gram-negative bacteria and poses a major chemotherapeutic problem particularly when such organisms can acquire R plasmids to, as it were, supplement their drug resistance [71]. The major mechanism of resistance is efflux although this may be accompanied by reduced drug uptake [41]. For instance, tetracycline accumulation in high-level Mar mutants is greatly reduced and an energy-dependent tetracycline efflux system can be demonstrated in whole cells and in everted membrane vesicles prepared by lysis of *E. coli* cells [41, 42].

The resistance of *N. gonorrhoeae* to hydrophobic antibiotics and to other hydrophobic agents such as detergent-like fatty acids is determined by the multiple transferable resistance (*mtr*) system [286]. This is mentioned here because the *mtr* system consists of the *mtr* gene and three tandemly linked genes, *mtrCDE*. The latter encode proteins that are analogous to a family of bacterial efflux/transport proteins such as the AcrAE and EnvCD proteins of *E. coli* mediating resistance to drugs and other compounds.

LINKED ANTIBIOTIC-BIOCIDE RESISTANCE

Various mechanisms of bacterial resistance to antibiotics and biocides have been discussed. The question as to whether there is a link between resistance to chemotherapeutic drugs on the one hand and non-antibiotics on the other remains to be addressed.

Antibiotics usually have a specific site or mode of action whereby they achieve a selective toxic effect against bacteria but not human host cells. In contrast, biocides frequently have multiple target sites in the bacterial cell and by their very nature are often toxic not only to bacteria and other micro-organisms but also to host cells [7, 23, 24]. Thus, mutation at, or absence of, a normal target site (or the presence of an additional target site) may be responsible for producing resistance to antibiotics but not to biocides.

Mechanisms of transport of antibiotics and biocides into a bacterial cell may also be different (Table 4.4). However, the cell envelope in Gram-negative bacteria often presents a barrier not only to hydrophobic and high molecular weight hydrophilic antibiotics but also to several biocides (Table 4.5). Impermeability mechanisms are also associated with the reduced biocide susceptibility of mycobacteria and bacterial spores [30, 44, 95], and a mucoexopolysaccharide (glycocalyx) with reduced antibiotic or biocide diffusion into Gram-positive bacteria (Table 4.9). Intrinsic resistance to both antibiotics and biocides may thus occur in both Gram-positive and -negative bacteria. This can, however, only be considered as being indirect rather than direct since drug/biocide molecules may be prevented from reaching a target site(s) solely as a result of the nature and composition of the outer cell layers. Efflux mechanisms must also be considered, however, as described below.

An association between antibiotic resistance and chlorhexidine and QAC resistance in *Providencia stuartii* and *Proteus* has been observed, but no evidence of a plasmid link obtained [25, 73, 287, 288]. Chlorhexidine hypersensitivity has been noted in ciprofloxacin-resistant variants of *Ps. aeruginosa* [289] and vancomycin- and gentamicin-resistant strains of *E. faecium* retained sensitivity to the *bis*biguanide [289, 290] and to other biocides [270–272]. Anderson *et al.* [272] studied the inactivation kinetics of VRE and vancomycin-sensitive enterococci (VSE) exposed to environmental disinfectants at concentrations well below (extended dilutions) the recommended use-dilutions and found no differences in susceptibility of VRE and VSE. This type of approach is much more relevant than the widespread usage of MICs to measure responses to biocides.

Streptococcus mutans has retained its chlorhexidine sensitivity [291] and although cationic biocide-induced resistance has been demonstrated by a 'training' method no attempts were made in these investigations to determine antibiotic resistance [292, 293]. Antibiotic resistance strains of Enterobacteriaceae were not more resistant to an amphoteric surfactant [294]. However, antibiotic-resistant strains of psychotrophic Gram-negative bacteria were more resistant to a QAC and to an anionic surfactant [295].

Numerous reports have linked the presence of plasmids conferring antibiotic resistance with resistance to cationic biocides, including chlorhexidine, QACs, diamidines, acridines and ethidium bromide. These reports have applied to MRSA strains [214–231, 296] with increased minimal inhibitory concentrations (MICs) of biocides being observed in MRSA strains encoding resistance to gentamicin (MGRSA strains). Although MRSA or MGRSA strains have not always been found to show greater biocide resistance [232–236], it is now clear that this could result from the inherent instability of the gentamicin-nucleic acid-binding (GNAB) plasmid and curing of this plasmid resulted in a decreased MIC of chlorhexidine. There is then a GNAB-related association with chlorhexidine in MIC terms although in the bactericidal context, MRSA strains were killed as readily as methicillin-sensitive (MSSA) strains [297]. MRSA strains with low-level chlorhexidine resistance are sensitive to triclosan but MRSA strains which express low level triclosan (49) resistance, but with no cross-resistance to NAB compounds, have been isolated from patients treated with nasal mupirocin and daily triclosan baths [298]. Furthermore, triclosan resistance could be co-transferred with mupirocin resistance to sensitive *S. aureus* recipients [298]. The mechanism of this triclosan resistance is unknown and there is no evidence to link it with the efflux mechanisms proposed for NAB compounds [298]. Multidrug resistance to antiseptics in coagulase-negative staphylococci has also been demonstrated [225].

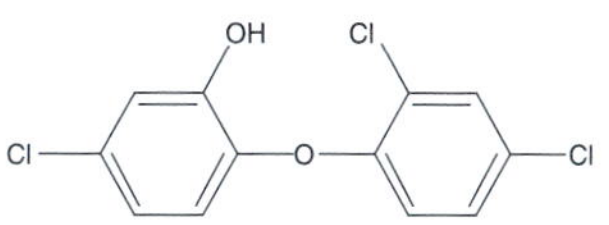

(49)

Inorganic mercury (Hg^{2+}) and organomercury resistance is a common property of strains of *S. aureus* which carry penicillinase plasmids [186, 204–212]. In Gram-negative bacteria, plasmids are known which carry genes specifying resistance to antibiotics and in some cases to cobalt, nickel, cadmium and arsenate [7]. Of the compounds listed here, however, only the organomercurials are of medical or pharmaceutical significance as biocidal agents.

Plasmid R124 alters the cell surface of *E. coli* cells such that they show enhanced resistance to the QAC, cetrimide, and other agents [299]. Generally, plasmids do not promote resistance in Gram-negative bacteria to biocidal agents [300], although hospital isolates may be more resistant to biocides than laboratory strains [301]. It is to be wondered whether an Mdr system is associated with this resistance.

Plasmid-mediated resistance to QACs and chlorhexidine in *S. aureus* has been cloned in *E. coli* [302] but the level of resistance is low and the mechanism not fully elucidated. The efflux-mediated antiseptic resistance gene *qac*A from *S. aureus* has a common ancestry with tetracycline- and sugar-transported proteins [227–229].

The mdr schemes in Gram-negative bacteria were described earlier (q.v.). For example, the Acr system in *E. coli* has been known for over 30 years [71] but has recently been considered in more detail. Mutations at an *acr* locus render *E. coli* more sensitive to hydrophobic antibiotics and to detergents and basic dyes [71]. In *N. gonorrhoeae*, the *mtr* (multiple transferable resistance) system determines levels of resistance to hydrophobic agents [286]; this system consists of the *mtrR* gene and three tandemly linked genes *mtrCDE*. The *mtrCDE*-encoded proteins were analogous to a family of bacterial efflux/transport proteins (a) encoded by the *mexAB* (multiple efflux) gene in *Ps. aeruginosa* [285] and (b) AcrA and EnvC in *E. coli* [303]. In all of these systems, efflux is involved (the *acrA* mutation referred to above inactivates a multidrug efflux complex [71]). Another transporter, EmrB, pumps out phenylmercuric nitrate and nalidixic acid [304].

Clearly, efflux is a major mechanism of multidrug resistance and several authors [41, 42, 71] are now reconsidering reduced drug accumulation in terms of efflux rather than as cellular impermability. It is, however, pertinent to query whether efflux of biocides is of relevance in relation to the concentrations employed in practice and whether biocides are responsible for selecting antibiotic-resistant strains as has been suggested. Clearly, much future experimentation is needed to establish a direct link between antibiotic and biocide resistances.

OVERCOMING BACTERIAL RESISTANCE

It is clear from the preceding sections that bacterial resistance to antibacterial agents, especially antibiotics, is of mounting concern. Increasingly, therefore, attention is being devoted to ways of overcoming this problem. This section will thus consider possible means of counteracting bacterial resistance.

ANTIBIOTICS

Several mechanisms of antibiotic resistance have been described, but there is no doubt that drug efflux, which may be plasmid-encoded or chromoso-mally-mediated, is now regarded as a major mechanism [41, 42, 71, 205, 305–311]. Thus, the possible design of inhibitors that are not extruded would be of considerable benefit [312, 313]. It must, of course, be remembered that one single efflux system may be responsible for multiple resistance [41, 42, 71, 314] and multiple drug-resistant strains of many species remain a major problem [237, 314]. An earlier finding [315] that a thiatetracycline, thia-cycline (50, a tetracycline with a sulphur atom at 6) was active against plas-mid-containing tetracycline-resistant bacteria was interesting and was probably due to inability of the cells to remove accumulated drug. Unfortu-nately, the thiatetracyclines are too toxic for clinical use. Much more re-cently, a newer generation of tetracycline analogues, the glycylcyclines, has been developed [316]. These are tetracyclines which are substituted with a di-methylglycylamido side-chain at C9. Two examples are N,N-dimethylglycyl-amido-6-demethyl-6-deoxytetracycline (51) and N,N-dimethylglycylamido-minocycline (52). These drugs are not recognised by the eight tetracycline efflux proteins [316].

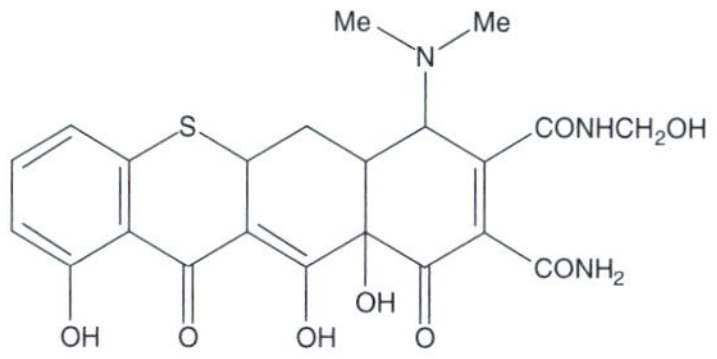

(50)

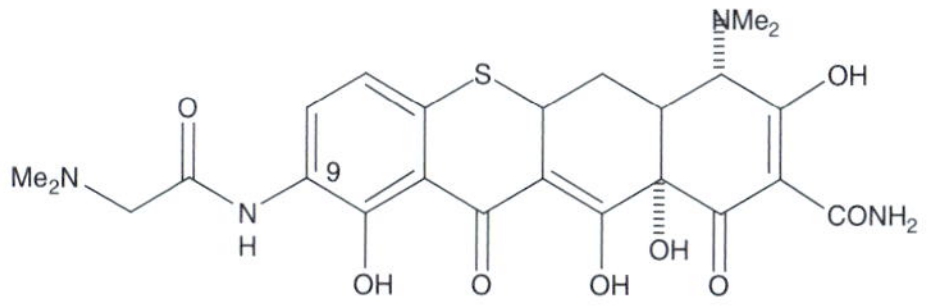

(51)

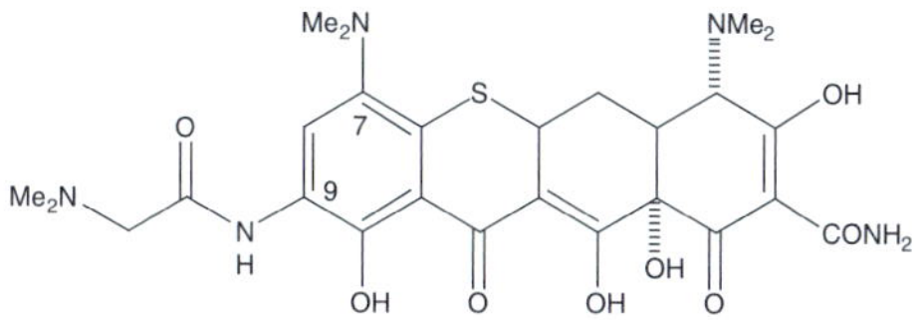

(52)

Enzymatic inactivation or modification of antibiotics has been discussed by many authors [179–182, 186–188]. As described earlier, β-lactams may be susceptible to β-lactamases. During the past 30-odd years, several β-lactams have been synthesized that are less susceptible to these enzymes. Such drugs include (i) newer types of β-lactam structures, e.g. carbapenems (37), cephamycins (40) and carbacephems (53), and (ii) modifications of the side-chains of existing penams (38) or cephems (39) [317]. Nevertheless, the wide diversity of β-lactamases [180, 181, 318] means that organisms producing enzymes with broad-spectrum activity may be able to resist some members of the β-lactam group. β-lactamase inhibitors such as clavulanic acid (36), and the penicillanic acid sulphone (54) derivatives, tazobactam (55) and sulbactam (56) have been combined with, and protect, appropriate β-lactamase-susceptible penicillins, with useful clinical results. Most extended-spectrum β-lactamases are susceptible to these inhibitors, but newer β-lactamase inhibitors may still be needed.

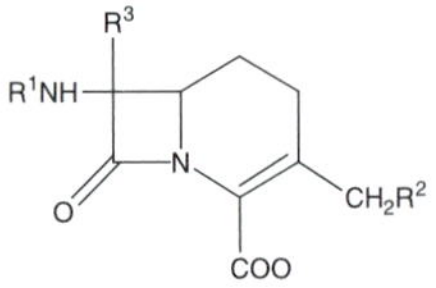

(53) 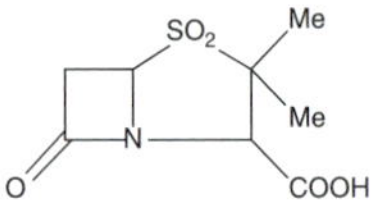   (54)

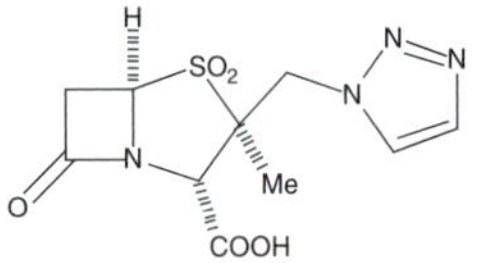

(55) 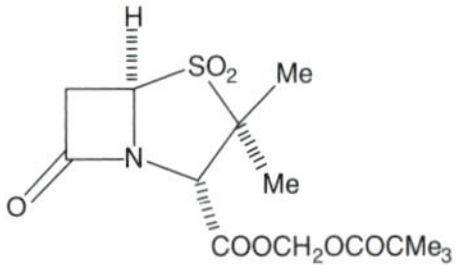   (56)

Chloramphenicol (9) is liable to breakdown by chloramphenicol acetyltransferases [185]. Fluoro derivatives (57, 58) resist enzymatic attack but little has been heard of these, apparently because of their toxicity [319]. Aminoglycoside antibiotics (AGACs) may be chemically modified by AMEs. Some derivatives (e.g. amikacin, 43) are more recalcitrant than others, e.g. kanamycin (42) (see Figure 4.2). Other enzyme-resistant AGACs of low toxicity are needed.

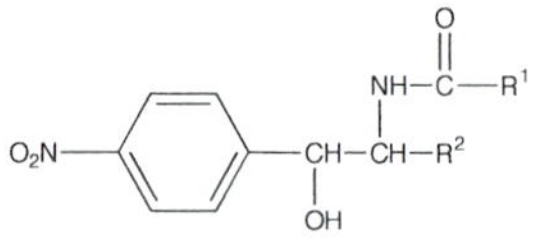

(57) R^1 = CHF_2 , R^2 = OH
(58) R^1 = OH, R^2 = CH_2F

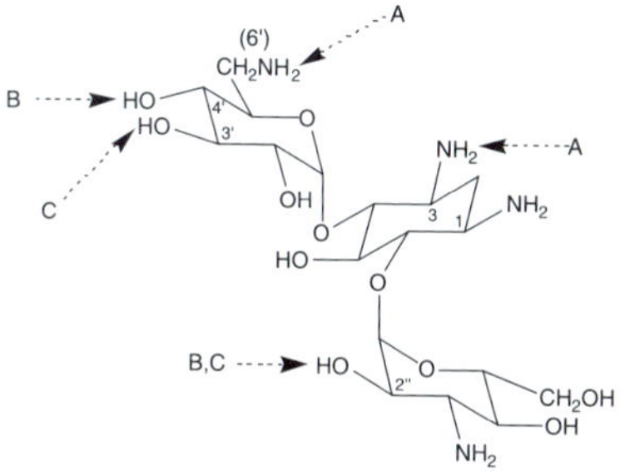

(a) Kanamycin A

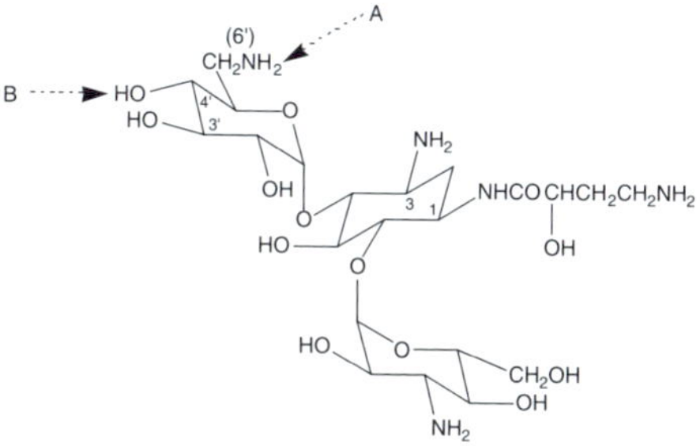

(b) Amikacin

Vancomycin-resistant enterococci (VRE) are of increasing concern. As pointed out previously, the VanA phenotype encompasses resistance to both vancomycin and teicoplanin, the VanB to vancomycin (moderate/high-level) and the VanC to low-level vancomycin (Table 4.16). Strains are thus not always resistant to teicoplanin and this might be used as the starting-point in the development of new glycopeptides. However, as shown above, strains with the VanA phenotype – the most important – are resistant to this antibiotic also. An investigational glycopeptide (LY333328) is bactericidal towards *vanA* VRE, its activity being dose-dependent; at its MIC, this compound had no lethal activity, whereas at a concentration of 10xMIC it produced a 3–4 log inactivation and at 100xMIC about 5 log reduction [320]. It is too early to predict the clinical efficacy of this compound, but it looks promising even if such high concentrations are unlikely to be achieved in practice.

Multidrug-resistant *M. tuberculosis* (MDRTB) strains have become a problem. As pointed out previously, the mycobacterial cell wall presents a major barrier to the entry of chemotherapeutic drugs. Possible solutions are (i) to design drugs of increasing hydrophobicity that will pass through the cell wall, as with isoniazid derivatives [89, 90], (ii) to use a second antibiotic in combination with a known inhibitor of arabinogalactan or mycolic acid biosynthesis [238].

BIOCIDES

There is not, perhaps, the same urgency to develop new biocidal agents, although resistance to biocides can take many forms (Table 4.19). Nevertheless, more active, less toxic compounds would always provide welcome additions to the current, rather limited range. This is particularly true for new agents with mycobactericidal activity and the replacement of glutaralde-

Table 4.19.   SUMMARY OF POSSIBLE MECHANISMS OF BIOCIDE RESISTANCE IN NON-SPORULATING BACTERIA

Biocide(s)	Intrinsic resistance		Acquired resistance		
	Impermeability	Enzyme inactivation	Efflux	Plasmid-encoded	Mutation
Acridines	OM?		Mdr	Efflux in MRSA	Mechanism unknown
Alcohols					Altered phospholipids in inner membrane
Anions					
Arsenate				Efflux by ATPase pump (ars)	
Chromate				Reduced uptake	
Chlorhexidine	OM, GCX	Possible in some bacteria		Efflux in MRSA?	Prov. stuartii: perplasmic protective proteins?
Chlorine	SL				
Chloroxylenol	OM				Mechanism unknown
Crystal violet	OM		Mdr	Efflux in MRSA	
Diamidines	OM			Efflux in MRSA	
Ethidium bromide	OM?		Mdr	Efflux in MRSA	
EDTA	Proteus? Providencia?				Altered HI OM protein in Ps. aeruginosa
Formaldehyde		Dehydrogenase		Dehydrogenase	
Hexachlorophane	OM				Mechanism unknown
Iodine	GCX				
Metals					
Cadmium				cadA system (S. aureus): efflux by ATPase pump	Cd^{2+} not accumulated
Mercury (inorganic)				Reduction to Hg^0 and vaporization	
Mercury (organic)			Mdr	Hydrolase then reductase	
Silver				Decreased uptake	
Parabens	OM	(+)			
Phenols	OM	(+)			
QACs	OM		Mdr(?)	Efflux in MRSA	Altered lipid content
Triclosan	OM			Mechanism unknown	

[a]OM, outer membrane of Gram-negative bacteria; GCX; glycocalyx; SL, slime layer; MRSA, methicillin-resistant S. aureus strains; Mdr, multidrug resistance
Where no comment is made, not described to date
(+), not conclusively established as a basis for resistance
See also Table 4.6 (resistance mechanisms in mycobacteria)

hyde, because of its toxicity, in endoscopy units is necessary. Moreover, the isolation of glutaraldehyde-resistant *M. chelonae* strains supports this view. The mycobactericidal activity of the dialdehyde itself is claimed to be enhanced when used with α, β-unsaturated and aromatic aldehydes [321], although the mechanism is unknown. Biocides with increased activity against Gram-negative organisms, notably *Ps. aeruginosa, Proteus* spp. and *Providencia* spp. need to be developed, as do agents with increased sporicidal activity.

Appropriate permeabilizers are probably the best way of potentiating activity against all of these organisms listed above, although biocide combinations should also be considered where relevant [319].

Finally, the question of staphylococci bearing *qac* genes needs to be assessed. The possession of such genes has been claimed to give these organisms the ability to survive in inimical conditions. It is still unclear, however, how important such genes are in the context of concentrations of QACs, chlorhexidine and other biocides used in clinical practice. There is an urgent need to provide information about the bactericidal, as opposed to bacteriostatic, activity [298, 322] of such compounds on MRSA and MRSE strains, as well as on Gram-negative bacteria possessing Mdr, where multidrug efflux exporters again provide the resistance mechanism. If there is a clinical problem with resistance to biocides being linked to antibiotic resistance then it will be necessary to devise new molecules to counteract efflux. In addition, in the hospital environment attention must be paid to suitable disinfection policies [323] and to the usage of skin antiseptics [324]. It is interesting to note that a recent review dealing with the reasons for the emergence of antibiotic resistance does not list biocide selection of resistant organisms as being a cause [325].

CONCLUDING REMARKS

Bacterial resistance to antibiotics and biocides is essentially of two types, intrinsic and acquired. Whilst the latter is of greater significance clinically with antibiotics, specific examples of intrinsic resistance to both antibiotics, e.g. mycobacteria, and biocides (e.g. mycobacteria, Gram-negative bacteria, spores) are also of importance.

Within these two broad types of resistance, several biochemical mechanisms are known, including reduced uptake, enzymatic inactivation, target site modification and enhanced efflux. The last mentioned is assuming greater importance as additional studies are undertaken. A direct link between antibiotic and biocide resistance remains to be established.

Antibiotic resistance is a major hospital problem with several types of bacteria, notably MRSA, VRE and MDRTB (with the possibility of VRSA becoming significant, also). Improved knowledge of resistance mechanisms is essential in the continuing fight to develop new ways of circumventing bacterial resistance [313, 326]

REFERENCES

1   Russell, A.D. (1998) in Pharmaceutical Microbiology (Hugo, W.B. and Russell, A.D., eds.) 6th Edn, pp. 91–129. Blackwell Science, Oxford.
2   Scott, E.M. and Gorman, S.P. in Ref. 1, pp. 201–228.
3   Favero, M.S. and Bond, W.W. (1991) in Manual of Clinical Microbiology (Balows, A., Hausler, W.J., Hermann, K.L. Isenberg, H.D. and Shadomy, H.J., eds.) 5th Edn, pp. 183–200 American Society for Microbiology, Washington, D.C.
4   Beveridge, E.G. in Ref. 1, pp. 355–373.
5   Beveridge, E.G. (1998) in Principles and Practice of Disinfection, Preservation and Sterilization (Russell, A.D., Hugo, W.B. and Ayliffe, G.A.J., eds.) 3rd Edn, (in press) Blackwell Science, Oxford.
6   Russell, A.D. and Chopra, I. (1996) Understanding Antibacterial Action and Resistance, pp. 172–206. Ellis Horwood, Chichester.
7   Russell, A.D. and Chopra, I. in Ref. 6, pp. 207–242.
8   Russell, A.D. and Chopra, I. in Ref. 6, pp. 243–256.
9   Williams, J.D. (1991) in Topley and Wilson's Principles of Bacteriology, Virology and Immunology (Linton, A.H. and Dick, H.M., eds.) 8th Edn, Vol. 1, pp. 105–51, Edward Arnold, London.
10 Cox, R.A., Conquest, C., Mallaghan, C. and Marples, R.R. (1995) J. Hosp. Infect. 29, 87–106.
11 Eltringham, I. (1997) J. Hosp. Infect. 35, 1–8.
12 Woodford, N., Johnson, A.P. and George, R.C. (1991) J. Antimicrob. Chemother. 28, 483–486.
13 Gray, J.W. and Pedler, S.J. (1992) J. Hosp. Infect. 21, 1–14.
14 Musser, J.M. (1995) Clin. Microbiol. Rev. 8, 496–514.
15 Cohn, D.L. (1995) J. Hosp. Infect. 30, Suppl. 322–328.
16 Bone, R.C. (1993) Clin. Microbiol. Rev. 6, 57–68.
17 Skolnick, A. (1991) J. Am. Med. Assoc. 265, 14–16.
18 Meyer, K.S., Urban, C., Eagan, J.A., Berger, B.J. and Rahal, J.J. (1993) Ann. Intern. Med. 119, 353–358.
19 Centers for Disease Control and Prevention (1993) J. Am. Med. Assoc. 271, 421–422.
20 McGowen, J.E., Jr. and Metchcock, B.G. (1995) J. Hosp. Infect. 30, Suppl. 472–482.
21 Spencer, R.C. (1995) J. Hosp. Infect. 30, Suppl. 453–464.
22 Sanders, W.E. and Sanders, C.C. (1997) Clin. Microbiol. Rev. 10, 220–241.
23 Russell, A.D. (1994) in Chemical Germicides in Health Care (Rutala, W.A., ed.) pp. 255–269, Polyscience Publications Inc., Morin Heights, Canada.
24 Russell, A.D. (1995) Int. Biodeterior. Biodeg. 36, 247–265.
25 Stickler, D.J. and King, J.B. in Ref. 5, (in press).
26 Gillespie, M.T. and Skurray, R.A. (1986) Microbiol. Sci. 3, 53–58.

27 Townsend, D.E., Ashdown, N., Momoh, M. and Grubb, W.B. (1985) J. Antimicrob. Chemother. 15, 417–434.
28 Tennent, J.M., Lyon, B.R., Midgley, M., Jones, I.G., Purewal, A.S. and Skurray, R.A. (1989) J. Gen. Microbiol. 135, 1–10.
29 Russell, A.D., Dancer, B.N. and Power, E.M.G. (1991) in Soc. Appl. Bacteriol. Techn. Ser. No. 27, pp. 23–44, Blackwell Scientific Publications, Oxford.
30 Russell, A.D. (1990) Clin. Microbiol. Rev. 3, 99–119.
31 Bloomfield, S.F. and Arthur, M. (1994) J. Appl. Bacteriol., Symp. Suppl., 91S–104S.
32 Courvalin, P. (1996) J. Antimicrob. Chemother. 37, 855–869.
33 Hugo, W.B. in Ref. 5, (in press).
34 Franklin, T.J., Snow, G.A., Barrett–Bee, K.J. and Nolen, R.D. (1989) Biochemistry of Antimicrobial Action, 4th Edn., Chapman and Hall, London.
35 Yao, J.D.C. and Moellering , R.J., Jr. (1995) in Manual of Clinical Microbiology (Murray, P.R., Baron, E.J., Pfaller, M.A., Tenover, F.C. and Yolken, R.H., eds.) 6th Edn. pp. 1281–1307, American Society of Microbiology, Washington, D.C.
36 Smith, J.T. and Lewis, C.S. (1988) in The Quinolones (Andriole, V.T., ed.) pp. 23–82, Academic Press, London.
37 Crumplin, G.C. (1990) Rev. Med. Microbiol. 1, 67–75.
38 Power, E.G.M. (1997) Prog. Med. Chem. 34, 149–201.
39 Russell, A.D., Furr, J.R. and Maillard, J.-Y. (1997) ASM News 63, 481–487.
40 Hancock, R.E.W. and Bellido, F, (1992) J. Antimicrob. Chemother. 29, 235–239.
41 Levy. S.B. (1992) Antimicrob. Agents Chemother. 36, 695–703.
42 Nikaido, H. (1994) Science (Washington, D.C.) 264, 382–388.
43 Gilbert, P. (1988) in Microbial Quality Assurance in Pharmaceuticals, Cosmetics and Toiletries (Bloomfied, S.F., Baird, R., Leak, R.E. and Leech, R., eds.) pp. 171–194, Ellis Horwood, Chichester.
44 Russell, A.D. (1997) in Topley & Wilson's Microbiology and Microbiol Infections (Duerden, B.I. and Balows, A., eds.) 9th edn., Vol. 2, pp. 149–188, Edward Arnold, London.
45 Sleytr, U.B., Messner, P., Minniken, D.E., Heckels, V.E. and Virj, M. (1988) in Bacterial Cell Surface Techniques (Hancock, I. and Poxton, I., eds.) pp. 1–31, John Wiley & Sons, Chichester.
46 Hugo, W.B. in Ref. 1, pp. 256–262.
47 Waxman, D.L. and Strominger, J.L. (1983) Annu. Rev. Biochem. 52, 825–869.
48 Tomasz, A. (1986) Rev. Infect. Dis. 8, S270–S278.
49 Nagarajan, R. (1991) Antimicrob. Agents Chemother. 35, 605–609.
50 Watanakunakorn, C. (1981) Rev. Infect. Dis. 3, S210–S215.
51 Davis, B.D. (1988) J. Antimicrob. Chemother. 22, 1–3.
52 von Aksen, U. and Noller, H.F. (1993) Science (Washington, D.C.) 260, 1500–1503.
53 Brisson–Noël, A., Trieu–Cuot, P. and Courvalin, P. (1988) J. Antimicrob. Chemother. 22, Suppl. B, 13–23.
54 Navarro, F. and Courvalin, P. (1994) Antimicrob. Agents Chemother. 38, 1788–1793.
55 Tai, P.C. and Davis, B.D. (1985) Symp. Soc. Gen. Microbiol. 38, 41–68.
56 Yamaguchi, A., Onmori, H., Kaneko–Ohdera, M., Nomura, T. and Sawai, T. (1991) Antimicrob. Agents Chemother. 35, 53–56.
57 Hodgson, J.E., Curnock, S.P., Dyke, K.G.H., Morris, R., Sylvester, D.R. and Gross, M.S. (1994) Antimicrob. Agents Chemother. 38, 1205–1208.
58 Abdel–Sayed, S. (1987) J. Antimicrob. Chemother. 19, 7–20.
59 Farr, B.M. (1995) in Principles and Practice of Infectious Diseases (Mandell, G.L., Bennett, J.E. and Dolin, R., eds.) 4th Edn, pp. 317–329, Churchill Livingstone, New York.

60 Beveridge, E.G., Boyd, I., Dew, I., Haswell, M. and Lowe, C.W.G. in Ref. 29, pp. 135–153.
61 Lambert, P.A. in Ref. 29, pp. 121–134.
62 Denyer, S.P. and Hugo, W.B. in Ref. 29, pp. 171–187.
63 Russell, A.D. and Chopra, I. in Ref. 6, pp. 96–149.
64 Falaha, B.M.A., Russell, A.D. and Furr, J.R. (1985) Lett. Appl. Microbiol. 1, 21–24.
65 Russell, A.D. and Furr, J.R. (1986) J. Hosp. Infect. 8, 47–56.
66 Russell, A.D. (1986) Infection 14, 212–215.
67 El–Moug, T., Rogers, D.T., Furr, J.R., El-Falaha, B.M.A. and Russell, A.D. (1985) J. Antimicrob. Chemother. 16, 685–689.
68 Ismael, N., Furr, J.R. and Russell, A.D. (1986) J. Appl. Bacteriol. 61, 373–381.
69 Russell A.D. and Furr, J.R. (1986) Int. J. Pharm. 34, 115–123.
70 Russell, A.D., Furr, J.R. and Pugh, W.J. (1985) Int. J. Pharm. 27, 163–173.
71 George, A.M. (1996) FEMS Microbiol. Lett. 139, 1–10.
72 Hugo, W.B. (1991) Int. Biodetetior. Bull. 27, 185–194.
73 Stickler, D.J. and Thomas, B. (1980) J. Clin. Pathol. 33, 288–296.
74 Reverdy, M.E., Bes, M., Nervi, C., Mastra, A. and Flewette, J. (1992) Med. Microbiol. Lett. 1, 56–63.
75 Russell, A.D. (1998) in Ref. 5, (in press).
76 Ogase, H., Nagai, I., Kemeda, K., Kame, S. and Ono, S. (1992) J. Appl. Bacteriol. 73, 71–78.
77 van Klingeren, B. and Pullen, W. (1993) J. Hosp. Infect. 25, 147–149.
78 Lynham, P.A., Babb, J.R. and Fraise, A.P. (1995) J. Hosp. Infect. 30, 237–239.
79 Griffiths, P.A., Babb, J.R., Bradley, C.R. and Fraise, A.P. (1997) J. Appl. Microbiol. 82, 519–526.
80 Bloom, B.R. and Murray, C.J.L. (1992) Science (Washington, D.C.) 257, 1055–1064.
81 Cole, S.T. (1994) in Tuberculosis: Back to the Future (Porter J.D.H. and McAdam, K.P.W., eds.) pp. 233–235, John Wiley, Chichester.
82 Uttley, A.H.C. and Pozniak, A. (1993) J. Hosp. Infect. 23, 249–253.
83 David, H.L. (1981) Rev. Infect. Dis. 3, 878–884.
84 Draper, P. (1984) Int. J. Leprosy, 52, 527–532.
85 Inderlied, C.B., Kemper, C.A. and Bermulez, L.E.M. (1993) Clin. Microbiol. Rev. 6, 266–310.
86 Nikaido, H., Kim, S.–H. and Rosenberg, E.Y. (1993) Mol. Microbiol. 8, 1025–1030.
87 Jarlier, V. and Nikaido, H. (1990) J. Bacteriol. 172, 1418–1423.
88 Emmerson, A.M. (1995) in Antimicrobial Chemotherapy (Greenwood, D., ed.), 3rd Edn. pp. 301–305, Oxford University Press, Oxford.
89 Zhang, Y. and Young, D. (1994) J. Antimicrob. Chemother. 34, 313–319.
90 Zhang, Y. and Young, D. (1993) Trends Microbiol. 1, 109–113.
91 Rastogi, N., Goh, K.S. and David, H.L. (1990) Antimicrob. Agents Chemother. 34, 759–764.
92 David, H.L., Rastogi, N., Clavel–Serës, S. and Clément, F. (1988) Clin. Microbiol. 17, 61–68.
93 Takayama, K. and Kilburn, J.O. (1988) Antimicrob. Agents Chemother. 33, 1493–1499.
94 McNeil, M.R. and Brennan, P.J. (1991) Res. Microbiol. 142, 451–463.
95 Russell, A.D. (1996) J. Appl. Bacteriol., Symp. Suppl. 81, 87 S–101S.
96 Gordon, S. and Andrew, P.W. (1996) J. Appl. Bacteriol., Symp. Suppl. 81, 10S–22S.
97 Russell, A.D. and Russell, N.J. (1995) Symp. Soc. Gen. Microbiol. 53, 327–365.
98 Favero, M.S. and Bond, W.W. (1991) in Disinfection, Sterilization and Preservation (Block, S.S., ed.) 4th Edn., pp. 617–641, Lea & Febiger, Philadelphia.
99 Marsik, F.J. and Denys, G.A. (1995) in Manual of Clinical Microbiology (Murray, P.R.,

Baron, E.J., Pfaller, M.A., Tenover, F.C. and Yolkien, R.H., eds.) 6th Edn., pp. 86–98, American Society for Microbiology, Washington, D.C.

100 Croshaw, B. (1971) in Inhibition and Destruction of the Microbiol Cell (Hugo, W.B., ed.) pp. 420–449, Academic Press, London.

101 Ascenzi, J.M. (1991) J. Hosp. Infect. 18, Supp. A, 256–263.

102 Rutala, W.A., Cole, E.C., Wannamaker, M.S. and Weber, D.J. (1991) Am. J. Med. 91, Suppl. B, 267S–271S.

103 Ayliffe, G.A.J., Coates, D. and Hoffman, P.N. (1993) Chemical Disinfection in Hospitals, 2nd Edn., Public Health Laboratory, London.

104 Broadley, S.J., Jenkins, P.A., Furr, J.R. and Russell, A.D. (1991) Lett. Appl. Microbiol. 13, 118–122

105 Shen, T.H. (1934) cited in Ref. 100.

106 Broadley, S.J., Jenkins, P.A., Furr, J.R. and Russell, A.D. (1995) J.Med. Microbiol. 43, 458–460.

107 Slayden, R.A., Lee, R.E., Armour, J.R., Cooper, A.M., Orme, I.M., Brennan, P.J. and Besra, G.S. (1996) Antimicrob. Agents Chemother. 40, 2813–2819.

108 Bardou, F., Quémard, A., Dupont, M.–A., Horn, C., Marchal, G. and Daffé, M. (1996) Antimicrob. Agents Chemother. 40, 2459–2467.

109 Gilbert, P. and Brown, M.R.W. (1980) J. Appl. Bacteriol. 48, 223–230.

110 Hugo, W.B. and Franklin, I. (1966) J. Gen. Microbiol. 52, 365–373.

111 Kolawole, D.O. (1984) FEMS Microbiol. Lett. 25, 205–209.

112 Berkeley, R.C.W. and Ali, N. (1994) J. Appl. Bacteriol., Symp. Suppl. 76, 1S–8S.

113 Foster, S.J. (1994) J. Appl. Bacteriol., Symp. Suppl. 76, 25S–39S.

114 Moir, A., Kemp, E.H., Robinson, C. and Corfe, B.M. (1994) J. Appl. Bacteriol., Symp. Suppl. 76, 9S–16S.

115 Johnstone, K. (1994) J. Appl. Bacteriol., Symp. Suppl. 76, 17S–24S.

116 Russell, A.D. (1982) in The Destruction of Bacterial Spores, pp. 169–231, Academic Press, London.

117 Gould, G.W. (1983) in The Bacterial Spore (Hurst, A. and Gould, G.W., eds.) Vol. 2, pp. 173–209, Academic Press, London.

118 Gould, G.W. (1984) Symp. Soc. Appl. Bacteriol. 12, 119–220.

119 Fowler, G.G. and Gasson, M.J. (1991) in Food Preservatives (Russell, N.J. and Gould, G.W., eds.) pp. 135–152, Blackie, London.

120 Hill, C. (1995) in New Methods of Food Preservation (Gould, G.W., ed.) pp. 22–57, Blackie, London.

121 Reisinger, P., Seidel, H., Tschesche, H. and Hammes, W.P. (1980) Arch. Microbiol. 127, 187–193.

122 Scott, V.N. and Taylor, S.L. (1981) J. Food Sci. 46, 117–120.

123 Denny, C.B. and Bohrer, C.W. (1959) Food Res. 24, 247–252.

124 Gould, G.W. (1964) 4th Int. Symp. Food Microbiol., pp. 17–24, SIK, Göteborg.

125 Greenberg, R.A. and Silliker, J.H. (1962) J. Food Sci, 27, 64–68.

126 Wilkins, K.M. and Board, R.G. in Ref. 119, pp. 285–362.

127 Bloomfield, S.F. in Ref. 5, (in press).

128 Power, E.G.M., Dancer, B.N. and Russell, A.D. (1988) FEMS Microbiol. Lett. 50, 223–226.

129 Power, E.G.M. and Russell, A.D. (1989) FEMS Microbiol. Lett. 66, 271–276.

130 Power, E.G.M. and Russell, A.D. (1989) J. Appl. Bacteriol. 67, 329–342.

131 Knott, A.G., Russell, A.D. and Dancer, B.N. (1995) J. Appl. Bacteriol. 79, 492–498.

132 Knott, A.G. and Russell, A.D. (1995) Lett. Appl. Microbiol. 21, 117–120.

133 Shaker, L.A., Furr, J.R. and Russell, A.D. (1988) J. Appl. Bacteriol. 64, 531–539.

134 Bloomfield, S.F. and Arthur, M. (1992) J. Appl. Bacteriol. 72, 166–172.
135 Gorman, S.P., Scott, E.M. and Hutchinson, E.P. (1984) J. Appl. Bacteriol. 57, 153–163.
136 Bloomfield, S.F. and Megid, R. (1994) J. Appl. Bacteriol. 76, 492–499.
137 Setlow, P. (1988) Annu. Rev. Microbiol. 42, 319–338.
138 Setlow, B. and Setlow, P. (1993) Appl. Environ. Microbiol. 59, 3418–3423.
139 Setlow, P. (1994) J. Appl. Bacteriol., Symp. Suppl. 76, 49S–60S.
140 Williams, N.D. and Russell, A.D. (1991) J. Appl. Bacteriol. 70, 427–436.
141 Williams, N.D. and Russell, A.D. (1992) FEMS Microbiol. Lett. 99, 277–280.
142 Williams, N.D. and Russell, A.D. (1993) J. Appl. Bacteriol. 75, 69–75.
143 Williams, N.D. and Russell, A.D. (1993) J. Appl. Bacteriol. 75, 76–81.
144 Costerton, J.W., Irwin, R.T. and Cheng, K.-J. (1981) Annu. Rev. Microbiol. 35, 399–424.
145 Costerton, J.W., Cheng, K.-J., Geesey, K.G., Ladd, P.I., Nickel, J.C., Dasgupta, M. and Marrie, T.J. (1988) Annu. Rev. Microbiol. 41, 435–464.
146 Poxton, I.R. (1993) J. Appl. Bacteriol., Symp. Suppl. 74, 1S–11S.
147 Brown, M.R.W. and Gilbert, P. (1993) J. Appl. Bacteriol., Symp. Suppl. 74, 87S–97S.
148 Carpentier, B. and Cerf, O. (1994) J. Appl. Bacteriol. 75, 499–511.
149 Wimpenny, J., Nichols, W., Stickler, D. and Lappin-Scott, H. (1994) eds., Bacterial Biofilms and their Control in Medicine and Industry. BioLine, Cardiff.
150 Lappin-Scott, H.M. and Costerton, J.W. (1995) eds., Microbiol Biofilms, Cambridge University Press, Cambridge.
151 Lawrence, J.R., Korber, D.R., Wolfaardt, G.M. and Caldwell, D.E. (1995) Adv. Microb. Ecol. 14, 1–75.
152 Wood, P., Jones, M., Bhakoo, M. and Gilbert, P. (1996) Appl. Environ. Microbiol. 62, 2598–2602.
153 Hiramatsu, K., Hanaki, H., Ino, T., Yabuta, K., Oguri, T. and Tenover, F.C. (1997) J. Antimicrob. Chemother. 40, 135–136.
154 Anderson, R.L., Bland, L.A., Favero, M.S., McNeil, M.M., Davis, B.J., Mackel, D.C. and Gravelle, C.R. (1985) Appl. Environ. Microbiol. 50, 1343–1348.
155 Anderson, R.L., Vess, R.W., Panlilio, A.L. and Favero, M.S. (1990) Appl. Environ. Microbiol. 56, 3598–3600.
156 Anderson, R.L., Holland, B.W., Carr, J.K., Bond, W.W. and Favero. M.S. (1990) Am. J. Public Health 80, 17–21.
157 Russell, A.D. and Day, M.J. (1993) J. Hosp. Infect. 25, 229–238.
158 Russell, A.D. and Day, M.J. (1996) Microbios 85, 45–65.
159 Spratt, B.G. (1994) Science (Washington, D.C.) 264, 388–393.
160 Piddock, L.J.V. (1991) J. Antimicrob. Chemother. 27, 399–403.
161 Maxwell, A. (1992) J. Antimicrob. Chemother. 30, 409–414.
162 Power, E.G.M., Munoz Bellido, J.L. and Phillips, I. (1992) J. Antimicrob. Chemother. 29, 9–17.
163 Cambau, E. and Gutman, L. (1993) Drugs 45, Suppl. 3, 15–23.
164 Lewin, C.S., Allen, R.A. and Amyes, S.G.B. (1990) J. Med. Microbiol. 31, 153–162.
165 Courvalin, P. (1992) ASM News 58, 368–375.
166 Hancock, R.E.W., Farmer, S.W., Li, Z. and Poole, K. (1991) Antimicrob. Agents Chemother. 35, 1309–1314.
167 Wehrli. W. (1983) Rev. Infect. Dis. 5, S407–411.
168 Lyon, B.R., Tennent, J.M., May, J.W. and Skurray, R.A. (1986) FEMS Microbiol. Lett. 33, 189–192.
169 Amyes, S.G.B. and Towner, K.J. (1990) J. Med. Microbiol. 31, 1–19.
170 Sundström, L, Radström, P., Swedberg, G. and Sköld, O. (1988) Mol. Gen. Genet. 213, 191–201.

171   Hughes, D.T.D. (1997) in Antibiotic and Chemotherapy (O'Grady, F.W., Lambert, H.P., Finch, R.G. and Greenwood, D., eds.) 7th Edn., pp. 460–468, Churchill Livingstone, Edinburgh.

172   Chaplin, C.E. (1952) J.Bacteriol. 63, 453–458.

173   Russell, A.D. and Furr, J.R. (1977) J. Appl. Bacteriol. 43, 253–260.

174   Fitzgerald, K.A., Davies, A. and Russell, A.D. (1992) Lett. Appl. Microbiol. 14, 91–95.

175   Brown, M.R.W. and Melling, J. (1969) J. Gen. Microbiol. 54, 439–444.

176   Nicas, T.I. and Hancock, R.E.W. (1980) J. Bacteriol. 143, 872–878.

177   Gilleland, H.E., Jr. and Murray, R.G.E. (1976) J. Bacteriol. 125, 267–281.

178   Gilleland, H.E., Jr. and Lyle., R.D. (1979) J. Bacteriol. 138, 839–845.

179   Davies, J. (1994) Science (Washington, D.C.) 264, 375–382.

180   Jacoby, G.A. and Medeiros, G.A. (1991) Antimicrob. Agents Chemother. 35, 1697–1704.

181   Bush, K. (1989) Antimicrob. Agents Chemother. 33, 259–263.

182   Richmond M.H. and Sykes, R.B. (1973) Adv. Microb. Physiol. 9, 31–88.

183   Sirot, D., Recule, C., Chaibi, E. B., Bret, L., Croize, J., Chanal–Claris, C., Labia, R. and Sirot, J. (1997) Antimicrob. Agents Chemother. 41, 1322–1325.

184   Henquell, C., Chanal, D., Sirot, D., Labia, R. and Sirot, J. (1995) Antimicrob. Agents Chemother. 39, 427–430.

185   Shaw, W.V. (1983) Crit. Rev. Biochem. 14, 1–46.

186   Lyon, B.R. and Skurray, R.A. (1987) Microbiol. Rev. 51, 88–134.

187   Bryan, L.E. (1989) in Microbial Resistance to Drugs (Bryan, L.E. ed.) pp. 35–57, Springer–Verlag, Berlin.

188   Davies, J. (1991) in Antibiotics in Laboratory Medicine (V. Lorian, ed.) pp. 691–713, Williams & Williams Co., Baltimore.

189   Murray, B.E. (1990) Clin. Microbiol. Rev. 3, 46–65.

190   Patterson, J.E. and Zervos, M.J. (1990) Rev. Infect. Dis. 12, 644–652.

191   Woodford, N., McNama, E., Smyth, E. and George, R.C. (1992) J. Antimicrob. Chemother. 28, 483–486.

192   Salyers, A.A., Speer, B.S. and Shoemaker, N.B. (1990) Mol. Microbiol. 4, 151–156.

193   Chopra, I., Hawkey, P.M. and Hinton, M. (1992) J. Antimicrob. Chemother. 29, 245–277.

194   Speer, B.S. and Salyers, A.A. (1989) J. Bacteriol. 171, 148–153.

195   Speer, B.S., Shoemaker, N.B. and Salyers, A.A. (1992) Clin. Microbiol. Rev. 5, 387–399.

196   Ball, P. (1991) J. Hosp. Infect. 19, Suppl. A, 47–59.

197   Ohara, K., Kanda, T., Ohmiga, K., Ebisn, T. and Kono, M. (1989) Antimicrob. Agents Chemother. 33, 1354–1357.

198   Leclerq, R. and Courvalin, P. (1991) Antimicrob. Agents Chemother. 35, 1267–1272.

199   Leclerq, R. and Courvalin, P. (1991) Antimicrob. Agents Chemother. 35, 1273–1276.

200   Cookson, B.D. (1989) Eur. J. Clin. Microbiol. Infect. Dis. 8,1038–1040.

201   Gilbert, J., Parry, C.R. and Slocombe, B. (1993) Antimicrob. Agents Chemother. 37, 32–38.

202   Rahman, M., Connolly, S., Noble, W.C., Cookson, B. and Phillips, I. (1990) J. Med. Microbiol. 33, 97–100.

203   Towner, K.J. in Ref. 88, pp. 147–158.

204   Huovinen, P., Sunström, L., Swedberg, G. and Sköld, O. (1995) Antimicrob. Agents Chemother. 39, 279–289.

205   Paulsen, I.T., Brown, M.H. and Skurray, R.A. (1996) Microbiol. Rev. 60. 575–608.

206   Eady, E.A., Ross, J.I. and Cove, J.H. (1991) J. Antimicrob. Chemother. 26, 461–465.

207   Russell, A.D. (1985) J. Hosp. Infect. 6, 9–19.

208   Russell, A.D. (1997) J. Appl. Microbiol. 82, 155–165.

209   Foster, T.J. (1983) Microbiol. Rev. 47, 361–409.

210   Silver, S. and Misra, A. (1988) Ann. Rev. Microbiol. 42, 717–743.
211   Misra, T.K. (1992) Plasmid 27, 17–28.
212   Beveridge, T.J., Hughes, M.N., Lee, H., Leung, K.T., Poole, R.K., Savvaidis I., Silver, S. and Trevors, J.T. (1997) Adv. Microb. Physiol. 38, 176–243.
213   Trevor, J.T. (1987) Enzyme Microb. Technol. 9, 331–333.
214   Townsend, D.E., Grubb, W.B. and Ashdown, N. (1983) J. Hosp. Infect. 4, 331–337.
215   Townsend, D.E., Greed, L.C., Ashdown, N. and Grubb, W.B. (1983) Med. J. Aust. 2, 310.
216   Emslie, K.R., Townsend, D.E., Bolton, S. and Grubb, W.B. (1985) FEMS Microbiol. Lett. 27, 61–64.
217   Emslie, K.R., Townsend, D.E. and Grubb, W.B. (1985) J. Med. Microbiol. 20, 139–145.
218   Gillespie, M.T., May, J.W. and Skurray, R.A. (1986) FEMS Microbiol. Lett. 34, 47–51.
219   Gillespie, M.T., Lyon, B.R. and Skurray, R.A. (1989) Lancet, 1, 503.
220   Gillespie, M.T., Lyon, B.R. and Skurray, R.A. (1990) J. Med. Microbiol. 31, 57–64.
221   Littlejohn, T.G., Diberardino, D., Messerotti, L.J., Spiers, S. J. and Skurray, R.A. (1991) Gene 101, 59–66.
222   Littlejohn, T.G., Paulsen, I.T., Gillespie, M.T., Tennent, J.M., Midgeley, M., Jones, I.G., Purewal, A.S. and Skurray, R.A. (1992) FEMS Microbiol. Lett. 95, 259–266.
223   Reverdy, M.E., Bes, M., Brun, Y. and Fleurette, J. (1993) Pathol. Biol. 41, 897–904,
224   Behr, H., Reverdy, M.E., Mabilat, C., Freney, J. and Fleurette, J. (1994) Pathol. Biol. 42, 438–444.
225   Leelaporn, A., Paulsen, I.T., Tennent, J.M., Littlejohn, T.G. and Skurray, R.A. (1994) J. Med. Microbiol. 40, 214–220.
226   Heir, E., Sundheim, G. and Holck, A.L. (1995) J. Appl. Bacteriol. 79, 149–156.
227   Paulsen, I.T., Brown, M.H., Littlejohn, T.G., Mitchell, B.A. and Skurray, R.A. (1996) Proc. Natl. Acad. Sci. U.S.A., 93, 3630–3635.
228   Paulsen, I.T., Skurray, R.A., Tam, R., Saier, M.H., Jr., Turner, R.J., Weiner, J.H., Goldberg, E.B. and Grinius, L.L. (1996) Mol. Microbiol. 19, 1167–1175.
229   Rouch, D.A., Cram, D.S., Dibaradino, O., Littlejohn, T.G. and Skurray, R.A. (1990) Mol. Microbiol. 4, 2051–2062.
230   Sasatsu, M., Shidata, Y., Noguichi, N. and Kono, M. (1994) Biol. Pharm. Bull. 17, 136–138.
231   Sasatsu, M., Shirai, Y., Hase, M., Noguchi, N., Kono, M., Behr, H., Freney, J. and Arai, T. (1995) Microbios 84, 161–169.
232   Al-Masaudi, S.B., Day, M.J. and Russell, A.D. (1988) J. Appl. Bacteriol. 65, 329–337.
233   Al-Masaudi, S B., Russell, A.D. and Day, M.J. (1991) J. Med. Microbiol. 34, 103–107.
234   Al-Masaudi, S.B., Day, M.J. and Russell, A.D. (1991) J. Appl. Bacteriol. 71, 239–243.
235   Al-Masaudi, S.B., Day, M.J. and Russell, A.D. (1991) J. Appl. Bacteriol. 70, 279–290.
236   Al-Masaudi, S.B., Russell, A.D. and Day, M.J. (1991) J. Appl. Bacteriol. 71, 331–338.
237   Tomasz, A. (1994) N. Engl. J. Med. 330, 1247–1251.
238   Barry, C.E. 3rd and Maluli, K. (1996) Trends Microbiol. 4, 275–281.
239   Minniken, D.E. (1991) Res. Microbiol. 142, 423–427.
240   Mitchison, D.A. (1996) J. Appl. Bacteriol., Symp. Suppl. 81, 72S–80S.
241   Banerjee, A., Dubnaue, E., Quemerd, A., Balasabramanian, V., Um, K.S., Wilson, T., Collins, D., de Lisle, G. and Jacobs, W.R. (1994) Science (Washington, D.C.) 263, 227–230.
242   Morris, S., Bai, G. H., Suffys, P., Portillo-Gomez, L., Fairchok, M. and Rouse, D. (1995) J. Infect. Dis. 171, 954–960.
243   Quemard, A., Laneelle, G. and Lacave, C. (1992) Antimicrob. Agents Chemother. 36, 1316–1321.
244   Williams, D.L., Waguespack, C., Eisenach, K., Crawford, J.T., Portaels, F., Salfinger, M.,

Nolan, C.M., Abe, C., Sticht–Groh, V. and Gillis, T.P. (1994) Antimicrob. Agents Chemother. 38, 2380–2386.

245   Takiff, H.E., Salazar, L., Guerrero, C., Phillip, W., Huang, W.M., Kreiswirth, B., Cole, S.T., Jacobs, W.R. and Telenti, A. (1994) Antimicrob. Agents Chemother. 38, 773–780.

246   Lejeune, B., Buzie–Losquim, F., Simitzis-Le Flohic, A.M., Le Bras, M.P. and Aliz, D. (1986) J. Hosp. Infect. 7, 21–25.

247   Hackbarth, C.J. and Chambers, H.F. (1989) Antimicrob. Agents Chemother. 33, 991–994.

248   Tenover, F.C., Poporic, T. and Olsvik, O. (1995) in ref. 35, pp. 1368–1378.

249   Report (1990) J. Hosp. Infect. 16, 351–377.

250   Shalita, Z., Murphy, E. and Novick, R.P. (1980) Plasmid 3, 291–311.

251   Parker, M.T. (1983) in Staphylococci and Staphylococcal Infection (Edsmon, F.S. and Adlam, C., eds.) Vol. 1, pp. 33–62, Academic Press, London.

252   Peters, G. (1988) J. Antimicrob. Chemother. 21, Suppl. C, 139–148.

253   Day, M.J. and Russell, A.D. in Ref. 5. (in press).

254   McDonnell, R.W., Sweeney, H.M. and Cohen, S. (1983) Antimicrob. Agents Chemother. 23, 151–160.

255   Coello, R., Jiménez, J., Garcia, M., Arroyo, P., Minguez, D., Fernández, C., Cruzet, F. and Gaspar, C. (1994) Eur. J. Clin. Microbiol. Infect. Dis. 13, 74–81.

256   Casewell, M.W. (1995) J. Hosp. Infect. 30. Suppl. 465–471.

257   Georgopapadakou, N.H. (1993) Antimicrob. Agents Chemother. 37, 2045–2053.

258   Murray, B.E. (1992) Antimicrob. Agents Chemother. 36, 2355–2359.

259   Jett, B.D., Huyoke, M.M. and Gilmore, M.S. (1994) Clin. Microbiol. Rev. 7, 462–478.

260   Courvalin, P. (1994) Antimicrob. Agents Chemother. 38, 1447–1451.

261   Wade, J.J. (1995) J. Hosp. Infect. 30 Suppl. 483–493.

262   Arthur, M., Reynolds, P.E., Depardieu, F., Evers, S., Durka–Malen, S., Quiniliani, R., Jr. and Courvalin, P. (1996) J. Infect. 32, 11–16.

263   Woodford, N., Johnson, A.P., Morrison, D. and Speller, D.C.E. (1995) Clin. Microbiol. Rev. 8, 585–615.

264   Rowe, R.M. (1996) Lancet 347, 252.

265   Noble, W.C., Viriani, Z. and Cree, R.G.A. (1992) FEMS Microbiol. Lett. 93, 195–198.

266   Milewski, W.M., Boyle-Vavra, S., Moreira, B., Eben, C.C. and Daum, R.S. (1966) Antimicrob. Agents Chemother. 40, 166–172.

267   Sieradzki, K. and Tomasz, A. (1997) J. Bacteriol. 179, 2557–2566.

268   Williams, D., Bergan, T. and Moosdeen, F. (1997) Newsletter Int. Soc. Chemother. 1, 1.

269   Arthur, M., Reynolds, P. and Courvalin, P. (1996) Trends Microbiol. 4, 401–407.

270   Alqurashi, A.M., Day, M.J. and Russell, A.D. (1996) J. Antimicrob. Chemother. 38, 745.

271   Bradley, C.R. and Fraise, A.P. (1996) J. Hosp. Infect. 34, 191–196.

272   Anderson, R.L., Carr, J.H., Bond, W.W. and Favero, M.S. (1997) Infect. Cont. Hosp. Epidem. 18, 195–199.

273   Cohen, S.P, McMurry, L.M.M., Hooper, D.C., Wolfson, J.S. and Levy, S.B. (1989) Antimicrob. Agents Chemother. 33, 1318–1325.

274   Cohen, S.P., Yan, W. and Levy, S.B. (1993) J. Infect. Dis. 168, 484–488.

275   Ariza, R.R., Cohen, S.P., Bachhawat, N., Levy, S.B. and Demple, B. (1994) J. Bacteriol. 176, 143–148.

276   Martin, R.G. and Rosner, J.L. (1995) Proc. Natl. Acad. Sci. U.S.A. 92, 5456–5460.

277   Seoane, A.S. and Levy, S.B. (1995) J. Bacteriol. 177, 3414–3419.

278   Goldman, J.D., White, D.G. and Levy, S.B. (1996) Antimicrob. Agents Chemother. 40, 1266–1269.

279   Maneewannakul, K. and Levy, S.B. (1996) Antimicrob. Agents Chemother. 40, 1695–1698.

280   Alonso, A. and Martínez, J.L. (1997) Antimicrob. Agents Chemother. 41, 1140–1142.
281   McMurry, L.M., George, A.M. and Levy, S.B. (1994) Antimicrob. Agents Chemother. 38, 542–546.
282   Asako, H., Nakeyima, K., Kobayashi, M. and Aono, R. (1997) Antimicrob. Agents Chemother. 63, 1428–1433.
283   Miller, P.F., Gambino, L.A., Sulavik, M.C. and Grachneck, S.J. (1994) Antimicrob. Agents Chemother. 38, 1773–1779.
284   Nakajima, H., Kobayashi, K., Kobayahi, M., Asako, H. and Aono, R. (1995) Antimicrob. Agents Chemother. 61, 2302–2307.
285   Poole, K., Krebes, K., McNally, C. and Neshat, S. (1993) J. Bacteriol. 175, 7363–7372.
286   Hagman, K.E., Pan, W., Spratt, B.G., Balthazar, J.T., Judd, R.C. and Shafer, W.M. (1995) Microbiology, 141, 611–622.
287   Stickler, D.J., Thomas, B., Clayton, C.L. and Chawla, J. (1993) Br. J. Clin. Pract., Symp. Suppl. 25, 23–30.
288   Dance, D.A.B., Pearson, A.D., Seal, D.V. and Lowes, J.A. (1987) J. Hosp. Infect. 10, 10–16.
289   Baillie, L.W.J., Power, E.G.M. and Phillips, I. (1993) J. Antimicrob. Chemother. 31, 219–225.
290   Baillie, L.W.J., Wade, J.J. and Casewell, M.W. (1992) J. Hosp. Infect. 20, 127–128.
291   Jarvínen, H., Temovuo, J. and Huovin, P. (1993) Antimicrob. Agents Chemother. 37, 1158–1159.
292   Jones, M.V., Herd, T.M. and Christie, H.J. (1989) Microbios 58, 49–61.
293   Nicoletti, G., Boghossien, V., Gurevitch, Y., Borland, R. and Morgenroth, P. (1993) J. Hosp. Infect. 23, 87–111.
294   Stecchini, M.L., Manzano, M. and Sarais, I. (1992) Int. J. Food Microbiol. 16, 79–85.
295   Fernández-Astorga, A., Hijarrubie, M.J., Hernandez, M., Arani, I. and Sunea, E. (1995) J. Appl. Bacteriol. 20, 308–311.
296   Eadon, H.J. and Pinney, R.J. (1991) J. Clin. Pharm. Ther. 16, 453–462.
297   Cookson, B.D., Bolton, M.C. and Platt, J.H. (1991) Antimicrob. Agents Chemother. 35, 1997–2002.
298   Cookson, B.D., Farrely, H., Stapleton, P., Garvey, R.R.J. and Price, M.R. (1991) Lancet 1, 1548–1549.
299   Roussow, F.T. and Rowbury, R.J. (1984) J. Appl. Bacteriol. 56, 63–74.
300   Nagai, I. and Ogase, H. (1990) J. Hosp. Infect. 15, 149–155.
301   Russell, A.D., Hammond, S.A. and Morgan J.R., (1986) J. Hosp. Infect. 7, 213–225.
302   Yamomoto, T., Tamura, Y. and Yokota, T. (1988) Antimicrob. Agents Chemother. 32, 932–935.
303   Ma, D., Cook, D.N., Alberti, M., Pon, N.G., Nikaido, H. and Hearst, J.E. (1993) J. Bacteriol. 175, 6299–6313.
304   Yerushalmi, H., Lebendiker, M. and Schyldiner, S. (1995) J. Biol. Chem. 270, 6856–6863.
305   Chopra, I. (1992) J. Antimicrob. Chemother. 30, 737–739.
306   Neyfakh, A.A., Borsch, C.M. and Kaat, G.W. (1993) Antimicrob. Agents Chemother. 37, 128–129.
307   Rothstein, D.M., McGlynn, M., Beman, V., McGahren, J., Zaccardi, J., Ceklemak, N. and Bertrand, K.P. (1993) Antimicrob. Agents Chemother. 37, 1624–1629.
308   Kaatz, G.W., Seo, S.M. and Ruble, C.A. (1993) Antimicrob. Agents Chemother. 37, 1086–1094.
309   Lewis, K. (1994) TIBS 19, 119–123.
310   Leelaporn, A., Paulsen, I.T., Tennent, J.M., Littlejohn, T.G. and Skurray, R.A. (1994) Antimicrob. Agents Chemother. 40, 214–220.

311 Kayser, F.H. (1996) Chemotherapy 42, S2, 2–12.
312 Coleman, K., Athalye, M., Clancey, A., Davison, M., Payne, D.J., Perry, C.R. and Chopra, I. (1994) J. Antimicrob. Chemother. 33, 1091–1116.
313 Chopra, I., Hodgson, J., Metcalf, B. and Poste, G. (1997) Antimicrob. Agents Chemother. 41, 497–503.
314 Murray, B.E. (1994) N. Engl. J. Med. 330, 1229–1230.
315 Russell, A.D. and Ahonkhai, I. (1982) J. Antimicrob. Chemother. 9, 445–449.
316 Tally, F.T., Ellestad, G.A. and Testa, R.T. (1995) J. Antimicrob. Chemother. 35, 449–452.
317 Rolinson, G.N. (1995) Symp. Soc. Gen. Microbiol. 53, 53–65.
318 Coulton, S. and François, I. (1994) Prog. Med. Chem. 31, 297–349.
319 Russell, A.D. and Chopra, I. in Ref. 6, pp. 257–275.
320 Zelenitsky, S.A., Karlowsky, J.A., Zhanel, G.G., Hoban, D.J. and Nicas, T. (1997) Antimicrob. Agents Chemother. 41, 1407–1408.
321 Gordon, M.D., Ezzell, R.J., Bruchner, N.I. and Ascenzi, J.M. (1994) J. Ind. Microbiol. 13, 77–82.
322 Cookson, B.D. (1994) Zentralbl. Bakteriol., Suppl. 26, 227–234.
323 Coates, D. and Hutchinson, D.N. (1994) J. Hosp. Infect. 26, 57–68.
324 Baquero, F., Patron, C., Patron, R. and Ferrer, M.M. (1991) J. Hosp. Infect. 18, Suppl. B, 5–11.
325 Tenover, F.C. and McGowen, J.E., Jr. (1996) Am. J. Med. Sci. 311, 9–16.
326 Silver, L. and Bostian, K. (1995) Antimicrob. Agents Chemother. 37, 377–383.

Progress in Medicinal Chemistry – Vol. 35
Edited by G.P. Ellis, D.K. Luscombe and A.W. Oxford
© 1998 Elsevier Science B.V. All rights reserved.

5 Towards cannabinoid drugs – revisited

R. MECHOULAM, Ph.D., L. HANUŠ, Ph.D. and ESTER FRIDE, Ph.D.

Brettler Medical Research Center, Faculty of Medicine, Hebrew University of Jerusalem, Ein Kerem, Jerusalem 91120, Israel

INTRODUCTION

Ten years ago we published a review in this series on cannabinoids as medicinal agents [1]. In the 1987 review, we discussed in considerable detail the pharmacohistory of *Cannabis sativa* and showed that many of the modern applications were well-known in the ancient world and that cannabis was quite extensively used until early in the 20th century. Over the last 70 years cannabis, as a crude drug, has not been officially used in medicine any more, although the active constituent, Δ^9-tetrahydrocannabinol (Δ^9-THC)* (1) found some use, as described below. In the last decade, there has been considerable popular pressure to allow the sale of marijuana as a medicinal agent. This has become a political problem and in two states in the USA, California and Arizona, voters have approved propositions allowing physicians to prescribe marijuana. Support for this change has come from various groups and individuals including the editor of the *New England Journal of Medicine*, J.P. Kassirer, who in a signed editorial has stated that 'a federal policy that prohibits physicians from alleviating suffering by prescribing marijuana for seriously ill patients is misguided, heavy-handed and inhumane' [2, 3]. Not surprisingly, the US Federal Government has opposed this grass-roots change of the law. However, it has promised funding of clinical studies on marijuana which until now has not been supported [4]. It will be of interest to see whether cannabis, smoked or eaten, is better than Δ^9-THC administered orally.

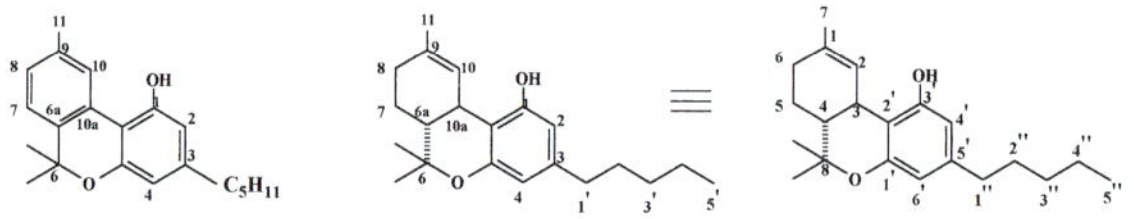

cannabinol (CBN)

(Chemical Abstract: 6,6,9-trimethyl-3-pentyl dibenzo[b,d]pyran-1-ol

Δ^9-tetrahydrocannabinol (Δ^9-THC)

Δ^1-THC (terpene numbering seldom used now)

Figure 5.1 Alignment and numbering of cannabinoid

Since our 1987 review, there has been enormous progress in our understanding of the biochemical basis of cannabinoid action – receptors were identified, endogenous cannabinoids were isolated, and their structures were elucidated. The signalling transduction mechanisms involved have been the object of numerous studies. In the present review, we shall try to

* For alignment and numbering of cannabinoids see *Figure 5.1*

summarize the major advances achieved in cannabinoid biochemistry, without going into details, as a book [5] and numerous review articles have been published recently [6–15].

In contrast to the important advances in biochemistry, medical progress in the cannabinoid field has been rather limited, possibly due to the legal situation. However, at least one new synthetic cannabinoid (HU-211) is in clinical trials (see below). We shall review the advances made since 1987 with emphasis on novel uses, on synthetic compounds which are being developed as drugs, on new synthetic leads and on chemical tools for further biochemical and pharmacological investigations.

CANNABINOID RECEPTORS AND ENDOGENOUS LIGANDS

THE MODE OF ACTION OF CANNABINOIDS

The active psychotropic constituent in *Cannabis sativa*, Δ^9-THC (1), was isolated in pure form and its structure was elucidated by our group in 1964 [16]. Yet the molecular basis of cannabinoid activity remained an enigma for several decades. This lack of knowledge made the rational development of cannabinoids as potential therapeutic agents difficult. During the 1970's and early 1980's it was generally assumed that the high lipophilicity of the cannabinoids is the basis of their pharmacological action. Paton, in a 1975 review pointed out that 'underlying much of the pharmacology of cannabis is the high lipophilicity of its active principles which is responsible for the slowness of its kinetics, its cumulation, [and] its persistence' [17]. Cannabis pharmacology parallels in many ways that of the general anaesthetics, yet it does not produce the expected surgical anaesthesia. Paton suggested that 'considering the strongly hydrophobic character of Δ^9-THC, it is possible that there is too great a physicochemical disparity between it and biological membranes into which it is inserted for a volume fraction to be achieved sufficient to produce the phenomenon of full anaesthesia' [17]. Hence Δ^9-THC should be considered, according to Paton, to belong to the group of biologically active lipophiles and its effects should be compared with the chronic effects of anaesthetics and solvents.

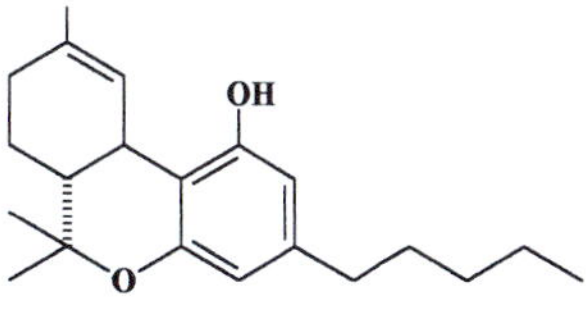

(1) (-)-Δ^9-THC 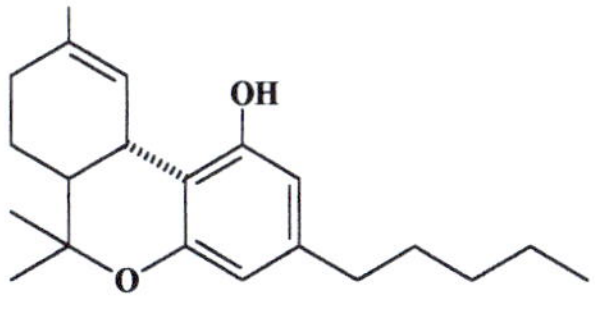   (2) (+)-Δ^9-THC

It was therefore possible to explain the action of cannabinoids without postulating the existence of a specific cannabinoid receptor and of an endogenous mediator of cannabinoid action. Gill and Lawrence found experimental evidence to support the above suggestion and concluded that: '..... as the liposomal membrane is apparently able to discriminate between the various cannabinoids in a way similar to the nerve cell, it is unnecessary to postulate the existence of a more complex macromolecular receptor substance to account for the observed structure-activity relationships' [18].

A further problem was the presumed lack of, or at least a low degree of, stereospecificity of cannabinoid action. As all receptors are asymmetric, it is conceivable that their interactions with asymmetric ligands will be limited to one enantiomer only. However, synthetic (+)-Δ^9-THC (2) showed cannabimimetic activity of 5–10% compared with that of the natural (−)-Δ^9-THC (1) [19]. This observation cast doubt over the existence of a specific cannabinoid receptor and hence of a cannabinoid mediator. This presumed low degree of stereoselectivity and the above described suggestions and data on the actions of cannabinoids on membranes, delayed research aimed at the identification of a receptor-mediator cannabinoid system.

In the 1980's, we showed that THC-type compounds exhibit very high stereospecificity of cannabinoid action [20–24]. Some of the previous observations regarding lack of stereospecificity were apparently due to separation problems. The (−)-enantiomer of Δ^8-11-hydroxy-THC-DMH (3) (HU-210) (DMH=1,1-dimethylheptyl) was shown to be several thousand times more potent than the (+)-enantiomer (HU-211) (4) in a series of animal tests. The synthesis of HU-211 is presented in *Figure 5.2* [25]. The intermediate (+)-4-oxo-myrtenyl pivalate (5), and its (−)-enantiomer are highly crystalline and can be obtained essentially stereochemically pure by recrystallization. This lucky observation made possible the ultimate synthesis of both enantiomers, HU-211 and HU-210, in e.e. higher than 99.8%, as determined by HPLC analysis of the respective diastereoisomeric bis-(S) (+)-α-methoxy-α-(trifluoromethyl)phenyl-acetyl (MTPA) esters [25]. Nonclassical cannabinoids were also shown to exhibit high stereospecificity of cannabinoid activity [21, 26].

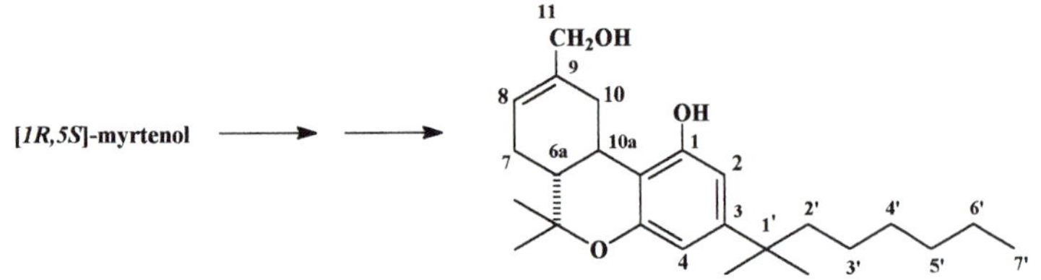

[1R,5S]-myrtenol ⟶ ⟶

(3) HU-210

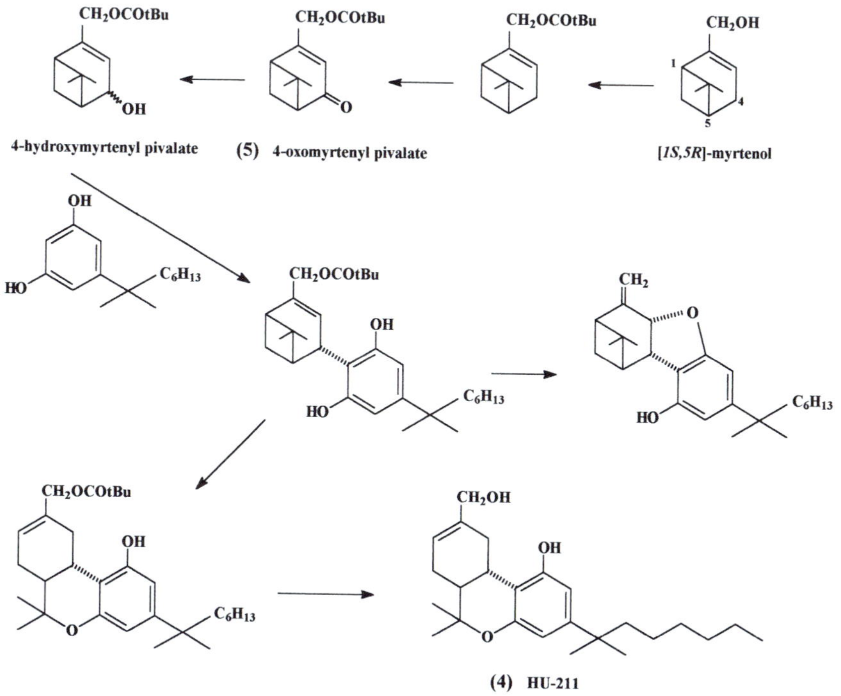

Figure 5.2 Synthesis of HU-211.

These developments led to the next crucial step, the identification of a cannabinoid receptor which was achieved by Howlett's group in 1988 [27]. A radiolabelled, potent synthetic cannabinoid was demonstrated to bind to brain membranes in a highly specific and selective manner, features that are characteristic of receptor binding. Excellent correlation was found between structure and pharmacological potencies of cannabinoids in several animal behavioural tests assessing sedation, body temperature and analgesia and their affinity for the cannabinoid brain receptor [9, 10, 28]. It was also shown that cannabinoid ligands inhibit the activity of adenylate-cyclase and N-type voltage-dependent calcium channels, and that these signal transduction pathways are mediated through the inhibitory guanosine triphosphate (GTP)-binding protein G_i/G_o [10].

Within two years, the receptor (named CB_1) was cloned from rat brain by a group at NIH [29]. It was shown to belong to the 7 transmembrane domain receptor family, members of which mediate their activity through GTP-binding proteins.

A peripheral cannabinoid receptor (CB_2) was later cloned from a human promyelocytic leukaemic line (HL60) [30]. This receptor was shown to be only partially homologous with the brain receptor, and is expressed in cells of the immune system, predominantly on B-cells [31].

These developments have been described and discussed in the reviews cited above which should be consulted for details, in particular as regards molecular biology, signal transduction mechanisms, pharmacological and physiological implications of the discovery of cannabinoid receptors [5–15].

ANANDAMIDE

The identification of a specific cannabinoid receptor in the brain suggested the existence of a brain cannabinoid ligand. It seemed to us quite unacceptable that the brain will waste its energy to synthesize a receptor (in high concentrations) in order to bind a constituent of a plant. The only reasonable assumption which could be made was that the brain produces a neuronal mediator, a specific compound which binds to and activates the cannabinoid receptor. The plant cannabinoid, Δ^9-THC, by structural coincidence happens to bind to the same receptor. In the late 1980's, several groups initiated work aimed at the discovery of such a brain constituent.

Our group proceeded by initially preparing a novel, highly potent probe for the cannabinoid receptor, which could be easily labelled with tritium [32]. This probe (tritiated HU-243) (6) was bound to the receptor and the

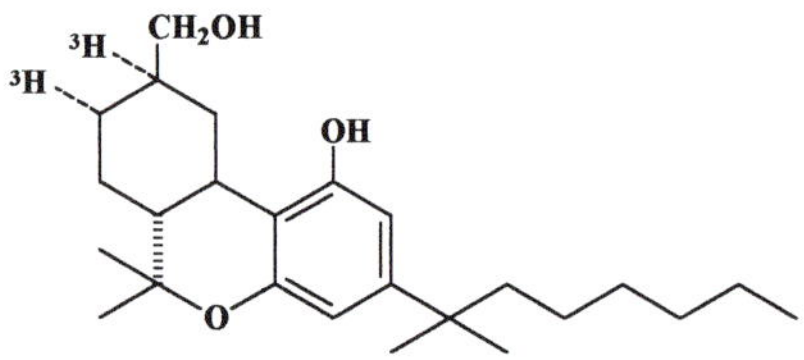

(6)   tritiated HU-243

complex formed was used in a bioassay-directed fractionation aimed at the isolation of components with cannabis-like activity. Porcine brain fractions were found to compete with this probe for binding to cannabinoid receptors. Chromatography of such brain fractions led to the identification of a family of unsaturated fatty acid ethanolamides, termed anandamides (see

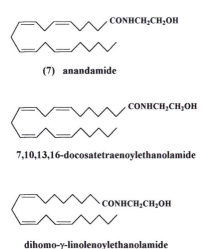

(7) **anandamide**

7,10,13,16-docosatetraenoylethanolamide

dihomo-γ-linolenoylethanolamide

Figure 5.3 Bioactive fatty acid ethanolamides in porcine brain.

Figure 5.3). The most abundant of these ligands is arachidonoylethanola-mide (anandamide, 7) [33, 34]. Two additional anandamides isolated from porcine brain are docosatetraenoylethanolamide and dihomo-γ-linolenoyl-ethanolamide [34] (*see Figure 5.3*).

Structurally there is little in common between $\varDelta^9$-THC and anandamide. The cannabinoids are terpenophenols, while the anandamides are fatty acid derivatives. Yet, pharmacologically they have much in common. Both $\varDelta^9$-THC and anandamide were shown to cause a typical tetrad of behavioural actions: hypothermia, hypomotility, antinociception and catalepsy. In most behavioural tests, anandamide is somewhat less potent than $\varDelta^9$-THC [35, 36]. Repeated injections of anandamide (i.p.) produced tolerance to a chal-lenge with THC or anandamide. This tolerance however was less persistent than that commonly seen with THC, lasting for only one week [37].

Anandamide parallels $\varDelta^9$-THC in competing with the binding of [^{3}H]HU-243 (6) to the brain cannabinoid receptor [33]. However, under most experi-mental conditions an amidase blocker has to be used in order to prevent anandamide hydrolysis during the binding experiments [38]. Using neuro-blastoma cells that express this receptor naturally, as well as cultured cell lines transfected with this receptor, it was shown that, like $\varDelta^9$-THC, anand-amide inhibits N-type voltage-dependent calcium channels [39] and adeny-late cyclase [40, 41].

Low doses of administered anandamide have been observed to produce effects opposite to those of high doses [42, 43]. The question may be raised as to whether the physiologically relevant anandamide functions are those observed after injection of commonly applied doses, or more in line with the opposite effects observed after low concentrations.

Anandamide is present in brain cells in very low concentrations [44–47]. Apparently it is released (presumably mostly when needed) through cleavage catalysed by phospholipase D from N-arachidonoyl phosphatidyl ethanolamine, which is formed by a calcium-dependent enzyme activity which catalyses the transfer of arachidonic acid (and other fatty acids) from various donor phospholipids to phosphatidylethanolamine (see *Figure 5.4*) [48, 49]. In animal post mortem brains kept at room temperature, the concentration of anandamide gradually increases [44, 45]. It would be of interest to find out whether anandamide concentrations also increase *in vivo* in brain after trauma, indicating a possible role for this fatty acid amide. Preliminary experiments by Dr. E. Shohami (unpublished) have shown that administration

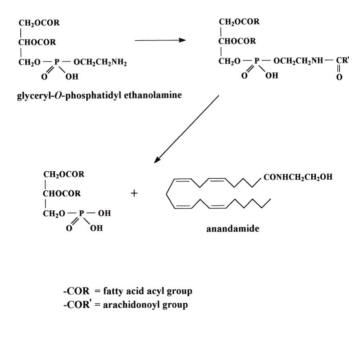

Figure 5.4 Biosynthesis of anandamide.

of anandamide intracerebrally to rats after brain trauma reduces the neurological deficit caused. If these results are confirmed, they may indicate not only a role for anandamide in brain but may point out a new direction for pharmacological treatment of brain injury.

Anandamide is rapidly hydrolysed enzymatically to arachidonic acid and ethanol-amine by a fatty acid amide hydrolase (FAAH) [50]. The molecular characterization, cloning and expression of FAAH have been reported in a recent study [51]. FAAH can be blocked with either the general serine protease inhibitor phenyl methylsulphonyl fluoride [38] or with the highly efficient methyl arachidonyl fluorophsphonate [52]. The non-steroidal antiinflammatory ibuprofen in therapeutic doses, but not aspirin, sulindac or acetaminophen, also inhibits anandamide metabolism [53]. This observation may have therapeutic implications.

It is outside the scope of this review to summarize the numerous publications dealing with anandamide activity *in vitro* and *in vivo*. Several recent reviews, cited above should be consulted. We would like to point out some results of potential pharmaceutical value.

Interaction between anandamide and several neuroreceptor systems has been recorded: (a) anandamide enhances GABA-ergic neurotransmission [54] and reduces GABA uptake in the globus pallidus [55]; (b) inhibition of motor behaviour by anandamide is paralleled by lowering the activity of nigrostriatal dopaminergic neurons [56–58]; (c) activation of cannabinoid receptors by anandamide inhibits the presynaptic release of glutamate [59]; (d) anandamide inhibits exocytotic noradrenaline release by activation of CB_1 receptors located on peripheral sympathetic nerve terminals [60]; (e) anandamide inhibits the activation of serotonin (5-HT_3) receptors in rat nodose ganglion neurons [61]. The relevance of these interactions, as well as a long list of other *in vitro* effects [10], to the pharmaceutical potential of anandamide is yet to be established.

Topical application of anandamide to rabbit cerebral arterioles at low concentrations caused a dose-dependent dilation [62]. This observation may be related to the use of cannabis in the past in migraine.

Intracerebroventricular (i.c.v.) administration of anandamide increased significantly the serum levels of adrenocorticotropic hormone (ACTH) and corticosterone and caused a pronounced depletion of corticosterone releasing factor (CRF-41) in the median eminence [63]. This observation may be of some importance in future development of cannabinoid anti-inflammatory drugs.

Anandamide is a potent inhibitor of gap junction conductance [64]. It blocks the propagation of astrocyte calcium waves and may thus control certain intercellular communication between neurons and astrocytes. Other cannabinoids did not cause this effect and the CB_1 antagonist SR141716A (8) does not prevent it. Apparently, the mechanism of action of anandamide in gap junctions is not *via* CB_1 or CB_2 and may follow the 'anaesthetic-type' of action put forward by Paton (see above). Gap junctions play important roles in the exchange of information and metabolites, in particular in the nervous system and changes in their stuctures and function may be associated with several diseases such as peroneal muscular atrophy (Charcot-Marie-Tooth disease) and intractable epilepsy.

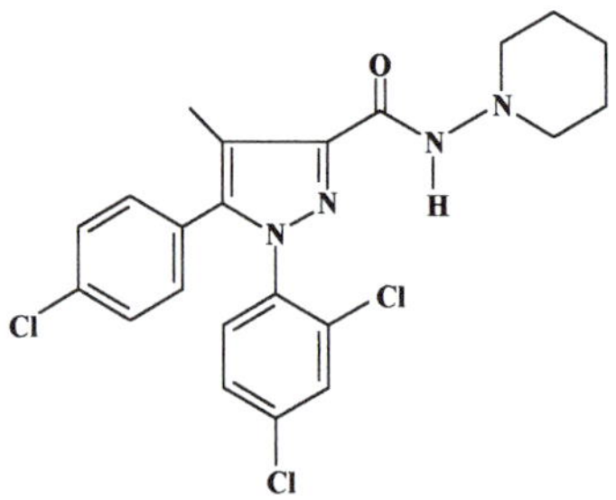

(8) SR141716A

Nitric oxide (NO) is a naturally occurring vasodilator (endothelium-derived relaxing factor). A further, lipid-soluble factor – endothelium-derived hyperpolarizing factor, EDHF – has been postulated. Randall *et al.* in a recent publication suggest that anandamide, or a closely related endogenous cannabinoid, is this endothelium-derived vasorelaxant [65]. Indeed they found that the NO-independent component of endothelium-dependent relaxation (which is presumed to be mediated *via* EDHF), is inhibited by SR141716A, and that anandamide is a potent vasorelaxant in the mesentery. These findings may have far-reaching implications in our understanding of several biological mechanisms and may represent a new lead to cardiovascular drugs.

A rather unexpected observation is that anandamide reduces sperm fertilizing capacity in sea urchins by inhibiting the acrosome reaction [66]. The acrosome is an organelle that surrounds part of the nucleus of the sperm. When the sperm approaches the egg a specific ligand in the jelly coat of the egg crosses over to the sperm and stimulates the transformation of a non-fertilizing sperm to a fertilizing one (the acrosome reaction). This observation may explain the presence of CB_1 receptors in sperm cells [67]. Whether the inhibition of the acrosome reaction by anandamide is of physiological relevance in mammals remains to be clarified. Anandamide may perhaps serve as a lead to novel local contraceptive drugs.

A presumably related observation is that CB_1 (but not CB_2) is expressed in mouse uterus [68] and that this organ produces anandamide which peaks when the uterus becomes resistant to implantation of the embryo [69]. These publications indicate that anandamide may be a regulator of fertility mechanisms. We have observed that although repeated injections of anandamide during the last trimester of gestation in the mouse fail to affect birth outcome (litter size, pup weights and early development), such treatment produces long-lasting effects in the offspring which are observed only at maturity [70, 71] and which may be similar but not necessarily identical to sequelae of prenatal THC administration [71, 72]. Thus on prenatal exposure to either anandamide or Δ^9-THC, the offspring after day 40 displayed a mild behavioural syndrome (as assessed by the tests for locomotion, catalepsy, analgesia and hypothermia described above), suggesting that the CB_1 receptor was mildly but permanently stimulated. Indeed, we found higher concentrations of CB_1 receptors in the forebrains of offsping of mothers which had been exposed to THC during pregnancy.

2-ARACHIDONOYLGLYCEROL (2-ARA-GL) (9)

The existence of a peripheral cannabinoid receptor (CB_2) led us to investigate the possibility that peripheral cannabinoid ligands, distinct from the

Table 5.1   INCREASE OF BINDING AFFINITY OF 2-Ara-Gl TO CB_2-CHO BY ENDOGENOUS MONOACYLGLYCEROLS

COMPOUNDS	RATIOS (molar)	Ki ± S.E. (nM)
2-Arachidonoylglycerol (2-Ara-Gl)		1640 ± 260
2-Linoleoylglycerol (2-Lino-Gl)		inactive
2-Palmitoylglycerol (2-Palm-Gl)		inactive
2-Lino-Gl + 2-Palm-Gl	2.4 : 1	inactive
2-Ara-Gl + 2-Lino-Gl	1 : 12	476 ± 23.5
2-Ara-Gl + 2-Palm-Gl	1 : 5	339 ± 5.6
2-Ara-Gl + 2-Lino-Gl + 2-Palm-Gl	1 : 12 : 5	273 ± 22

Data are the means ± S.E. of 3 experiments performed in triplicate.
The ratios in the mixture of the monoacylglycerols were as present in the spleen on the basis of the CG spectra.

anandamides, are present in such organs. Following the procedures described by us for the isolation of anandamide, we isolated a lipid-soluble constituent from canine gut. Its structure was elucidated as 2-arachidonoyl-glycerol (2-Ara-Gl) (9) [73] on the basis of chromatographic, NMR and MS data, and ultimately by direct comparison with a synthetic sample. Later, 2-Ara-Gl was also located in brain by Sugiura *et al.* [74]. 2-Ara-Gl binds to both CB_1 and CB_2. In pancreas and spleen, 2-Ara-Gl is accompanied by additional 2-acylglycerols, such as 2-palmitoylglycerol (10) and 2-linoleoyl-glycerol (11), which considerably potentiate its binding to CB_2 (see *Table 5.1*) [75].

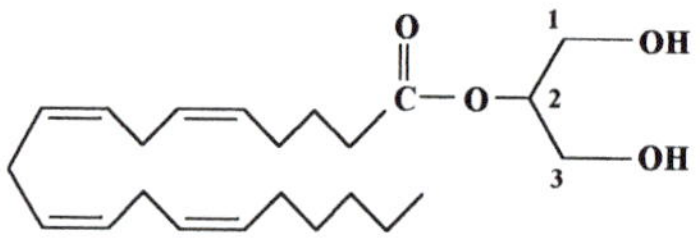

(9) 2-arachidonoylglycerol

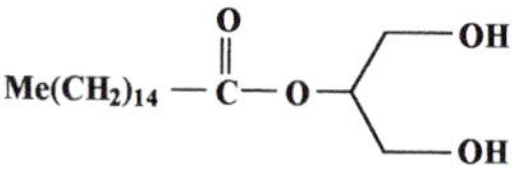

(10) 2-palmitoylglycerol

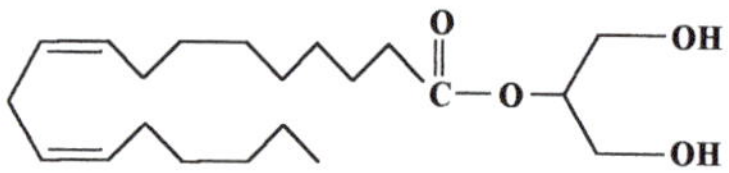

(11) 2-linoleoylglycerol

Lee *et al.* have found that 2-Ara-Gl produced a potent, dose-related inhibition of T and B cell proliferation. 2-Ara-Gl also inhibited forskolin-stimulated adenylyl cyclase [76]. In contrast, anandamide did not inhibit forskolin-stimulated adenylyl cyclase activity in splenocytes, and did not inhibit T or B cell proliferation. Anandamide and 2-Ara-Gl apparently differ considerably in their biochemical profiles in the immune system.

Sugiura *et al.* have reported that 2-Ara-Gl activity is not confined to immune cells. In neuroblastoma cells, 2-Ara-Gl in low doses (nM) induced a rapid, transient elevation of intracellular Ca^{++} [77]. At high doses (μM), the opposite effect was seen [78]. The implications for medicinal chemistry of the presence and effects of 2-Ara-Gl in both the brain and periphery are not yet clear.

STRUCTURE-ACTIVITY RELATIONSHIPS (SAR)

THE ANANDAMIDE SERIES

Anandamide was first reported late in 1992. Since then a sizeable list of publications has appeared that clarify the SAR mostly as regards binding [79–89]. These are summarized below. As K_i values are very sensitive to the experimental conditions, they vary in the different publications. However, a trend seems to persist. The following tentative relationships can be formulated [86]:

1. In the 20:4, n-6 series [90], the unsubstituted amide is inactive; *N*-monoalkylation, at least up to a branched pentyl group, leads to significant binding. The following regularities in binding were noted for anandamide-type compounds with the indicated *N*-alkyl moieties: n-C_5H_{11} < branched C_5H_{11} < CHMeCH$_2$Me (either *R* or *S*) < n-Bu < t-Bu < Me < Et < iPr <Pr. The last two compounds were the most active in these homologous series, with K_i values about three times lower than that of anandamide.

2. *N,N*-Dialkylation, with or without hydroxylation on one of the alkyl groups, leads to loss of activity.

3. Hydroxylation of the *N*-monoalkyl group at the ω-carbon atom (as in anandamide itself) retains activity, as compared to the parent *N*-alkyl group. However, in most cases this activity is slightly lower, at least in the relatively potent compounds.

4. The methyl ether and the phosphate are less active than the parent alcohol. The carboxylic acid derivatives are inactive.

5. In the 20:x, n-6 series, x has to be 3 or 4; two double bonds only leads to inactivation.

6. In the n-3 series, the limited data suggest that the derived ethanolamides are either inactive or less active than related compounds in the n-6 series.

7. Alkylation or dialkylation of the α-carbon adjacent to the carbonyl group retains the level of binding in the case of anandamide. α-Monomethylation or α,α-dimethylation of *N*-propyl derivatives potentiated binding and led to highly active compounds.

8. The presence of a chiral centre on the *N*-alkyl substituent leads to enantiomers with significantly different levels of binding. One of these compounds *R*-(+)-arachidonoyl-1′-hydroxy-2′-propylamide [(*R*)-methandamide] (12) has a 4-fold lower K_i than anandamide and has been shown to possess stability to enzyme hydrolysis [80].

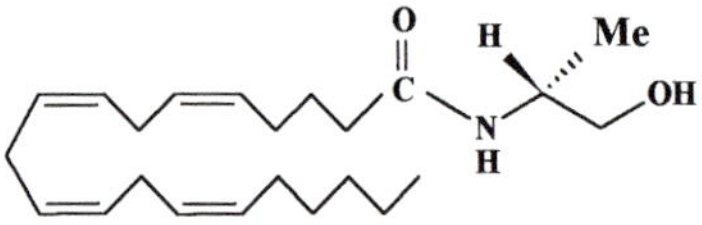

(12) (R)-methanandamide

9. The OH group in anandamide can be replaced by a fluorine with about 10 times increase in specific binding to CB_1 but not to CB_2 [87].
10. Conjugation of the double bonds leads to reduced activity [89].

THE 2-ARACHIDONOYLGLYCEROL (2-ARA-GL) SERIES

No SAR studies have been reported so far; however, unpublished results from our laboratory indicate that, as expected, 1-arachidonoylglycerol is as potent as 2-arachidonoylglycerol (9) on binding to both CB_1 and CB_2 and that 2-palmitoyl-glycerol (2-Palm-Gl) (10) and 2-linoleoylglycerol (2-Lino-Gl) (11), which accompany 2-Ara-Gl in the gut, spleen and pancreas do not bind to the receptors. However, as mentioned above, 2-Palm-Gl (10) and 2-Lino-Gl (11) administered together significantly potentiated the binding of 2-Ara-Gl. 2-Lino-Gl and 2-Palm-Gl together also significantly potentiated inhibition of motor behaviour, immobility on a ring and hypothermia caused by 2-Ara-Gl in mice. 2-Lino-Gl, but not 2-Palm-Gl, significantly inhibited both the hydrolysis and the uptake of 2-Ara-Gl by RBL-2H3 cells [75]. All these effects indicate that the potency of 2-Ara-Gl can be increased by related 2-acylglycerols, which alone show no significant activity in any of the tests employed. This effect ('entourage effect') may represent a novel route for molecular regulation of endogenous cannabinoid activity.

CLASSICAL CANNABINOIDS

The term classical cannabinoids refers to tricyclic THC-type cannabinoids. The SAR in these series have been quite extensively investigated in the past and several reviews have appeared [21, 91, 92]. The structure-activity requirements formulated have withstood the erosion of time. Most of the published work refers to *in vivo* tests for effects now assumed to be due to CB_1 activation. Compton *et al.* have recently correlated binding to CB_1 with *in vivo* activity and have concluded that the requirements for binding to the cannabinoid receptor correlate well with activity across different species, and that receptor binding mediates most of the pharmacological effects of cannabinoids [93].

Martin *et al.* [92] have critically summarized the molecular modelling studies in the cannabinoid field, much of it based on the excellent analyses of Reggio [94–97] which have thrown light on the binding pocket for cannabinoids at the cannabinoid receptor. A receptor steric (RS) map was calculated. The usefulness of this map was demonstrated by an examination of the interaction of cannabinol (CBN) (13) (a natural cannabinoid with low activity) with the CB_1 receptor. The results implied that 'one reason for the reduced affinity of CBN may be that only 53.2% of CBN molecules are shaped properly to fit in the binding pocket for cannabinoids at the CB_1 receptor' [97].

It is generally accepted that three regions are of basic importance in defining the psychotropic activity of tricyclic cannabinoids: (i) the northern part of the terpene ring (carbons 9 and 11); (ii) the free phenolic group, and (iii) the lipid side-chain [21].

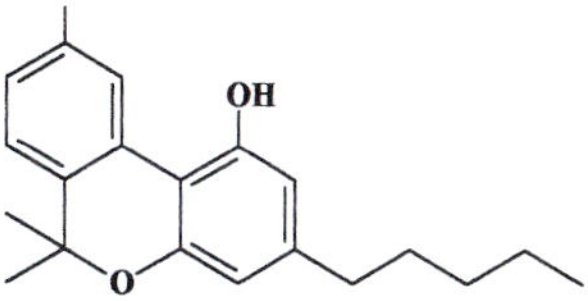

(13) CBN

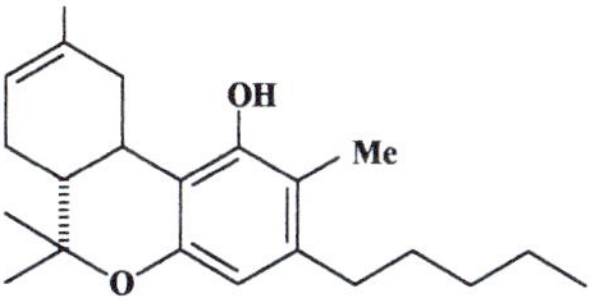

(14) 2-methyl-Δ^8-THC

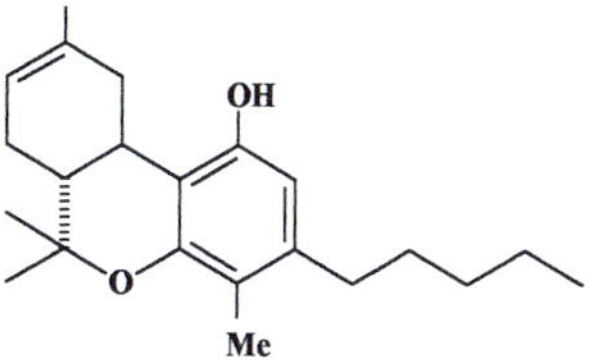

(15) 4-methyl-Δ^8-THC

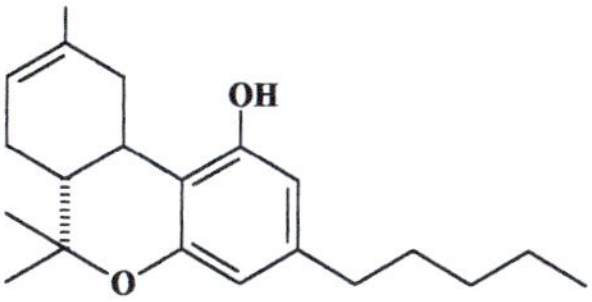

(16) Δ^8-THC

A region where substitution may cause inactivation is the C-4 position (next to the carbon carrying the pyran oxygen). While 2-methyl-Δ^8-THC (14) is a potent cannabimimetic, both on binding to CB_1 and *in vivo*, the isomeric 4-methyl-Δ^8-THC (15) is inactive. It was shown that the active Δ^8-THC (16) and 2-methyl-Δ^8-THC (14) share common structural features that are not exhibited by the inactive 4-methyl-Δ^8-THC (16) [98].

While the SAR of the classical cannabinoids for CB_1 are well known, those for CB_2 have not yet been thoroughly exploited. The few reports published so far indicate that the structural requirements for binding to CB_2 differ very much from those of CB_1. Gareau *et al.* reported that the dimethylheptyl homologue (17a) of Δ^8-THC binds to both CB_1 and CB_2 (K_i values of 0.83 and 0.49 nM, respectively); the methyl ether (17b) of this compound binds weakly to CB_1 (K_i 1585 nM), which is compatible with data for other methyl ethers of cannabinoids. However, (17b) binds to CB_2 with a K_i of 20 nM [99]. A further example is the methyl ether (18) of $\Delta^{9(11)}$-THC-DMH in which the ratio of K_i values of CB_1/CB_2 binding is > 1000. This selectivity of binding activity, if shown to hold for *in vivo* tests, may represent an opening to new cannabinoid drugs which act on the periphery only, without CNS side-effects.

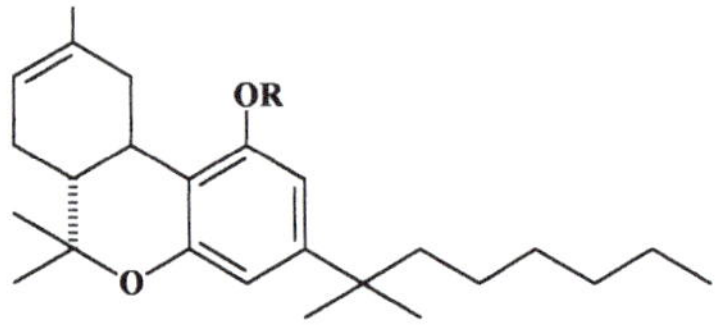

(17) a, R = H
b, R = Me

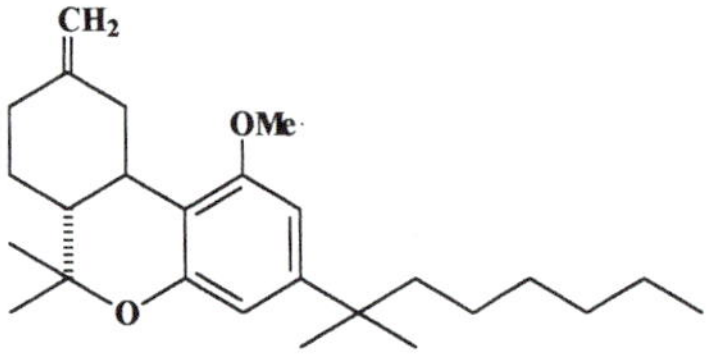

(18)

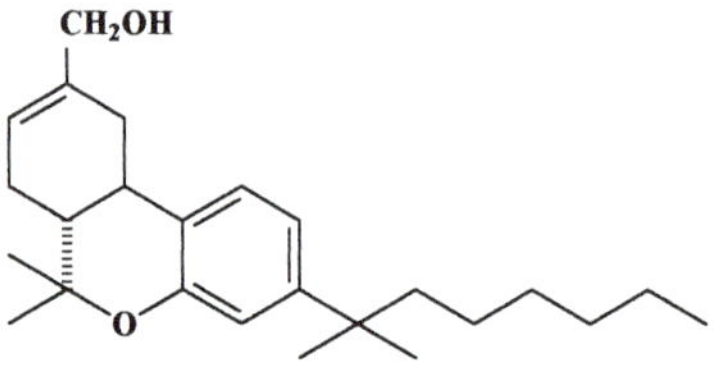

(19)   1-deoxy-11-hydroxy-Δ^8-THC-DMH

The same group has also found indole derivatives with K_i ratios of CB_1 to CB_2 binding of up to 160 [100]. Huffman *et al.* have reported related observations. With racemic 1-deoxy-11-hydroxy-Δ^8-THC-DMH (19) the ratio of the K_i for CB_1 and CB_2 is 37.5, which, though considerably less than the values found by Gareau *et al.*, points in the same direction [101].

Numerous cannabinoids, anandamides and heterocyclics have been compared for CB_1 and CB_2 binding [87]. Most compounds did not differ significantly in their binding values for the two receptors except the specific CB_1 antagonist SR141716A (see below) which binds weakly to CB_2 and the indole compound JWH-015 (20) which showed a CB_1/CB_2 binding ratio of 27.75. Busch-Petersen *et al.* synthesized a series of cannabinoids in which rotation around the C_1'-C_2' bond was blocked [102]. The compound with the cis-heptene side-chain (21) had the highest affinity for CB_1 (0.89 μM) with the widest separation between CB_1 and CB_2 affinities. Pertwee *et al.* showed that 6'-cyano-2'-yne-Δ^8-THC (22) (K_i = 0.77 μM) has agonist-antagonist properties [103]; 6'-cyano-Δ^8-THC-DMH was a potent agonist [104].

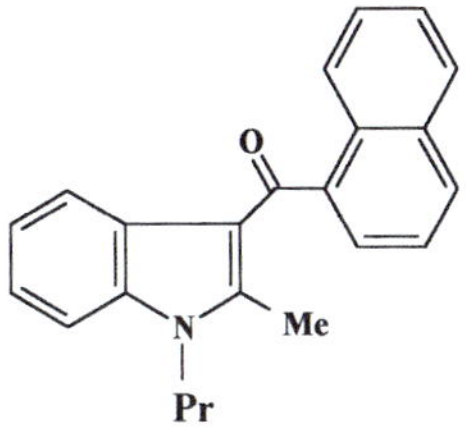

(20) JWH-015 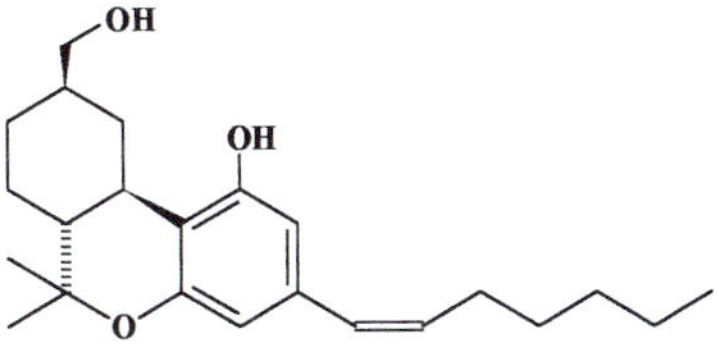   (21)

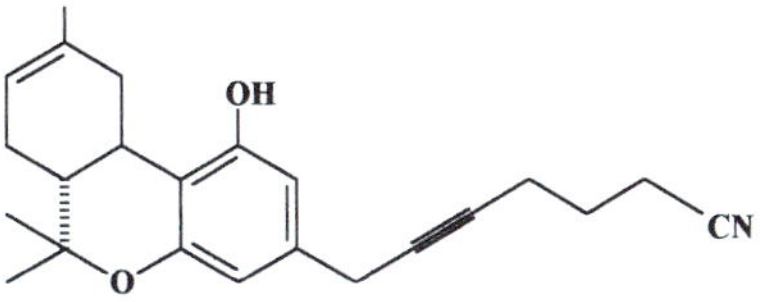

(22)

NON-CLASSICAL CANNABINOIDS

There are two major groups of synthetic compounds which have cannabinoid activity, but which differ chemically from the tricyclic THC-like cannabinoids: the bicyclic cannabinoids, exemplified by compound CP55940 (23), and the (aminoalkyl)indoles exemplified by pravadoline (24a). A detailed SAR analysis of these groups of compounds is beyond the scope of this review. The bicyclic cannabinoids and derivatives have been reviewed previously [105]; recent publications deal mainly with related tricyclic non-classical cannabinoids [106] and with the (aminoalkyl)indoles [92, 107]. It is of interest to note that while the bicyclic cannabinoids were originally prepared as 'simplified' cannabinoids, the cannabinoid-type activity of the (aminoalkyl)indoles was discovered by serendipity. These compounds were synthesized in a project aimed at the discovery of novel nonsteroidal anti-inflammatory agents presumably based on the indomethacin structure. However, while they did not possess anti-inflammatory properties, they were found to be antinociceptive, and to inhibit the electrically evoked contractions in a mouse vas deferens muscle preparation. This led to binding experiments: these (aminoalkyl)indole derivatives were able to displace [^{3}H]CP55940 (23) binding to the cannabinoid receptor in brain membranes, but did not displace radioligands selective for 20 other neuroreceptors.

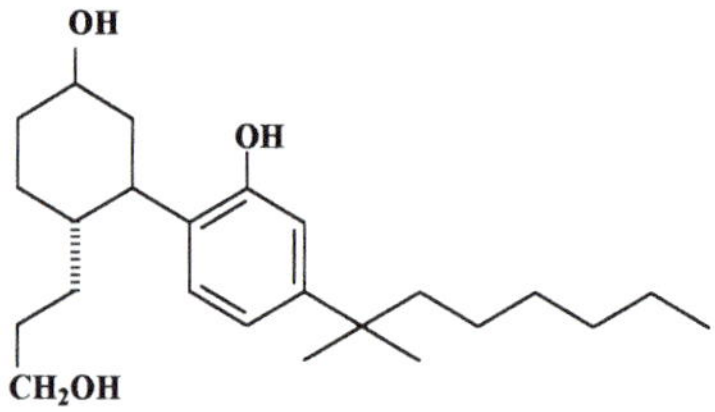

(23) CP-55940

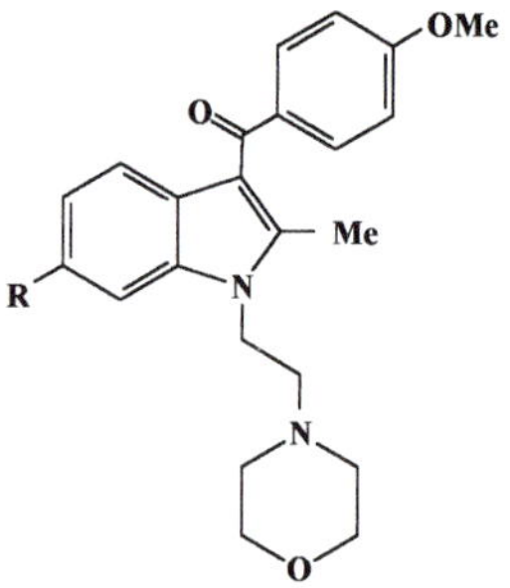

(24) a, R = H, pravadoline
b, R = I, AM 630

ANTAGONISTS OF CANNABINOIDS

A striking example of the possible difference in binding to CB_1 and CB_2 is the highly specific antagonist to CB_1, the heterocyclic SR141716A (8), which has become a very useful tool in cannabinoid research [108]. It binds weakly to CB_2 but its K_i for CB_1 is in the low nM region. Δ^9-THC interacts competitively with the binding sites labelled with SR141716A; by contrast, anandamide displaced the antagonist non-competitively, emphasizing that THC and anandamide differ in their binding characteristics [109].

SR141716A produces a 2-fold potentiation of electrically stimulated acetylcholine (ACh) release in the hippocampus [110]. As cannabinoid receptor activation can produce a strong inhibition of ACh release in the hippocampus, the potentiation of ACh release in the hippocampus by SR141716A suggests that this compound may actually be an inverse agonist at cannabinoid receptors or that it antagonizes the inhibitory actions of an endogenous ligand acting on these receptors. Similar conclusions have been drawn by Compton et al. [111], who found that while SR141716A completely inhibited THC-induced activity in a locomotor test at doses between 0.3 and 3 mg/kg (i.v.) in mice, at high doses (between 3 and 20 mg/kg i.v.) locomotion was simulated. It was suggested that SR141716A produced this effect through mechanisms not yet identified which would infer some lack of specificity for the CB_1 receptor. The possibility that SR141716A was an inverse agonist at the CB_1 receptor was also raised.

Terranova et al. [112] found that SR141716A reverses inhibition of long-term potentiation caused by anandamide and concluded that anandamide could be a candidate for an endogenous neuromessenger involved in memory processes. Indeed, SR141716A at low doses was found to improve short-term olfactory memory in a social recognition test in rodents [113]. A role for anandamide in forgetting was tentatively proposed.

A second antagonist, AM 630 (24b), a novel aminoalkylindole, was found to attenuate the ability of some cannabinoids to inhibit electrically-evoked twitches of the mouse isolated vas deferens [114]. AM 630 was a more potent antagonist of Δ^9-THC than of anandamide (K_d of 14.0 and 278.8 nM, respectively). It was suggested that the receptors for which AM 630 has the highest activity may not be CB_1 cannabinoid receptors. This is supported by the observation that AM 630 is actually a cannabinoid agonist in the myenteric plexus – muscle preparation [115]. Yamada *et al.* [116] showed that isothiocyanate derivatives of pravadoline can serve as potential electrophilic affinity ligands for CB_1.

MEDICINAL PROPERTIES OF CANNABINOIDS

ANTIEMETIC ACTION OF CANNABINOIDS

In our 1987 review, we summarized the research and clinical experience in this area [1]. Surprisingly, in spite of the enormous public interest in 'medical marijuana' and countless articles in the daily press and magazines focused predominantly on this aspect of marijuana use, little progress has been reported on the antiemetic activity of cannabinoids in the last decade. Plasse *et al.* have reviewed the clinical experience gained over 7 years with dronabinol (Δ^9-THC) in antiemetic treatment [117]. With doses of 7 mg/m^2 or below, complete response was noted in 36% of the patients, 32% showed partial response and 32% showed no response. However, 65% displayed drowsiness and dizziness and 12% had dysphoric effects. Combination treatment of dronabinol with prochlorperazine (a dopamine receptor blocker widely used as an antipsychotic drug with antiemetic effects) was more effective than each drug alone [118].

The antiemetic activity of cannabinoids is not mediated, at least not with all cannabinoids, *via* either CB_1 or CB_2. We found that HU-211 (4), a nonpsychotropic cannabinoid, which does not bind significantly to either CB_1 or CB_2, is a potent antiemetic agent in pigeons administered cisplatin [119]. Unexpectedly, addition of low doses of cupric ions potentiated the activity of HU-211. The activity of HU-211 was U shaped: 2.5 mg/kg HU-211 (with 0.8 mg/kg $CuCl_2$) reduced the amount of vomitus by 97%; at higher or lower doses the activity was significantly lowered. The antiemetic activity of Δ^9-THC was compared with that of HU-211: at 5 mg/kg Δ^9-THC (without Cu^{++}) reduced vomiting by 24%, while HU-211 (2.5 mg/kg, no Cu^{++}) showed a 89% reduction. So far, HU-211 has not been tested as an antiemetic in patients. As HU-211 is now in Phase II clinical trials against brain trauma (see below), it may be evaluated later for this type of activity.

The mechanism of cannabinoid antiemetic activity is not well understood. As pointed out above, it does not seem to involve CB_1 or CB_2. Fan has shown that anandamide and some synthetic cannabinoid agonists inhibit the activation of $5\text{-}HT_3$ receptors in rat nodose ganglion neurons [61]. These receptors are known to mediate emetic activity. However, we have found that HU-211 does not bind to $5\text{-}HT_3$ receptors (unpublished observations). Hence, although cannabinoids may act, in part at least, through this serotonin receptor, this mode of action does not account for the total activity.

We have noted that in mice the response to THC develops gradually and does not reach maximal potency until adulthood [120]. THC administered to mice shortly after birth exhibits essentially no activity. Significant activity is observed only after day 20. We assumed that if this observation is relevant for humans, children will not experience THC-type cannabimimetic effects until about the begining of puberty. As the antiemetic effects are apparently not mediated by the cannabinoid receptors, it was reasonable to expect that antiemetic effects of THC in young children could be observed while cannabimimetic effects would be absent. We chose Δ^8-THC, rather than the marketed Δ^9-THC, as it is less psychotropic, much less expensive than Δ^9-THC to produce and is much more stable than Δ^9-THC. We started a clinical trial with a low dose (5 mg/m^2) which in some adults already causes psychotropic effects. None was observed in young children. Hence, the antiemetic dose was increased from 5 mg/m^2 up to 18 mg/m^2. At his very high dose, which cannot be administered to adults due to side-effects, children showed no such effect and no vomiting or nausea were noted [121]. Although this clinical trial was an open one, we believe that the results are highly significant as, almost without exception, young children in the same hospital undergoing similar treatments vomited during cancer chemotherapy. The number of children was not large (n = 14); however, together they were administered Δ^8-THC about 400 times over a period of 2–3 years. No tolerance to the drug was noted [121].

A survey of oncologists on the use of marijuana (on smoking) as an antiemetic drug has indicated that about 50% are willing to recommend it to their patients, and considered it somewhat more effective than dronabinol (oral administration) [122]. It is therefore surprising that Δ^9-THC has not been administered so far by inhalation.

Serotonin (5-HT$_3$) antagonists are the mainstay today of drug treatment of acute chemotherapy-induced emesis [123]. However, the problem of delayed emesis has yet to be solved. Apparently dronabinol has not been tested so far as an add-on drug to HT$_3$ antagonists to prevent this condition.

APPETITE STIMULATION

Δ^9-THC (dronabinol) was introduced in the clinic as an antiemetic drug. However, it was soon realized that numerous AIDS patients use it primarily as an appetite stimulant. One of the typical symptoms of AIDS is cachexia. Indeed, in Central Africa, AIDS is known as the 'thin disease'. Cannabis has actually been used in Indian popular medicine for appetite stimulation especially by young women [124].

Dronabinol is now an FDA-approved drug for the treatment of cachexia and is prescribed for cancer and AIDS patients. The clinical literature however is limited. Plasse *et al.* reported the results of two multicentre, open studies on the effects of dronabinol as an appetite stimulant in cancer patients [117]. Various doses and dose schedules were followed and side-effects *versus* dose levels were evaluated. Their conclusion was that 'overall, no group gained a significant amount of weight while on the study, though some individual patients did actually gain weight. However, there was a reduction in the rate of weight loss in all groups'. The reduction in the rate of weight loss was significant for the groups administered 2.5 mg and 5 mg four times a day. Psychotropic effects were not noted at the lower dose; weakness and fatigue, dizziness, drowsiness and memory or concentration difficulties were recorded at the higher dose. Preliminary results of an open study with ten symptomatic AIDS patients who received 2.5 mg dronabinol three times a day for 1–5 months were also reported [117]. The patients were able to adjust the dose to avoid side-effects. Before treatment, the patients were reported to be losing 0.93 kg/month (median). On therapy, a weight gain of 0.54 kg/month (median) was recorded.

A second study gave comparable results. Twelve AIDS patients were administered dronabinol, 5 mg twice a day [125]. An improved appetite score was noted, body fat increased and symptom stress decreased.

Based on the results of these pilot studies, the effects of dronabinol (2.5 mg, twice a day) on appetite, weight, nausea and mood were examined in 139 AIDS patients over a six-week period [126]. After 4 weeks, significant differences in anorexia were seen and these were sustained throughout the trial. Weight was stable in the treated patients but was lower in the placebo recipients. The data indicated that dronabinol caused increased appetite (38%), decreased nausea (20%) and some improvement in mood. The side-effects noted were minor and were not severe enough to require intervention. A small number of patients, that received only 2.5 mg once a day due to the side-effects, showed an appetite increase similar to that of patients treated with dronabinol twice a day. This low dose of THC is not expected to produce any significant side-effects. The authors of this study conclude that 'dronabinol is a safe and effective treatment for anorexia in patients with weight loss due to AIDS. By improving appetite and mood, decreasing nausea, and stabilizing weight, dronabinol may significantly improve the quality of life of patients infected with HIV.'

A 3-day appetite stimulation study (with dronabinol 2.5 mg twice a day orally) with healthy subjects showed no appetite increase, but food intake was significantly increased when the drug was administered by a rectal suppository [127].

The use of dronabinol and marijuana in AIDS patients has been correlated with their drug therapy [128]. Azidothymidine (AZT) and AZT-dideoxycytidine therapy was not compatible with Δ^9-THC or marihjuana as a declining health status was noted. However, all clinical indicators of pancreatitis (a common AIDS related disease) improved in dideoxyinosine-treated patients who were administered THC or marijuana.

The physiological basis of the effects of cannabinoids on feeding behaviour is not known. A French group has reported that neurons in the solitary tract nucleus, which respond to increases in glucose concentrations, are sensitive to Δ^9-THC and have suggested that such neurons may mediate cannabinoid effects on feeding behaviour [129].

Dronabinol is a rather expensive drug for chronic, long-term disease and may cost a patient several hundred dollars a month. The use of the much less expensive Δ^8-THC should be evaluated. This compound, which parallels Δ^9-THC in its antiemetic effects in children (see above), is considerably less expensive to prepare and is much more stable. While Δ^9-THC is easily oxidized to cannabinol by air, Δ^8-THC does not undergo this reaction.

MULTIPLE SCLEROSIS (MS) AND SPINAL INJURY

The Assyrians apparently used cannabis for neurological diseases. Azallu (an Assyrian term for cannabis used for medicinal purposes) was a drug against 'poison of all limbs' and against 'arimtu', which was probably a neurological disease of the legs [130]. Was one of these two disorders multiple sclerosis? We shall, of course, never know.

Multiple sclerosis is a disorder of the brain and the spinal cord that attacks the myelin sheath of nerve fibres. The cause is unknown but it is presumably an autoimmune disease triggered by an environmental factor. There is a genetic susceptibility. The disease is usually slowly progressive with exacerbations and remissions over many years. The symptoms vary; they include tiredness, spasticity, weakness in the legs and inability to walk, lack of balance, muscle pain, tremor, vision problems, urinary hesitancy and incontinence, slurred speech and memory loss. Anxiety and depression are common. A variety of partially effective drug treatments are in use. Corticosteroids are the cornerstone of therapy. However, their efficacy is limited and their chronic use entails considerable risk. Carbamazepine and phenytoin are used for pain (including trigeminal neuralgia) and numerous other drugs are employed for specific symptoms. Several new drugs have been approved for use in the last year, e.g., copexone. However, apparently they are far from ideal. The lack of an efficient drug treatment has led many patients to search for additional therapies, one of which is cannabis. Today, cannabis, is used by an apparently large number of multiple sclerosis patients, although it is not an approved drug.

Two animal studies have been described with cannabinoids in a model of MS. This model, experimental autoimmune encephalomyelitis (EAE), is based on the generalized atonia, quadriplegia and death of rats or guinea-pigs administered CNS tissue or myelin basic protein. The animals are observed over a period of several weeks. In one of these studies, Lyman *et al.* noted that Δ^9-THC decreased EAE inflammation, and led to much reduced effects and that the time of the appearance of the MS effects was delayed [131]. In the second study, Δ^8-THC was employed. Again the drug significantly reduced the incidence and severity of neurological deficit in two strains of rats [132]. Serum corticosterone levels were elevated two-fold in rats with EAE chronically treated with Δ^8-THC. These results suggest that suppression of EAE by cannabinoids may be related to their effect on corticosterone secretion. A related publication reports the use of WIN55212–5, an (aminoalkyl)indole derivative, in Syrian hamsters with primary dystonia [133]. Again, improvement was noted. Diazepam acted synergistically with WIN55212–5. This observation parallels previous work in which the activity of cannabinoids was found to be synergized by benzodiazepines [134].

Consroe *et al.* have reported and analysed the answers to a questionnaire mailed to MS patients who use cannabis [135]. The 112 patients who responded to the questionnaire were about equally divided between USA and UK and between male and female patients. Most of the patients reported strong improvement after cannabis in spasticity at sleep onset, and on awakening at night, as well as reduction of leg pain at night and of tremor. The patients also reported improvement in anxiety and depression as well as in spasticity when awaking in the morning and on walking. There was only minor improvement in memory loss, in faecal incontinence and in constipation. This report should be considered a good basis for the initiation of clinical trials.

There are data from several small clinical trials with Δ^9-THC or with nabilone (25), a synthetic cannabinoid agonist. The first trial (double blind, placebo-controlled) was by Petro and Ellenberger who recorded that in a group of nine patients, administered up to 10 mg Δ^9-THC, three felt that they were better able to walk [136]. The authors measured deep tendon reflexes, muscular resistance to stretch of the legs and abnormal reflexes, and found improvements.

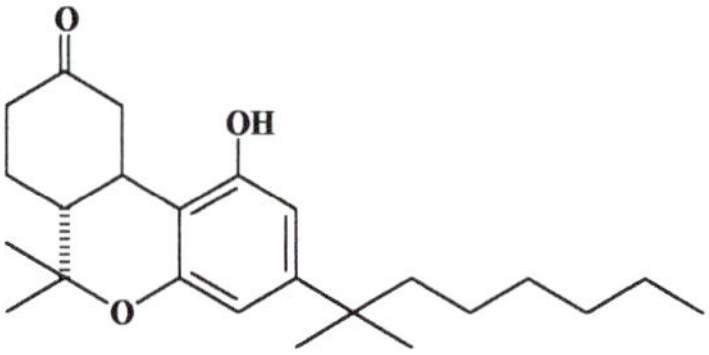

(25) nabilone

Clifford observed objective improvement in two patients in a single blind, placebo-controlled study with Δ^9-THC in eight patients with disabling tremors and ataxia. The rest of the patients reported subjective, though minor improvement in tremor, and in sense of well-being [137]. The therapeutic effects of THC in the two patients were reproducible. In one of them, a long-lasting improvement of tremor was noted, while in the other such an improvement was observed in a handwriting test.

A discrepancy between subjective improvement of spasticity *versus* objectively determined measurements was also noted by Ungerleider *et al.* who treated 13 MS patients (in a double blind, placebo controlled, cross-over study) who had serious side-effects on treatment with conventional drugs [138]. With a dose of 7.5 mg Δ^9-THC, all patients reported a decrease in spasticity; however, in objective functional tests (limb spasticity, co-ordination and weakness and reflexes) no improvement was noted. In contrast, Meinch *et al.* found that, on smoking marijuana in an open label trial, an MS patient showed improved mobility, reduced spasticity and tremor measured by objective methods [139].

A single report on nabilone has appeared. Martyn *et al.* (in a double blind, placebo-controlled trial) administered this drug to a single patient who reported improvement in severity of muscle spasms [140]. In a recent report, Brenneisen *et al.* administered orally Δ^9-THC (10–15 mg) to two patients, and compared its effects with those of Δ^9-THC hemisuccinate administered by suppositories (5 mg) [141]. Both treatments reduced spasticity and rigidity, and improved walking in objective measurements.

The effects of Δ^9-THC (5 mg) on the neurological symptoms of spinal cord injury have been studied in one patient [142]. Numerous treatments over several months resulted in reduction in subjectively-rated spasticity for periods over 12 h, in improvement of bladder control and pain as well as in the quality of mood and sleep. In contrast to the above reports, Greenberg *et al.* found no improvement in 10 MS patients who smoked marijuana on a single occasion [143]. Actually, an impairement of balance was noted.

The human data, as well as the animal studies indicate that cannabinoids may ameliorate some symptoms of MS. Unfortunately, a definitive large-scale study has yet to be performed. Such a study should take into account that (a) the placebo effect in MS is very high; (b) the effective doses recorded are close to, or identical with the doses that cause cannabimimetic effects; (c) cannabis (marijuana, hashish) is obviously not identical to Δ^9-THC, and that other constituents may synergize THC action; (d) smoking cannot be compared with oral administration, and (e) the anal route of administration [141] may be a preferred one.

ANALGESIA

Cannabis and some cannabinoids relieve pain. However, the therapeutic doses are essentially equivalent to the doses that cause CNS effects and, except in very specific conditions (possibly migraine) they are of limited use. Although several companies have produced a large number of derivatives, as previously reviewed in this series [1] and elsewhere [105,107] the situation has not changed. No practical separation of activity has been achieved, except with the synthetic cannabinoid HU-211 (see below) which does not bind to either CB_1 or CB_2. Hence most of the work reported in the last decade deals mainly with the pharmacology of pain reduction by cannabinoids, rather than with drug discovery and development.

Hollister has noted that 'the apparent paradox is that THC both increases and decreases pain' [144]. Indeed, Ames in one of the first modern clinical studies with cannabis reported that in one experimental subject, who had been admistered cannabis, a venepuncture was painful, while in another subject the drug caused marked analgesia [145].

THC augments opioid activity. Thus, administration to mice of Δ^8-THC (3 mg/kg) together with morphine (1 mg/kg) or codeine phosphate (3 mg/kg) potentiated the activity of the opioids 4–5 times [146]. Intrathecal administration of Δ^9-THC (3.13 μg and 6.26 μg) to mice caused a 4- and 12-fold shift in the dose response curve of morphine [147]. While the administration of naloxone does not affect cannabinoid analgesia, it completely blocked the effect of the drug combination. Pretreatment with morphine enhanced the effects of Δ^9-THC to a minor extent only. Cannabinoid-induced antinociception therefore is not due to direct interaction with the opiate receptor. Morphine antinociception is mediated predominantly by μ receptors. Its enhancement by Δ^9-THC, however is through activation of κ and δ receptors [148]. Indeed Δ^9-THC is cross-tolerant with κ opioid receptor agonists [149]. The κ antagonist norbinaltorphimine blocks Δ^9-THC and CP55940 (23) (i.v.) antinociception in mice, but intrathecal CP55940 was not blocked. Apparently κ receptors are involved in cannabinoid-produced analgesia. A Spanish group has published related results [150]. Δ^9-THC antinociception was significantly potentiated by morphine in the tail-flick and hot plate tests. This synergistic effect was blocked by the CB_1 antagonist SR141716A, by naloxone and by norbinaltorphimine. It was concluded that the synergism between Δ^9-THC and morphine involves μ supraspinal and κ spinal opioid as well as cannabinoid receptors [150]. It is of interest that Δ^9-THC also potentiated analgesia produced by acupuncture, a procedure assumed to be mediated by opioid receptors [151].

Lichtman and Martin have shown that cannabinoid-induced antinociception has both spinal and supraspinal components [152]. A spinal α_2-noradrenergic mechanism is involved in cannabinoid antinociception as yohimbine and/or methysergide altered Δ^9-THC induced antinociceptive effects in rats [153]. A supraspinal mechanism is also involved as cannabinoid analgesia can be produced in spinally transected rats [152]. Both similarities and differences were noted on comparison of the antinociceptive effects produced by anandamide (and the more potent fluroanandamide) and Δ^9-THC [154]. Anandamide was cross-tolerant to Δ^9-THC, but in contrast to THC, it did not alter opioid-induced antinociceptive effects nor was its action blocked by a κ antagonist.

Our group has reported that HU-211 (in combination with cupric chloride) caused suppression in a rat model of neuropathic pain [155]. As HU-211 produces no cannabimimetic effects these observations represent a separation of analgesic from psychotropic effects with a cannabinoid. A separation between antinociceptive effects and cannabimimetic effects has been noted with the metabolite, 11-nor-9-carboxy-Δ^9-THC (26) and with its 1,1-dimethylheptyl homologue [156]. A British group has reported that cannabidiol (CBD, 27), and cannabigerol (CBG, 28), which are natural, non-psychotropic cannabinoids, although more potent than Δ^9-THC, are less maximally effective. In the induced writhing test, the ED_{50} of cannabidiol was 0.042 mg/kg, while that of Δ^9-THC was 25 mg/kg [157].

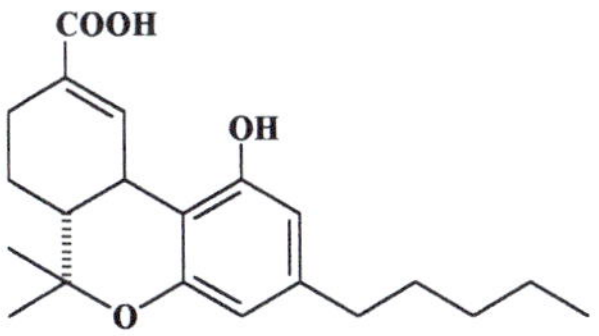

(26)   11-nor-9-carboxy-Δ9-THC

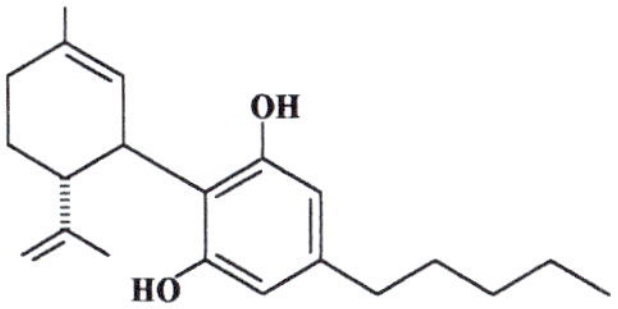

(27)   CBD

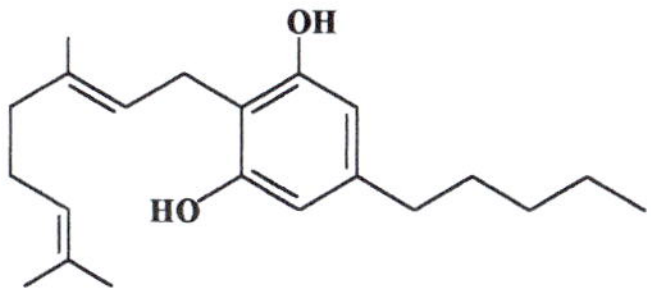

(28)   CBG

The analgesic effects of (+)-WIN55212–2 mesylate, an (aminoalkyl)indole derivative, in a rat model of neuropathic pain have been recorded [158]. The (−)-enantiomer was inactive. It was shown that 'a moderate dose (2.14 mg/ kg) completely alleviated the thermal and mechanical hyperalgesia, and mechanical allodynia without side-effects.' The results also suggested that changes in cannabinoid receptors occur in nerve injured animals.

It has been suggested that one function of endogenous cannabinoids may be to modulate pain sensitivity [159]. Indeed, a recent publication has indicated that the antagonist SR141716A produces hyperalgesia in untreated mice [160].

HYPOTENSIVE ACTIVITY

Cannabis and cannabinoids are known to reduce orthostatic blood pressure [144]. The literature up to 1986 on the cardiovascular effects has been reviewed by Graham [161]. Δ^9-THC (up to 5 mg/kg) administered i.v. to anaesthetized rats caused a profound decrease in mean arterial blood pressure in a dose-dependent manner [162]. Yet very little has been published on this topic. One of the reasons may be that the acute cardiovascular changes, including decreased standing blood pressure or increased supine blood pressure are of 'little consequence for users without cardiovascular disease' [163]. From the point of view of the medicinal chemist, this field apparently did not look promising as it was assumed that the lowering of blood pressure was of central origin and therefore it would be accompanied by the THC-type side-effects, although it was shown 20 years ago that a synthetic isomer of CBD, namely (29), which is not psychotropic, is hypotensive [164]. Several 9-azacannabinoids [such as (30) and (31)] were prepared in order to test their antihypertensive effects [165]. The most potent compound, the benzopyran (30), was less psychoactive than Δ^9-THC. However, on chronic administration to dogs, rapid tachyphylaxis was noted and apparently no further work was undertaken. It should be of interest to take a second look at these azacannabinoids.

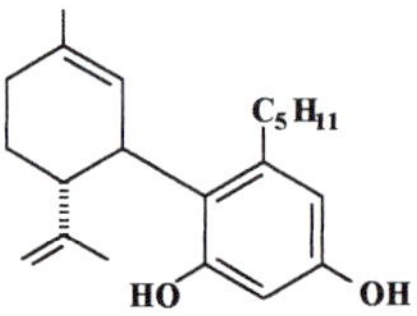

(29)

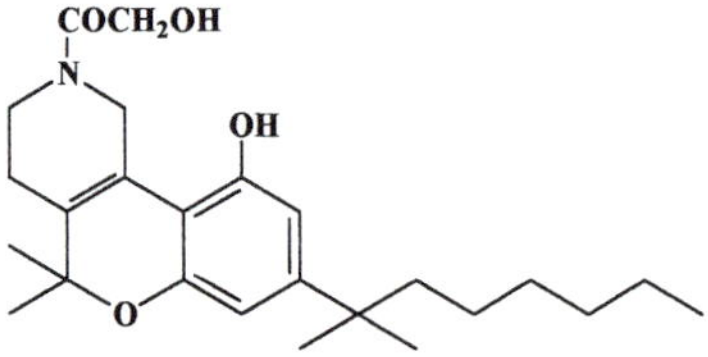

(30)

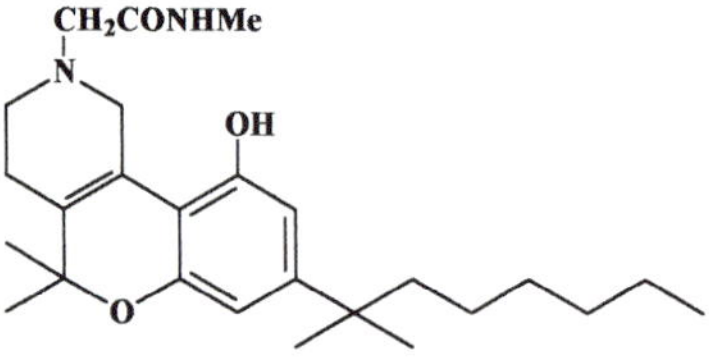

(31)

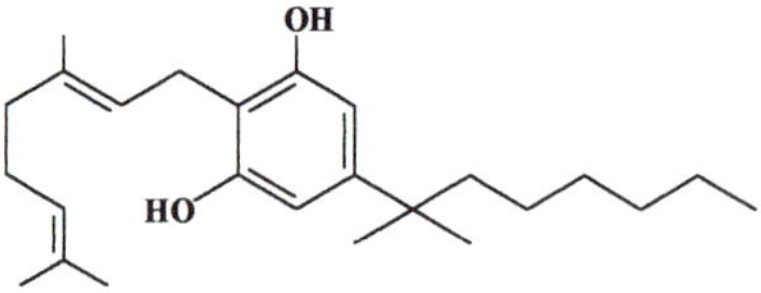

(32) CBG-DMH

In a recent publication, the effects of HU-210 on blood pressure are discussed [166]. It was noted that this very potent psychotropic agent also caused long-lasting hypotension (as well as bradycardia) in rats in doses between 10 and 1000 μg/kg. With Δ^9-THC, an initial pressor effect was found, but with HU-210, this effect was not observed.

In several papers the Kunos group has reported observations that may represent a starting point for novel medicinal chemistry research in this area [167, 168]. Anandamide (i.v. bolus; 4 mg/kg) caused a triphasic blood pressure response, brief hypotension, followed by a transient pressor and then a prolonged depressor phase. The hypotensive effect was not initiated in the CNS, but was due to a presynaptic action that inhibited norepinephrine release from sympathetic nerve terminals in the periphery (heart and vasculature). The inhibitory effect (but not the pressor effect) was antagonized by SR141716A, indicating that this peripheral action was mediated by CB_1 receptors.

In a late 1997 paper, the Kunos group reported that activation of peripheral CB_1 receptors contributes to haemorrhagic hypotension, and that the anandamide produced by macrophages may be a mediator of this effect [168a]. In a series of publications, Randall has proposed that anandamide (or a related cannabinoid) is a long-looked for endothelium-derived hyperpolarizing factor, EDHF [65, 168b]. EDHF relaxes vascular smooth muscle through activation of ATP-dependent potassium channels. As mentioned above, anandamide, like Δ^9-THC, induces vasodilation in cerebral arteries [62]. Randall has now shown that in rat mesenteric arteries (treated with inhibitors of NO synthase and cyclo-oxygenase), the CB_1 antagonist SR141716A inhibits endothellum-dependent relaxations evoked by carbachol. Such relaxations are produced by anandamide. As in most types of hypertension, endothellum-dependent relaxations are curtailed due to lowered production or activity of endothellum-derived NO and EDHF. The above observations may be of central importance in the development of anandamide-type drugs against hypertension [168c].

GLAUCOMA

Whenever 'medical marijuana' is discussed, glaucoma is mentioned alongside vomiting as the principal disease condition for which the drug is helpful. However, in spite of its prominence in the marijuana controversy, very little research has been done on the effects of cannabis on glaucoma in the last decade.

In our 1986 review [1] we summarized our views as follows:

'The cannabinoids tested so far appear to be of limited use in the treatment of glaucoma. They appear to act only against a primary (but not sole) symptom of the disease (i.e., ocular hypertension), rather than against the underlying disease process, which remains uncertain. The side-effects of those cannabinoids particularly effective in lowering intraocular pressure (IOP) restrict their clinical usefulness (with some exceptions such as cannabigerol). Cannabinoids administered intraocularly to humans cause no IOP reduction.'

The situation as regards the terpenophenolic cannabinoids has not changed much. Although cannabigerol, (28) was considered the most promising compound, no SAR studies with it have been reported. As the 1,1-dimethylheptyl THC's are much more active that the THC's, we prepared the same homologue (32) of cannabigerol. Colasanti, who originally had reported that cannabigerol is an active compound [169], tested this new compound and compared it with cannabigerol in a cat model, but found that it was inactive [unpublished results].

It is unfortunate that work on CBG has been discontinued as Colasanti has found that after systemic administration of Δ^9-THC to rats, polyspike discharges appeared in the cortical electroencephalogram, while CBG was devoid of this undesirable side-effect [170]. Additional light has been thrown on the mechanism of cannabinoid action on IOP. The effect of THC is retained after removal of either sympathetic or parasympathetic input in the eyes of cats; neither Δ^9-THC nor cannabigerol altered the rate of aqueous humor formation. However, both compounds increased aqueous outflow [170].

In order to determine whether the mechanism of action of the cannabinoids originates in the CNS, various cannabinoids were administered into the cerebral ventricles of New Zealand albino rabbits [171]. No change in the intraocular pressure was noted, although on intravenous administration Δ^9-THC produced dose-dependent ocular hypotension and miosis. Apparently, the action of the cannabinoids on IOP does not originate in the CNS, and the authors surmise that alterations in blood pressure may be involved in the ocular hypotension. It is surprising that Δ^8-THC did not cause ocular hypotension on i.v. administration as Δ^8-THC and Δ^9-THC are generally assumed to have the same pharmacological profile.

Both our group [172] and Pate *et al.* [173] found that anandamide in normotensive rabbits, on topical application, reduces IOP. In the report by Pate *et al.* anandamide was dissolved in an aqueous solution of a cyclodextrin, single eyedrops were instilled in one of the eyes and the changes in IOP were compared to those of the untreated eye. Topical application of 31.25 μg caused an immediate reduction of IOP. A dose of 62.5 μg induced an initial increase and a subsequent decrease of IOP below baseline, while a dose of 125 μg caused a significant increase without the subsequent decrease of IOP.

The initial increase followed by a prolonged decrease has not been observed previously with cannabinoids and may represent a serious drawback in the eventual clinical use of anandamides as antiglaucoma agents. This biphasic effect is reminiscent of the same phenomenon seen with anandamide in blood pressure measurements (see above).

A series of anandamide-type compounds were compared with anandamide in their action on IOP following the above described methodology [174]. None of the compounds had a better profile than anandamide. The limited SAR recorded did not parallel the SAR established for cannabimimetic activity. Thus, arachidonoyl propylamide, which is several times more potent on binding to CB_1 than anandamide, was inactive in the ocular test; arachidonoyl propanolamide which is as active as anandamide on binding to CB_1 caused only elevation at a dose of 62.5 μM, while anandamide at this dose caused elevation followed by reduction (see above). The observations by Pate *et al.*, in particular the technique of administration in a cyclo-dextrin, may ultimately lead to an active anandamide-type compound which does not cause initial IOP elevation.

SLEEP

Drowsiness or sleepiness are well-known effects in the later stages of intoxi-cation by marijuana. Hollister has reviewed the field [144]. The most signifi-cant finding up to 1986 was that THC reduces rapid eye movement (REM) sleep, in this respect not differing from conventional benzodiazepine hypno-tics. When THC was administered orally to human volunteers (10–30 mg dose), the subjects were reported to experience the initial 'high', but later fell asleep faster. The higher doses caused some 'hangover' the next day [144]. Cannabinoids are well-known to prolong barbiturate sleeping time [175]. This effect is due, in part at least, to a decrease in the disappearance rate of the barbiturates, which appears to be the result of increased distribu-tion volume and decreased rate of metabolism. However, the magnitude of the inhibitory effect of THC appears to be small even on prolonged adminis-tration [175a]. Apparently, cannabidiol is a more potent inhibitor of drug metabolism and it seems that prolongation of barbiturate sleeping time by cannabis extracts is due to their cannabidiol content and is mediated by a di-rect effect on the microsomal enzyme systems [175a]. Pretreatment of mice with THC prolonged sodium thiopental sleeping time. This effect was poten-tiated by α-fluoromethylhistidine, a specific inhibitor of histidine decarbox-ylase. The mechanism of this synergistic action is obscure [176].

Recently, the identification and sleep-inducing properties of *cis*-9-octade-cenamide (oleamide, 33), a lipid found in the cerebrospinal fluid of sleep-de-prived cats, were reported [177]. In order to determine which brain receptors might be associated with this activity, oleamide was tested on a range of re-ceptors. It was found that oleamide potentiated the action of serotonin on $5\text{-}HT_{2a}$ and $5\text{-}HT_{2c}$ receptors [178]. It was suggested that this effect 'may represent a novel mechanism for regulation of receptors that activate G pro-teins and thereby play a role in alertness, sleep, and mood as well as disturb-ances of these states.'

(33) oleamide

Our group has looked at the problem from a different angle. We assumed that oleamide and anandamide may have common pharmacological features as they possess related chemical structures – both are fatty acid amides, and as mentioned above, drowsiness or sleepiness are well-known effects in the later stages of intoxication by marijuana. This path of investigation received support from observations made by Santucci *et al.* who found that the CB_1 antagonist SR 141716A increased the time spent in wakefulness at the expense of both slow-wave and rapid eye movement sleep and suggested that the endogenous cannabimimetic system may regulate the organization of the sleep-waking cycle [179].

To establish a possible relationship between anandamide and oleamide we compared some *in vivo* and *in vitro* effects caused by them. We first examined oleamide in a tetrad of assays commonly used for the determination of cannabinoid activity, including that of anandamide: (a) the ring immobility (catalepsy) test, which measures the percent of time mice remain motionless on a ring, (*b*) the open field test, which measures locomotor activity, (c) hypothermia and (d) antinociception. This tetrad of tests, when evaluated together has proven to be highly predictive of cannabinoid action. The ED_{50} values obtained for oleamide were compared to those of anandamide and Δ^9-THC. It was found that oleamide has essentially the same profile as anandamide, although it is less potent in most tests. It was also found that oleamide increased the effect of anandamide in all four tests [179a]. However, oleamide does not bind to either CB_1 or CB_2 [86, 180].

A fatty acid amide hydrolase (FAAH) causes hydrolysis of both oleamide and anandamide [50,51]. We investigated the possibility that oleamide raises anandamide levels by inhibition of FAAH. Indeed, we found that oleamide dose dependently inhibited the hydrolysis of $[^{14}C]$-anandamide by mouse neuroblastoma $N_{18}TG_2$ cells. The ability of oleamide to affect the K_i of anandamide in competition binding experiments to CB_1 in transfected COS-7 cells was also investigated. In the presence of the amidase inhibitor PMSF, the observed K_i of anandamide decreased by approximately one order of magnitude. A similar enhancement in affinity was observed upon addition of oleamide [179a]. When PMSF and oleamide were present together, the affinity did not increase beyond the single order of magnitude. These observations, raise the possibility that the induction of sleep caused by oleamide may be mediated by anandamide.

'MEDICAL MARIJUANA' – IS IT BETTER THAN THC?

Over the last few years, and in particular, during the months preceding the 1996 USA general election, a heated public discussion took place over the legalization of marijuana for medical use – in particular for amelioration of nausea and vomiting in cancer patients, for appetite stimulation in AIDS patients, and against glaucoma as well as spasticity and pain in patients with multiple sclerosis and spinal cord injury.

A marijuana pub, selling the drug to patients was closed in San Francisco. This resulted in a public outcry which ultimately led to a referendum on the issue. California and Arizona residents voted to legalize 'medical marijuana', but the Federal Government declared that it will take legal action against physicians that prescribe an illegal drug. This unprecedented turmoil is in fact part of a much bigger social issue – should the USA legalize the use of cannabis (and later perhaps the use of all other illegal drugs) under certain restrictions, as was done with alcoholic drinks in the 1930's. The supporters of this step believe that it will reduce sharply the crime rate, without significantly increasing the number of addicts. The US Federal agencies are strongly opposed and claim that addiction, in particular within certain social groups, may become catastrophic. In their view the legalization of 'medical marijuana' is just the thin edge of the wedge and should be opposed. From the point of view of the medicinal chemist, this fierce debate, at least as regards cannabis, misses several central points. Most plant products of medical value are not usually administered as a crude plant extract but as a pure substance. In the Western world, morphine, not opium, is used against pain. The same is true for a long list of important drugs found in plants or micro-organisms. Why should not patients use THC rather than 'medical marijuana'? Even if we assume that the groups supporting the use of cannabis may be interested in the crude drug, rather than in THC, for 'political reasons', there are several advantages of 'medical marijuana' as compared with THC. Marijuana is usually smoked which, in this case at least, represents a very efficient way of administration and the effects are noted almost immediately; many users claim that the effects are milder than those of THC. Besides, THC (dronabinol) is considerably more expensive than marijuana. THC however has also significant advantages: it can be administered at exact dose levels. Cannabis contains additional cannabinoids which are difficult to quantify in each batch. THC and crude cannabis have not been directly compared in patients. THC in an aerosol form presumably can be compared with smoked marijuana. Such experiments have yet to be attempted.

A South American group has shown that CBD (27) has anxiolytic and antipsychotic properties [181,182] and decreases the anxiety effects of THC [183]. Thus, it is quite possible that CBD (and other constituents?) in marijuana 'mellow' the THC effect. Most of the clinical trials published so far are with THC or nabilone. Is it possible that marijuana is a better drug than THC, as is often claimed, because of the presence in it of both CBD and THC? If this is correct, is it also possible that other inactive constituents also modify THC activity? This possibility has not been evaluated so far except for the cannabimimetic effect in the rhesus monkey. Over 25 years ago we reported that the general sedative effect of THC on rhesus monkeys is not modified by the presence of several of the major constituents in Lebanese hashish [184]. However, this assay is a rather gross one and it is quite possible that it does not fully correspond to the effects in humans which may be more subtle, in particular as regards anxiety and mood changes.

Surprisingly, in spite of the furore over 'medical marijuana' no comparative studies have been published, and we do not know whether (a) medical marijuana is indeed better than THC, as claimed; (b) whether the additional cannabinoids in crude cannabis (particularly CBD) impart desirable qualities to the drug and (c) whether better formulations of THC (with or without additional constituents) can be obtained.

Synthetic Δ^9-THC is indeed quite expensive to prepare. In particular, the conversion of Δ^8-THC to Δ^9-THC is a difficult and expensive process [1]. However, Δ^8-THC can certainly be used in place of Δ^9-THC. In all tests reported, Δ^8-THC is qualitatively equivalent to Δ^9-THC, although it may be slightly less potent [91]. Both compounds are essentially nontoxic in animals and humans [185, 186]. However, Δ^8-THC is much more stable than Δ^9-THC. The latter oxidizes easily to cannabinol; the former is stable. As a matter of fact, it was found in an archeological dig nearly 1600 years after it was formed! [187]. Δ^8-THC is much less expensive to prepare. So far however, there has been only one clinical study with Δ^8-THC [121].

HU-211 (DEXANABINOL) – A NEUROPROTECTIVE AGENT

As discussed in some detail on p. 202 of this review, the enantiomers (−)-HU-210 (which retains the stereochemistry present in THC) and (+)-HU-211 were originally prepared in order to establish whether the activity of the cannabinoids is stereospecific. The synthetic path is shown in *Figure 5.2* [25]. The intermediate ketones (5 and enantiomer) can be easily crystallized to absolute purity and therefore the final products HU-210 (3) and HU-211 (4) are obtained with e.e. higher than 98.8%. This was a central aim of our synthetic approach in order to make possible the eventual therapeutic use of the [3S,4S] enantiomer (4), as the presence of traces of the highly psychotropic [3R, 4R] enantiomer (3) could lead to undesirable side-effects. The enantiomeric purity of thrice recrystallized (3) and (4) was established by h.p.l.c. analysis of the diastereoisomeric bis(MIPA) esters obtained by reaction with (S)-(+)-α-methoxy-α-(trifluoromethyl)phenylacetyl (MTPA) chloride. As expected, we found that HU-211 (4) has no cannabimimetic activity [20–24]. However, unexpectedly, we observed that it exhibits pharmacologi-

cal properties typical of *N*-methyl-D-aspartate (NMDA) receptor antagonists [188]. Binding studies with [³H]TCP and [³H]MK-801 showed that
HU-211 blocks NMDA receptors stereospecifically by interacting with a
site close to, but distinct from, that of noncompetitive NMDA receptor antagonists and from the recognition sites of glutamate, glycine, and poly-
amines. The NMDA receptor is one of the subreceptors of glutamate, which
is now well-established as the transmitter at most excitatory synapses in the
mammalian CNS. In numerous disease conditions (neuronal necrosis from
cerebral ischaemia for example) or on brain trauma, overactivation of the
NMDA receptor is noted. It causes increased influx of Ca^{2+} ions into cells
leading to their death. While numerous NMDA antagonists have been
tested, none has as yet passed the clinical tests because of various side-
effects.

IN VITRO STUDIES

The interaction of HU-211 with the NMDA receptor led to the investigation
of several of its effects on various physiological systems that could be of therapeutic interest. It was noted that HU-211 blocks the NMDA-induced
[⁴⁵Ca] uptake by primary neuronal cultures of rat forebrain [189]. While the
NMDA glycine-operated channel was blocked, that of kainate was not.
Furthermore, it was shown that HU-211 protects rat neuronal cultures
against NMDA-mediated glutamate toxicity [190, 191]. HU-211 was found
to inhibit the binding of [³H] MK-801, a widely investigated experimental
non-competitive NMDA antagonist, to rat forebrain membranes, with a K_i
of 11.0 μM. HU-211 is not identical with MK-801 in its effects: massive cell
damage produced by nitric oxide (NO) is attenuated by HU-211 (1–10 μM)
but not by MK-801 (up to 30 μM). HU-211 is an effective scavenger of peroxy radicals *in vitro*. Cultured neurons were protected from the toxic effect
of several free radical generators, or anoxia [192].

IN VIVO STUDIES

The above encouraging results led to numerous *in vivo* investigations. We
had already reported that HU-211 blocked NMDA-induced tremor, convul-
sions and death in mice [188]. As numerous effects of brain trauma, ischaemia and bacterial infections in the CNS are associated with delayed mechanisms of damage, mediated by glutamate, the action of HU-211 on such
conditions was investigated. In a series of papers, Shohami and collaborators reported that some effects of closed head injuries in rats could be significantly improved by HU-211 administration [193–196]. Thus HU-211,

5 mg/kg i.v. administered up to 4 h after closed head injury resulted in significant reduction of oedema formation, improvement of clinical status (established by a standard neurological severity score), and of spatial memory (water maze measurements). The integrity of the blood brain barrier (BBB) is severely broken down on brain injury. HU-211 was found to improve BBB integrity. A significant accumulation of Ca^{2+} in several brain regions takes place on closed head injury. HU-211 attenuated such accumulation, indicating that the proposed mechanism of action – blockage of the NMDA-operated Ca^{++} channel – is most probably correct.

Tumour necrosis factor-α (TNFα) accumulates in the brain after trauma. This cytokine is known to be an important factor in delayed CNS damage. It was found that, in addition to its anti-NMDA effect, HU-211 causes up to 90% inhibition of the TNFα surge after closed head injury in rats [195]. Bacterial and viral infections of the CNS are known to cause secretion of the TNFα as well as interleukin-1 and other cytokines which are involved in the inflammatory process and may cause secondary damage. Such infections may result in high mortality. It was found that rats infected with *Streptococcus pneumoniae* suffered less cerebral oedema on treatment with a combination of a suitable antibiotic with HU-211 than the antibiotic alone [196].

In two experimental models of ischaemia, it has been shown that HU-211 significantly increases cell survival. It was seen that after forebrain ischaemia produced by 20 min of carotid occlusion, the number of viable neurons in the hippocampal CA1 region of HU-211-treated rats was significantly higher than in controls. The same effect was seen in gerbils after 10 min of bilateral carotid occlusion on treatment with HU-211 [197, 198]. A related effect has been noted after rat optic nerve crush injury. Administration of HU-211 improved recovery of the nerve, with the visual evoked response amplitude increasing significantly [199].

These encouraging results have led to an examination of the physiological and toxicological effects of HU-211, which were compared with those produced by MK-801. Doses of 8–20 mg/kg HU-211 administered to rats (i.v. or i.p.) produced no ataxia, tremors, hyperactivity or coma as seen with MK-801. The latter compound and other NMDA antagonists are known to produce vacuolization of neurons in the posterior cingulate cortex. No such effect was observed with HU-211 [200]. Oral toxicity was studied in rats and cynomogous monkeys (28 day study; 5–50 mg/kg per day). An i.v. toxicity trial was undertaken with rats and rabbits (14 day study; 1–8 mg/kg per day). No signs of systemic toxicity were noted and no lethality was observed [200]. A phase I trial in volunteers (doses up to 100 mg) has been completed. No undesirable CNS effects or changes in blood pressure were noted. HU-211 is now in Phase II trials in several hospitals in Israel in cases of

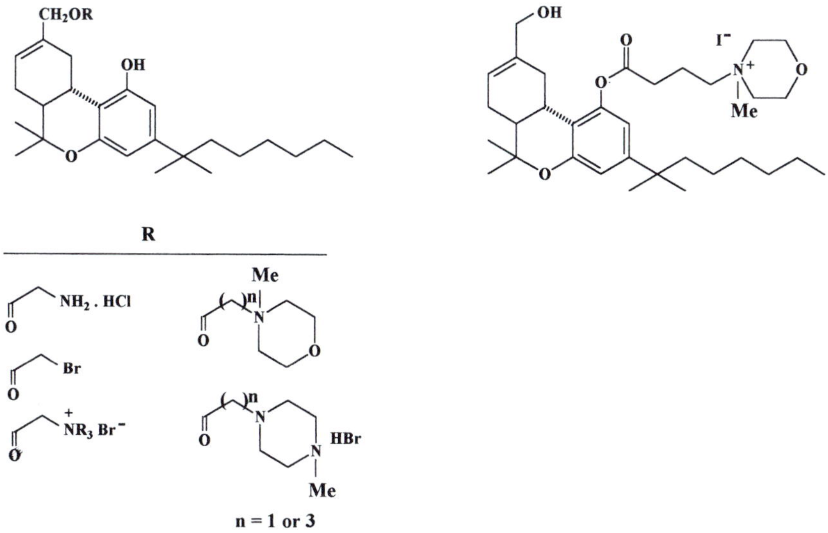

Figure 5.5 *Water-soluble derivatives of HU-211*

CNS trauma. The biomedical research on HU-211 has been reviewed [14, 200].

HU-211 is highly liposoluble, which makes it readily accessible to the central nervous system since it readily crosses the blood brain barrier. However, its poor solubility in water hampers development of formulations suitable for i.v. administration. In order to overcome this drawback, Popp *et al.* have prepared a series of water-soluble salts of glycinate esters (attached to the allylic hydroxyl) and salts of amino acid esters containing tertiary and quaternary nitrogen heterocyclics (attached to the phenolic hydroxyl) (see *Figure 5.5*) [201, 202]. Most of the new compounds were relatively soluble in water or 10% aqueous ethanol, and showed neuroprotective properties, attributed to the parent compound, formed on hydrolysis of the esters *in vivo*.

CONCLUSIONS

We have presented a somewhat subjective overview of the impressive developments achieved in the biochemistry of cannabinoids in the last decade.

From being a limited, specific topic in phytochemistry and psychobiology, cannabinoid research has become one of central importance in neurobiology, with implications in various areas. The newly discovered endogenous ligands – in particular anandamide – are rapidly gaining an important place in our understanding of signal transduction in neuronal cells, and in expanding our view in basic processes, stretching from reproduction to motor coordination, sleep, memory and presumably emotions. Anandamide seems to be, alongside GABA, a central inhibitory molecule in the CNS.

These impressive advances in biology have not been paralleled by developments in the therapeutic area. The psychotropic effects of Δ^9-THC, and the stigma attached to cannabis as an abused drug, has resulted in a pronounced lack of enthusiasm within the pharmaceutical companies. However, the recent development of cannabinoids that do not cause a psychotropic effect, and yet have therapeutically important features (HU-211, for example), the discovery of antagonists and of cannabinoids that bind preferentially to the peripheral CB_2, may bring about enhanced pharmaceutical research.

REFERENCES

1   Mechoulam, R. and Feigenbaum, J.J. (1987) Prog. Med. Chem. 24, 159–207.
2   Kassirer, J.P. (1997) N. Engl. J. Med. 336, 366.
3   Kassirer, J.P. (1997) N. Engl. J. Med. 336, 1186–1187.
4   Gopinath, L. (1997) Chem. Br. 33, 41–42.
5   Pertwee, R. (ed) (1995) Cannabinoid Receptors, Harcourt Brace, London.
6   Mechoulam, R., Hanuš, L. and Martin, B.R. (1993) Biochem. Pharmacol. 48, 1537–1544.
7   Mechoulam, R., Vogel, Z. and Barg, J. (1994) CNS Drugs 2, 255–260.
8   Fride E., Hanuš, L. and Mechoulam, R. (1994) in Lipid Mediators in Health and Disease (U. Zor, ed.) pp. 1–10, Freund Publ. London.
9   Martin, B.R., Welch, S.P. and Abood, M. (1994) Adv. Pharmacol. 25, 341–397.
10 Howlett, A.C. (1995) Annu. Rev. Pharmacol. Toxicol. 35, 607–634.
11 Di Marzo, V. and Fontana A. (1995) Prostaglandins Leukotrienes Essent. Fatty Acids 53, 1–11.
12 Piomelli, D. (1996) Arachidonic Acid in Cell Signaling, pp. 167–195, Chapman & Hall, New York.
13 Onaivi, E.S. Chakrabarti A. and Chaudhuri G. (1996) Progr. Neurobiol. 48, 275–305.
14 Shohami, E., Weidenfeld, J., Ovadia, H., Vogel, Z., Hanuš, L., Fride, E., Breuer, A., Ben Shabat, S., Sheskin, T. and Mechoulam, R. (1996) CNS Drug Rev. 2, 429–451.
15 Howlett, A.C., Berglund, B. and Melvin, L.S. (1998) Curr. Pharm. Design 1, 343–354.
16 Gaoni, Y. and Mechoulam, R. (1964) J. Am. Chem. Soc. 86, 1646.
17 Paton, W.D.M. (1975) Annu. Rev. Pharmacol. 15, 191–220.
18 Gill, E.W. and Lawrence, D.K. (1976) in Pharmacology of Marihuana (Braude, M.C. and Szara, S., eds.), pp. 147–155, Raven Press, New York.

19 Dewey, W.L., Martin, B.R. and May, E.L. (1984) in Handbook of Stereoisomers: Drugs in Psychopharmacology (Smith, D.F., ed.), pp. 317–326, CRC Press, Boca Raton, FL.

20 Mechoulam, R., Lander, N., Srebnik, M., Breuer, A., Segal, M., Feigenbaum, J.J., Järbe, T.U.C. and Consroe, P. (1987) in Structure Activity Relationships of the Cannabinoids (Rapaka, R.S. and Makriyannis, A., eds.), pp. 15–30, National Institute on Drug Abuse. Monograph 79 Washington, D.C.

21 Mechoulam, R., Devane, W.A. and Glaser, R. (1992) in Marijuana/Cannabinoids: Neurobiology and Neurophysiology (Murphy, L. and Bartke A., eds.), pp. 1–33, CRC Press, Boca Raton, FL.

22 Little, P.J., Compton, D.R., Mechoulam, R. and Martin, B. (1989) Pharmacol. Biochem. Behavior 32, 661–666.

23 Järbe, T.U.C., Hiltunen, A.J. and Mechoulam, R. (1989) J. Pharmacol. Exp. Ther. 250, 1000–1005.

24 Howlett, A.C., Champion, T.M., Wilken, G.H. and Mechoulam, R. (1990) Neuropharmacology 29, 161–165.

25 Mechoulam, R., Lander, N., Breuer, A. and Zahalka, J. (1990) Tetrahedron: Asymmetry 1, 315–319.

26 Melvin, L.S., Milne, G.M., Johnson, M.R., Wilken, G.H. and Howlett, A.C. (1995) Drug Design Discov. 13, 155–166.

27 Devane, W.A., Dysarz, F.A., Johnson, M.R., Melvin, L.S. and Howlett, A.C. (1988) Mol. Pharmacol. 34, 605–613.

28 Martin, B.R., Compton, D.R., Thomas, B.F., Prescott, W.R., Little, P.J., Razdan, R.K., Johnson, M.R., Melvin, L.S., Mechoulam, R. and Ward, S.J. (1991) Pharmacol. Biochem. Behav. 40, 471–478.

29 Matsuda, L.A., Lolait, S.J., Brownstein, M.J., Young, A.C. and Bonner, I.I. (1990) Nature (London) 346, 561–564.

30 Munro, S., Thomas, K.L. and Abu-Shaar, M. (1993) Nature (London) 365, 61–65.

31 Galiegue, S., Mary, S., Marchand, J., Dussossoy, D., Carriere, D., Carayon, P., Bouaboula, M., Shire, D., Le Fur, G. and Casellas, P. (1995) Eur. J. Biochem. 232, 54–61.

32 Devane, W.A., Breuer, A., Sheskin, T., Järbe, T.U.C., Eisen, M. and Mechoulam, R. (1992) J. Med. Chem. 35, 2065–2069.

33 Devane, W.A., Hanuš, L., Breuer, A., Pertwee, R.G., Stevenson, L.A., Griffin, G., Gibson, D., Mandelbaum, A., Etinger, A. and Mechoulam, R. (1992) Science (Washington, D.C.) 258, 1946–1949.

34 Hanuš, L., Gopher, A., Almog, S. and Mechoulam, R. (1993) J. Med. Chem. 36, 3032–3034.

35 Fride, E. and Mechoulam, R. (1993) Eur. J. Pharmacol. 231, 313–314.

36 Smith, P.B., Compton, D.R., Welch, S.P., Razdan, R.K., Mechoulam, R. and Martin, B.R. (1994) J. Pharmacol. Exp. Ther. 270, 219–227.

37 Fride, E. (1995) Brain Res. 697, 83–90.

38 Childers, S.R., Sexton, T. and Roy, M.B. (1994) Biochem. Pharmacol. 47, 711–715.

39 Mackie, K., Devane, W. and Hille, B. (1993) Mol. Pharmacol. 44, 498–503.

40 Vogel, Z., Barg, J., Levy, R., Saya, D., Heldman, E. and Mechoulam, R. (1993) J. Neurochem. 61, 352–355.

41 Felder, C.C., Briley, E.M., Axelrod, J., Simpson, J.T., Mackie, K. and Devane, W.A. (1993) Proc. Natl. Acad. Sci. U.S.A. 90, 7656–7660.

42 Sulcova, A., Mechoulam, R. and Fride, E. (1998) Pharmacol. Biochem. Behav. (in press).

43 Fride, E., Barg, J., Levy, R., Saya, D., Heldman, E., Mechoulam, R. and Vogel, Z. (1995) J. Pharmacol. Exp. Ther. 272, 699–707.

44   Schmid, P.C., Krebsbach, R.J., Perry, S.R., Dettmer, T.M., Maasson, J.L. and Schmid, H.H.O. (1995) FEBS Lett. 375, 117–120.
45   Kempe, K., Hsu, F.-F., Bohrer, A. and Turk, J. (1996) J. Biol. Chem. 17287–17295.
46   Sugiura, T., Kondo, S., Sukagawa, A., Tonegawa, T., Nakane, S., Yamashita, A., Ishima,Y. and Waku, Y. (1996) Eur. J. Biochem. 240, 53–60.
47   Cadas, H., di Tomaso, E. and Piomelli, D. (1997) J. Neurosci. 17, 1226–1242.
48   Di Marzo, V., Fontana, A., Cadas, H., Schinelli, S., Cimino, G., Schwartz, J.C. and Piomelli, D. (1994) Nature (London) 372, 686–691.
49   Sugiura, T., Kondo, S., Sukagawa, A., Tonegawa, T., Nakane, S., Yamashita, A. and Waku, K. (1996) Biochem. Biophys. Res. Commun. 218, 113–117.
50   Deutsch, D.G. and Chin, S.A. (1993) Biochem. Pharmacol. 46, 791–796.
51   Cravatt, B.F., Giang, D.K., Mayfield, S.P., Boger, D.L., Lerner, R.A. and Gilula, N.B. (1996) Nature (London) 384, 83–87.
52   Deutsch, G.D., Omeir, R., Arreaza, G., Salehani, D., Prestwich, G.D., Huang, Z. and Howlett, A. (1997) Biochem. Pharmacol. 53, 255–260.
53   Fowler, C.J., Stenstrom, A. and Tiger, G. (1997) Pharmacol. Toxicol. 80, 103–107.
54   Wickens, A.P. and Pertwee, R.G. (1993) Eur. J. Pharmacol. 250, 205–208.
55   Maneuf, Y.P., Nash, J.E., Crossman, A.R. and Brotchie, J.M. (1996) Eur. J. Pharmacol. 308, 161–164.
56   Mailleux, P. and Vanderhaeghen, J.J. (1993) J. Neurochem. 61, 1705–1712.
57   Rodriguez de Fonseca, F.R., Villanua, M.A., Munoz, R.M., San-Martin-Clark, O. and Navarro, M. (1995) Neuroendocrinology 61, 714–721.
58   Romero, J., Garcia, L., Cebeira, M., Zadrozny, D., Fernandez-Ruiz, J.J. and Ramos, J.A. (1995) Life Sci. 56, 2033–2040.
59   Shen, M., Piser, T.M., Seybold, V.S. and Thayer, S.A. (1996) J. Neurosci. 16, 4322–4334.
60   Ishac, E.J.N., Jiang, L., Lake, K.D., Varga, K., Abood, M.E. and Kunos, G. (1996) Br. J. Pharmacol. 1181, 2023–2028.
61   Fan, P. (1995) J. Neurophysiol. 73, 907–910.
62   Ellis, E.F., Moore, S.F. and Willoughby, K.A. (1995) Am. J. Physiol – Heart Circ. Physiol. 381, H1859-H1864.
63   Weidenfeld, J., Feldman, S. and Mechoulam, R. (1994) Neuroendocrinology 59, 110–112.
64   Venance, L., Piomelli, D., Glowinski, J. and Giaume, C. (1995) Nature (London) 376, 590–594.
65   Randall, M.D., Alexander, S.P., Bennett, T., Boyd, E.A., Fry, J.R., Gardiner, S.M., Kemp, P.A., McCulloch, A.I. and Kendall, D.A. (1996) Biochem. Biophys. Res. Commun. 229, 114–120.
66   Schuel, H., Goldstein, E., Mechoulam, R., Zimmerman, A.M. and Zimmerman, S. (1994) Proc. Natl. Acad. Sci., U.S.A. 91, 7678–7682.
67   Chang, M.C., Berkery, D., Schuel, R., Laychock, S.G., Zimmerman, A.M., Zimmerman, S. and Schuel, H. (1993) Mol. Reprod. Dev. 36, 507–516.
68   Das, S.K., Paria, B.C., Chakraborty, I. and Dey, S.K. (1995) Proc. Natl. Acad. Sci. U.S.A. 92, 4332–4336.
69   Schmid, P.G., Paria, B.C., Frebsbach, R.J., Schmid, H.H.O. and Dey, S.K. (1997) Proc. Natl. Acad. Sci. U.S.A. 94, 4188–4192.
70   Fried, P.A. (1974) Psychopharmacology 50, 285–291.
71   Fride, E. and Mechoulam R. (1996) Psychoneuroendocrinology 21, 157–172.
72   Fride, E., Ben-Shabat, S. and Mechoulam, R. (1996) Soc. Neurosci. Abs. 22, 1683.
73   Mechoulam, R., Ben-Shabat, S., Hanuš, L., Ligumsky, M., Kaminski, N.E., Schatz, A.R., Gopher, A., Almog, S., Martin, B.R., Compton, D.R., Pertwee, R.G., Griffin, G., Bayewitch, M., Barg, J. and Vogel, Z. (1995) Biochem. Pharmacol. 50, 83–90.

74 Sugiura, T., Kondo, S., Sukagawa, A., Nakane, S., Shinoda, A., Itoh, K., Yamashita, A. and Waku, K. (1995) Biochem. Biophys. Res. Commun. 215, 89–97.

75 Ben-Shabat, S., Fride, E., Di Marzo, V. and Mechoulam, R. (unpublished observations).

76 Lee, M., Yang, K.H. and Kaminski, N.E. (1995) J. Pharmacol. Exp. Ther. 275, 529–536.

77 Sugiura, T., Kodaka, T., Kondo, S., Tonegawa, T., Nakane, S., Kisimoto, S., Yamashita, A. and Waku, K. (1996) Biochem. Biophys. Res. Commun. 229, 58–64.

78 Sugiura, T., Kodaka, T., Kondo, S., Tonegawa, T., Nakane, S., Kishimoto, S., Yamashita, A. and Waku, K. (1997) Biochem. Biophys. Res. Commun. 233, 207–210.

79 Pinto, J.C., Potie, F., Rice, K.C., Boring, D., Johnson, M.R., Evans, D.M., Wilken, G.H., Cantrell, C.H. and Howlett, A.C. (1994) Mol. Pharmacol. 46, 516–522.

80 Abadji, V., Lin, S., Taha, G., Griffin, G., Stevenson, A., Pertwee, R. and Makriyannis, A. (1994) J. Med. Chem. 37, 1889–1893.

81 Adams, I.B., Ryan, W., Singer, M., Razdan, R.K., Compton, D.R. and Martin, B.R. (1995) Life Sci. 56, 2041–2048.

82 Fontana, A., Di Marzo, V., Cadas, H. and Piomelli, D. (1995) Prostaglandins Leukotrienes Essent. Fatty Acids, 53, 301–308.

83 Edgemond, W.S., Campbell, W.B. and Hillard, C.J. (1995) Prostaglandins Leukotrienes Essent. Fatty Acids 52, 83–86.

84 Adams, I.B., Ryan, W., Singer, M., Thomas, B.F., Compton, D.R., Razdan, R.K. and Martin, B.R. (1995) J. Pharmacol. Exp. Ther. 273, 1172–1181.

85 Thomas, B.F., Adams, I.B., Mascarella, W., Martin, B.R. and Razdan, R.K. (1996) J. Med. Chem. 39, 471–479.

86 Sheskin, T., Hanuš, L., Slager, J. and Mechoulam, R. (1997) J. Med. Chem. 40, 659–667.

87 Showalter, V.M., Compton, D.R., Martin, B.R. and Abood, M.E. (1996) J. Pharmacol. Exp. Ther. 278, 989–999.

88 Khanolkar, A.D., Abadji, V., Lin, S., Hill, W.A., Taha, G., Abouzid, K., Meng, Z., Fan, P. and Makriyannis, A. (1996) J. Med. Chem. 39, 4515–4519.

89 Wise, M.L., Soderstrom, K., Murray, T.F. and Gerwick, W.H. (1996) Experientia 52, 88–92.

90 The nomenclature of the anandamides uses a widely accepted fatty acid numbering: the first number indicates the number of carbon atoms of the fatty acid; the second number indicates the number of double bonds (all of them skipped); the third number indicates the carbon atom (counting from the end of the fatty chain) with the first double bond.

91 Razdan, R.K. (1986) Pharmacol. Rev. 38, 75–149.

92 Martin, B.R., Thomas, B.F. and Razdan, R.K. in Ref. 5, 36–85.

93 Compton, D.R., Rice, K.C., De-Costa, B.R., Razdan, R.K., Melvin, L.S., Johnson, M.R. and Martin, B.R. (1993) J. Pharmacol. Exp. Ther. 265, 218–226.

94 Reggio, P.H., McGaughey, G.B., Odear, D.F., Seltzman, H.H., Compton, D.R. and Martin, B.R. (1991) Pharmacol. Biochem. Behav. 40, 479–486.

95 Reggio, P.H., Panu, A.M. and Miles, S. (1993) J. Med. Chem. 36, 1761–1771.

96 Reggio, P.H., Bramblett, R.D., Yuknavich, H., Seltzman, H.H., Fleming, D.N., Fernando, S.R., Stevenson, L.A. and Pertwee, R.G. (1995) Life Sci. 56, 2025–2032.

97 Lagu, S.G., Varona, A., Chambers, J.D. and Reggio, P.H. (1995) Drug Des. Discovery 12, 179–192.

98 Glaser, R., Adin, I., Mechoulam, R. and Hanuš, L. (1995) Heterocycles 39, 867–877.

99 Gareau, Y., Dufresne, C., Gallant, M., Rochette, C., Sawyer, N., Slipetz, D.M., Tremblay, N., Weech, P.K., Metters, K.M. and Labelle, M. (1996) Bioorg. Med. Chem. Lett. 6, 189–194.

100 Gallant, M., Dufresne, C., Gareau, Y., Guay, D., Leblanc, Y., Prasit, P., Rochette, C.,

Sawyer, N., Slipetz, D.M., Tremblay, N., Metters, K.M. and Lab, M. (1996) Bioorg. Med. Chem. Lett. 6, 2263–2268.

101 Huffman, J.W., Yu, S., Showalter, V., Abood, M.E., Wiley, J.L., Compton, D.R., Martin, B.R., Bramblett, R.D. and Reggio, P.H. (1996) J. Med. Chem. 39, 3875–3877.

102 Busch-Petersen, J., Hill, W.A., Fan, P., Khanolkar, A., Xie, X-Q., Tius, M.A. and Makriyannis, A. (1996) J. Med. Chem. 39, 3790–3796.

103 Pertwee, R.G., Fernando, S.R., Griffin, G., Ryan, W., Razdan, R.K., Compton, D.R. and Martin, B.R. (1996) Eur. J. Pharmacol. 315, 195–201.

104 Wiley, J.L., Compton, D.R., Gordon, P.M., Siegel, C., Singer, M., Dutta, A., Lichtman, A.H., Balster, R.L., Razdan, R.K. and Martin, B.R. (1996) Neuropharmacology 35, 1793–1804.

105 Johnson, M.R. and Melvin, L.S. (1986) in Cannabinoids as Therapeutic Agents (Mechoulam, R., ed.), 121–145, CRC Press, Boca Raton.

106 Melvin, L.S., Milne, G.M., Johnson, M.R., Subramaniam, B., Wilken, G.H. and Howlett, A.C. (1993) Mol. Pharmacol. 44, 1008–1015.

107 D'Ambra, T.E., Eissenstat, M.A., Abt, J., Ackerman, J.H., Bacon, E.R., Bell, M.R., Carabateas, P.M., Josef, K.A., Kumar, V., Weaver, J.D., Arnold, R., Casiano, F.M., Chippari, S.M., Haycock, D.A., Kuster, J.E., Luttinger, D.A., Stevenson, J.I., Ward, S.J., Hill, W.A., Khanolkar, A. and Makriyannis, A. (1996) Bioorg. Med. Chem. Lett. 6, 17–22.

108 Rinaldi-Carmona, M., Barth, F., Heaulme, M., Shire, D., Calandra, B., Congy, C., Martinez, S., Maruani, J., Neiliat, G., Caput. D., Ferrura, P., Soubrie, P., Breliere, J.G. and Le Fur, G. (1994) FEBS Lett. 350, 240–244.

109 Rinaldi-Carmona, M., Pialot, F., Congy, C., Redon, E., Barth, F., Bachy, A., Breliere, J.G., Soubrie, P. and Le Fur, G. (1996) Life Sci. 58, 1239–1247.

110 Gifford, A.N. and Ashby Jr., C.R. (1996) J. Pharmacol. Exp. Ther. 277, 1431–1436.

111 Compton, D.R., Aceto, M.D., Lowe, J. and Martin, B.R. (1996) J. Pharmacol. Exp. Ther. 277, 586–594.

112 Terranova, J.P., Michaud, J.C., Le Fur, G. and Soubrie, P. (1995) Naunyn-Schmiedebergs Arch. Pharmacol. 352, 576–579.

113 Terranova, J.P., Storme, J.J., Lafon, N., Perio, A., Rinaldi-Carmona, M., Le Fur, G. and Soubrie, P. (1996) Psychopharmacology 126, 165–172.

114 Pertwee, R., Griffin, G., Fernando, S., Li, X., Hill, A. and Makriyannis, A. (1995) Life Sci. 56, 1949–1955.

115 Pertwee, R.G., Fernando, S.R., Nash, J.E. and Coutts, A.A. (1996) Br. J. Pharmacol. 118, 2199–2205.

116 Yamada, K., Rice, K.C., Flippen-Anderson, J.L., Eissenstat, M.A., Ward, S.J., Johnson, M.R. and Howlett, A.C. (1996) J. Med. Chem. 39, 1967–1974.

117 Plasse, T.F., Gorter, R.W., Krasnow, S.H., Lane, M., Shepard, K.V. and Wadleigh, R.G. (1991) Pharmacol. Biochem. Behav. 40, 695–700.

118 Lane, M., Vogel, C.L., Ferguson, J., Krasnow, S., Saiers, J.L., Hamm, J., Salva, K., Wiernik, P.H., Holroyde, C.P., Hammill, S., Shepard, K. and Plasse, T. (1991) J. Pain Symptom Manage. 6, 352–359.

119 Feigenbaum, J.J., Richmond, S.A., Weissman, Y. and Mechoulam, R. (1989) Eur. J. Pharmacol. 169, 159–165.

120 Fride, E. and Mechoulam, R. (1996) Dev. Brain Res. 95, 131–134.

121 Abrahamov, A., Abrahamov A. and Mechoulam, R. (1995) Life Sci. 56, 2097–2102.

122 Doblin, R.E. and Kleiman, M.A. (1991) J. Clin. Oncol. 9, 1314–1319.

123 Tavorath, R. and P.J. Hesketh. (1996) Drugs 52, 639–648.

124 Snyder, S.H. (1971) in Uses of Marijuana, pp. 6–7, Oxford University Press, New York.

125 Struwe, M., Kaempfer, S.H., Geiger, C.J., Pavia, A.T., Plasse, T.F., Shepard, K.V., Ries, K. and Evans, T.G. (1993) Ann. Pharmacother. 27, 827–831.

126 Beal, J.E., Olson, R., Laubenstein, L., Morales, J.O., Bellman, P., Yangco, B., Lefkowitz, L., Plasse, T.F. and Shepard, K.V. (1995) J. Pain Symptom Manage. 10, 89–97.

127 Mattes, R.D., Engelman, K., Shaw, L.M. and Elsohly, M.A. (1994) Pharmacol. Biochem. Behav. 49, 187–195.

128 Whitfield, R.M., Bechtel, L.M. and Starich,G.H. (1997) Alcohol. Clin. Exp. Res. 21, 122–127.

129 Himmi, T., Dallaporta, M., Perrin, J. and Orsini, J.C. (1996) Eur. J. Pharmacol. 312, 273–279.

130 Mechoulam, R. in Ref. 105, pp. 1–19.

131 Lyman, W.D., Sonett, J.R., Brosnan, C.F., Elkin, R. and Bornstein, M.B. (1989) Neuroimmunology 23, 73–81.

132 Wirgiun, I., Mechoulam, R., Breuer, A., Schezen, E., Weidenfeld, J. and Brenner, T. (1994). Immunopharmacology 28, 209–214.

133 Richter, A. and Loscher, W. (1994) Eur. J. Pharmacol. 264, 371–377.

134 Pertwee, R.G. and Greentree, S.G. (1988) Neuropharmacology 27, 485–491.

135 Consroe, P., Musty, R., Rein, J., Tillery, W. and Pertwee, R.G. (1997). Eur. Neurol. 38, 44–48.

136 Petro, D.J. and Ellenberger, C. (1981) J. Clin. Pharmacol. 21, 413s-416s.

137 Clifford, B.D. (1983) Ann. Neurol. 13, 669–671.

138 Ungerleider, J.T., Andyrsiak, T., Fairbanks, L., Ellison, G.W. and Myers, L.W. (1987) Adv. Alcohol Subst. Abuse 7, 39–50.

139 Meinck, H.-M., Schonle, P.W. and Conrad, B. (1989) J. Neurol. 236, 120–122.

140 Martyn, C.N., Illis, L.S. and Thom, J. (1995) Lancet 345, 579.

141 Brenneisen, R., Egli, A., Elsohly, M.A., Henn, V. and Spiess, Y. (1996) Int. J. Clin. Pharmacol. Ther. 34, 446–452.

142 Maurer, M., Henn, V., Dittrich, A. and Hofmann A. (1990) Eur. Arch. Psychiat. Clin. Neurosci. 240, 1–4.

143 Greenberg, H.S., Werness, S.A.S., Pugh, J.E., Andrus, R.O., Anderson, D.J. and Domino, E.F. (1994) Clin. Pharmacol. Ther. 55, 324–328.

144 Hollister, L.E. (1986) Pharmacol. Rev. 38, 1–20.

145 Ames, F. (1958) J. Ment. Sci. 104, 972–999.

146 Mechoulam, R., Lander, N., Srebnik, M., Zamir, I., Breuer, A., Shalita, B., Dikstein, S., Carlini, E.A., Leite, J.R., Edery, H. and Porath, G. (1984) in The Cannabinoids, Chemical, Pharmacologic and Therapeutic Aspects (Agurell, S., Dewey, W.L. and Willette, R.E., eds.), pp. 777–793, Academic Press, New York.

147 Welch, S.P. and Stevens, D.L. (1992) J. Pharmacol. Exp. Ther. 262, 10–18.

148 Pugh, G., Smith, P.B., Dombrowski, D.S. and Welch, S.P. (1996) J. Pharmacol. Exp. Ther. 279, 608–616.

149 Smith, P.B., Welch, S.P. and Martin, B.R. (1994) J. Pharmacol. Exp. Ther. 268, 1381–1387.

150 Reche, I., Fuentes, J.A. and Ruizgayo, M. (1996) Eur. J. Pharmacol. 318, 11–16.

151 Xu, S., Cao, X., Mo, W., Xu, Z. and Pan, Y. (1989) Acupunct. Electrother. Res. 14, 103–113.

152 Lichtman, A.H. and Martin, B.R. (1991) J. Pharmacol. Exp. Ther. 258, 517–523.

153 Lichtman, A.H. and Martin, B.R. (1991) Brain Res. 559, 309–314.

154 Welch, S.P., Dunlow, L.D., Patrick, G.S. and Razdan, R.K. (1995) J. Pharmacol. Exp. Ther. 273, 1235–1244.

155 Seltzer, R., Zeltser, Z., Eisen, A., Feigenbaum, J.J. and Mechoulam, R. (1991) Pain 47, 95–103.

156 Burstein, S.H., Audette, C.A., Breuer, A., Devane, W.A., Colodner, S., Doyle, S.A. and Mechoulam, R. (1992) J. Med. Chem. 35, 3135–3141.
157 Evans, F.J. (1991) Planta Med. 57 (Suppl. 1), S60–S67.
158 Herzberg, U., Eliav, E., Bennett, G.J. and Kopin, I.J. (1997) Neurosci. Lett. 221, 157–160.
159 Martin, W.J., Hohmann, A.G. and Walker, J.M. (1996) J. Neurosci. 16, 6601–6611.
160 Richardson, J.D., Aanonsen, L. and Hargreaves, K.M. (1997) Eur. J. Pharmacol. 319, R3–R4.
161 Graham, J.P.D. in Ref. 105, pp. 159–166.
162 Estrada, U., Brase, D.A., Martin, B.R. and Dewey, W.L. (1987) Life Sci. 41, 79–87.
163 Jones, R.T. (1983) Annu. Rev. Med. 34, 247–258.
164 Adams, M.D., Earnhardt, J.T., Martin, B.R., Harris, L.S., Dewey, W.L. and Razdan, R.K. (1977) Experientia 33, 1204–1205.
165 Zaugg, H.E. and Kyncl, J. (1983) J. Med. Chem. 26, 214–217.
166 Vidrio, H., SanchezSalvatori, M.A. and Medina, M. (1996) J. Cardiovasc. Pharmacol. 28, 332–336.
167 Varga, K., Lake, K., Martin, B.R. and Kunos, G. (1995) Eur. J. Pharmacol. 278, 279–283.
168 Varga, K., Lake, K.D., Huangfu, D., Guyenet, P.G. and Kunos, G. (1996) Hypertension 28, 682–686.
168a Wagner, J.A., Varga, K., Ellis, E.F., Rzigalinski, B.A., Martin B.R. and Kunos, G. (1997) Nature (London), 390, 518–521.
168b Randall, M.D. and Kendall, D.A. (1997) Eur. J. Pharmacol. 331, 205–209.
168c Vanhoutte, P.M. (1996) J. Hypertension Suppl. 14 S83–93.
169 Colasanti, B.K., Graig, C.R. and Allara, R.D. (1989) Exp. Eye Res. 39, 251–255.
170 Colasanti, B.K. (1990) J. Ocul. Pharmacol. 6, 259–269.
171 Liu, J.H. and Dacus, A.C. (1987) Arch. Ophthalmol. 105, 245–248.
172 Devane, W., Mechoulam, R., Breuer, A. and Hanus, L. (1994) PCT/US93/11625.
173 Pate, D.W., Jarvinen, K., Uritti, A., Jarho, P. and Jarvinen, T. (1995) Curr. Eye Res. 14, 791–797.
174 Pate, D.W., Jarvinen, K., Urtti, A., Jarho, P., Fich, M., Mahadevan, V. and Jarvinen, T. (1996) Life Sci. 58, 1849–1860.
175 Benowitz, N.L. and Jones, R.T. (1977) Clin. Pharmacol. Ther. 22, 259–268.
175a Paton, W.D.M. and Pertwee, R. (1972) Br. J. Pharmacol. 44, 250–261.
176 Oishi, R., Itoh, Y., Nishibori, M., Saeki, K. and Ueki, S. (1988) Psychopharmacology 95, 77–81.
177 Cravatt, B.F., Prospero-Garcia, O., Siuzdak, G., Gilula, N.B., Henriksen, S.J., Boger, D.L. and Lerner, R.A. (1995) Science (Washington, D.C.) 268, 1506–1509.
178 Huidobro-Toro, J.P. and Harris, R.A. (1996) Proc. Natl. Acad. Sci, U.S.A. 93, 8078–8082.
179 Santucci, V., Storme, J.J., Soubrie, P. and Le Fur, G. (1996) Life Sci. 58, PL103–110.
179a Mechoulam, R., Fride, E., Hanuš, L., Sheskin, T., Bisogno, T., DiMarzo, V., Buyewitch, M. and Vogel, Z. (1997) Nature (London) 389, 25–26.
180 Boring, D.L., Berglund, B.A. and Howlett, A.C. (1996) Prostaglandins Leukotrienes Essent. Fatty Acids 55, 207–210.
181 Guimaraes, F.S., Chiaretti, T.M., Graeff, F.G. and Zuardi, A.W. (1990) Psychopharmacology 100, 558–559.
182 Zuardi, A.W., Morais, S.L., Guimaraes, F.S. and Mechoulam, R. (1995) J. Clin. Psychiatr. 56, 485–486.
183 Zuardi, A.W., Shirakawa, I., Finkelfarb, E. and Karniol, I.G. (1982) Psychopharmacology 76, 245–250.
184 Mechoulam, R., Shani, A., Edery, H. and Grunfeld, Y. (1970) Science, (Washington, D.C.) 169, 611–612.

185   Thompson, G.R., Rosenkrantz, H., Schaeppi, U.H. and Braude, M.C. (1973) Toxicol. Appl. Pharmacol. 25, 363–372.
186   Hollister, L.E. and Gillespie, H.K. (1973) Clin. Pharmacol. Ther. 14, 353–357.
187   Zias, J., Stark, H., Seligman, J., Levy, R., Werker, E., Breuer, A. and Mechoulam, R. (1993) Nature (London) 363, 315.
188   Feigenbaum, J.J., Bergmann, F., Richmond, S.A., Mechoulam, R., Nadler, V., Kloog, Y. and Sokolovsky, M. (1989) Proc. Natl. Acad. Sci. U.S.A. 86, 9584–9587.
189   Nadler, V., Mechoulam, R. and Sokolovsky, M. (1993) Brain Res. 622, 79–85.
190   Nadler, V., Mechoulam, R. and Sokolovsky, M. (1993) Neurosci. Lett. 162, 43–45.
191   Eshhar, N., Streim, S. and Biegon, A. (1993). NeuroRep. 5, 237–240.
192   Eshhar, N., Streim, S., Kohen, R., Tirosh, O. and Biegon, A. (1995) Eur. J. Pharmacol. 283, 1–3.
193   Shohami, E., Novikov, M. and Mechoulam, R. (1993) J. Neurotrauma 10, 109–119.
194   Shohami, E., Novikov, M. and Bass, R. (1995) Brain Res. 674, 55–62.
195   Shohami, E., Gallily, R., Mechoulam, R., Bass, R. and Ben-Hur, T. (1997) J. Neuroimmunol. 72, 169–177.
196   Bass, R., Trembovler, V., Engelhard, D. and Shohami, E. (1996) J. Infect. Dis. 173, 735–738.
197   Belayev, L., Bar-Joseph, A., Adamchik, J. and Biegon, A. (1995) Mol. Chem. Neurophathol. 25, 19–33.
198   Bar-Joseph, A., Berkovitch, Y., Adamchik, J. and Biegon, A. (1994) Mol. Chem. Neuropathol. 23, 125–135.
199   Yoles, E., Belkin, M. and Schwartz, M. (1995) J. Neurotrauma 13, 49–57.
200   Biegon, A. and Bar-Joseph, A. (1995) Neurol. Res. 17, 275–280.
201   Pop, E., Liu, Z.Z., Brewster, M.E., Barenholz, Y., Korablyov, V., Mechoulam, R., Nadler, V. and Biegon, A. (1996) Pharm. Res. 13, 62–69.
202   Pop, E., Soti, F., Brewster, M.E., Barenholz, Y., Korablyov, V., Mechoulam, R., Nadler, V. and Biegon, A. (1996) Pharm. Res. 13, 469–475.

Subject Index

Cumulative Index of Authors for Volumes 1–35

The volume number, (year of publication) and page number are given in that order.

Cumulative Index of Subjects for Volumes 1–35

The volume number, (year of publication) and page number are given in that order.